THE FUNDAMENTALS OF
AEROSOL DYNAMICS

C S WEN

Nankai University

THE FUNDAMENTALS OF AEROSOL DYNAMICS

World Scientific

Singapore • New Jersey • London • Hong Kong

Published by

World Scientific Publishing Co. Pte. Ltd.

P O Box 128, Farrer Road, Singapore 912805

USA office: Suite 1B, 1060 Main Street, River Edge, NJ 07661

UK office: 57 Shelton Street, Covent Garden, London WC2H 9HE

Library of Congress Cataloging-in-Publication Data
Wen, C. S.
    The fundamentals of aerosol dynamics / C. S. Wen.
      p.   cm.
    Includes bibliographical references and index.
    ISBN 9810226616
    1. Aerosols.   2. Dynamics of a particle.   I. Title.
QD549.W427   1996
541.3'4515--dc20   96-1975
                                                    CIP

British Library Cataloguing-in-Publication Data
A catalogue record for this book is available from the British Library.

This book is printed on acid-free paper.

Printed in Singapore by Uto-Print

PREFACE

Aerosol dynamics is a new interdisciplinary science. It emerges and develops in the intersecting front of the fluid mechanics and aerosol science, and has become an independent discipline since Fuchs' foundation work ("The mechanics of aerosols", Science Press, 1960 (in Chinese), Pergamon Press, 1964). The aerosol system widely exists in the natural world, the engineerings, and the living beings, such as the respiratory of human body. Thus, the discipline has vast vistas in practice. The aerosol system is one of the two main parts of the fluid suspensions. Therefore, the aerosol dynamics is a branch of the mechanics of fluid suspensions, and the rapid developments in the mechanics of fluid suspensions during the last 20 years have carried the aerosol dynamics forward from the isolated-particle stage to the multi-particle interaction stage. The time now seems appropriate for a review and re-assimilation of the ideas and informations available in the multi-particle dynamics.

Up to now Fuchs' book has been an important reference book of aerosol science. However, some of his view-points are not correct. For example, in his book he alleged that the interparticle potential force is a short-range force and is much smaller than the other two forces exerting on aerosol particles—the external forces and the hydrodynamic resistance—and can be neglected in most cases. Therefore, he determined that the motions of aerosol particles become irrelevant to each other, and thus the mechanics of aerosols reduces to theoretically and experimentally investigating an isolated particle moving under external forces in a damping medium. In fact, the multi-particle interactions are not confined only to the interparticle potential. The hydrodynamic interactions among particles are long-range forces. Then, they cannot be neglected even when the aerosol dispersion is as dilute as $\phi \sim 10^{-2}$ where ϕ is the volume concentration of the particles. The enhanced settling of a particle cloud falling in an unbounded space, the hindered settling of particles dispersed in a bounded space and the coagulation processes of particles cannot be understood correctly without correct understanding of the hydrodynamic interactions. And the breakthrough advances in the investigation of multi-particle hydrodynamic interactions are obtained only in the last 30 years. The lack of a solid understanding of these particle interactions causes some shortcomings (for example, the exact theory of hindered settling) and errors (for example, the underestimation of the role of van der Waals attraction on Brownian coagulation) in Fuchs' book, and these problems have been solved only by the investigations in the last 20 years. Thus, it is really very necessary to give a new summary about the present status of the multi-particle interaction investigations. It is the purpose of the present book. The first three chapters describe the fundamental knowledge of the aerosols and the mechanics of viscous fluid, especially the low-Reynolds-number hydrodynamics with pair-interactions. Chapter 4 gives the kinetics of two aerosol particles at low Reynolds number and low Stokes number. The last three chapters present the basic theories of sedimentation, coagulation, mass or heat transfer, effective viscosity of aerosol dispersions and the evolution of the size distribution of aerosol particles.

In writing the book I have aimed to be critical rather than encyclopedic, and therefore the treatment reflects, to some extent, my personal view point. Of course, there is room for different opinions and interpretations. It is indeed difficult and perhaps undesirable to remain unbiased for a hot dispute. The book is fairly heavily weighted with theory, but not with mathematics. I shall be very pleased if it proves to be of some use to related research workers, and if it draws attention of students to a field which offers enormous scope for research into a wide range of phenomena, both in the free atmosphere, and in the engineerings and in the medical science .

I am grateful to the National Natural Science Foundation of China for its support of this research. I gratefully acknowledge Professor Lin Hai and Professor Jin Ya-qiu for their recommending the book to World Scientific.

I am also grateful to Professor Li Xue-qian for reading and revising the language of the manuscripts, and to Professor Shi Shu-zhong for his help with editing the manuscripts in its final camera-ready forms.

Finally, I should say a few words about my close comrade-in-arms Zhu Zhen-hua. The book is actually a summarization of my researches on the mechanics of fluid suspension and aerosol dynamics in the last fifteen years. During the long years, I have experienced her encouragement, kindness and support all the time and everywhere. Without this support, it is impossible for me to obtain all my achievements in this area. In addition to the above general support, she undertook most of the strenuous and endless work of preparing the final camera-ready manuscripts using her sparetime, and corrected some mistakes in the manuscripts. It may be thus said that this book is a fruit of our common painstaking labour. Therefore, I wish to express my heartfelt thanks to her, and dedicate this book to her.

C. S. Wen
Dept. of Physics
Nankai University
Tianjin, China
December, 1995

CONTENTS

CHAPTER 1
A SURVEY OF AEROSOLS

1. The General Properties and Classifications of Aerosols

Small particles either in solid state or in liquid state are suspended in gaseous medium. The suspension system is called the aerosol. Exactly speaking, aerosol means both the dispersed phase and the gaseous medium. However, in ordinary life it usually means the aerosol particles only, which have different terms in different disciplines, like "atmospheric float dust" in environment science, "atmospheric condensation nuclei" in could physics etc. In fact, cloud, fog and smoke, are themselves specific aerosols. We often encounter them in some atmospheric physical processes, in ordinary life, in industry, agriculture, in some military activities and in some scientific experiments.

The aerosol particles are very small and the disperse degree of them is very large. The smallest particles are of molecular scale. Several neutral molecules adhere to a charged molecule, and form a molecular aggregate. It may be called a small ion, or a light ion. The size of it may be of order $10^{-3}\mu$m. The upper limit of the particle size may be of several microns to 10μm in general, or even larger in some particular systems. For instance the size of large cloud drops and large hailstones in some severely convective clouds may be of order $10^3\mu$m to $10^4\mu$m. This shows that the disperse range of the aerosol particle size may span five to seven orders.

The aerosol particles may be either in solid state or in liquid state, either inorganic substance or organic matter. Besides, they may also be bacterium, microorganisms, pollens and spores, and the aerosol systems which are formed by such 'particles' are called bioaerosol.

The aerosol particles have electric charges in general. The situations are different from those of hydrosol particles. In hydrosol systems the particles have usually like-charges. However, in aerosol system the particles may have either like charges or unlike charges. For example, the particles of atmospheric aerosols include either positive or negative ions. The total amount of the positive ions is not very different from that of the negative ions, and is generally a little bit larger. Besides, during the operation of atomic reactor and the detonation of an atomic bomb, a great amount of radioactive dust is produced.

There are some relations between the particle size and the composition of the aerosols. Table 1 of Chapter 1 shows different characters of the particle size of aerosols having different composition (Batchelor[1]).

Obviously aerosol systems are very complicated colloidal dispersions. There is not

Table 1. The particle size s of different aerosol systems

0.01	0.1	1.0	10	100 μm
smog	smoke		dust	
	mist	fog		spray
carbon black	paint pigment		pulverized coal	
viruses		bacteria		
	pollen		spore	

a generally recognized classification of the dispersions. Different disciplines, different situations and different people have different classification ways.

In could physics, they are divided into three categories according to the particle size , viz.

1 Aitken nucleus : the particle radius a is smaller than 0.1μm;
2 Large nucleus: the particle radius a is larger than 0.1μm and smaller than 1μm;
3 Giant nucleus: the particle radius a is larger than 1μm.

According to the formation of the aerosol dispersions, they fall into two categories, viz.

1 Dispersed aerosols: material either in solid state or in liquid state are pulverized into granular state due to mechanical pulverization or natural weathering, then suspended in air due to the wind-force raising;
2 Condensed aerosols: the formation of the particle or the droplet is due to condensation of super-saturated gas on condensation nucleus or the particle droplet is formed in a mixture state of different gases through the photochemical reaction.

In his famous book "The mechanics of aerosols" Fuchs divided the aerosols into four categories[2]:

1 Fog: aerosol dispersion in liquid state (including both dispersed and condensed systems);
2 Smoke: condensed aerosol dispersions in solid state;
3 Dust: dispersed aerosol dispersions in solid state;
4 Smog: the mixture of aerosols both in solid state and in liquid state (smoke + fog).

According to in what situations the aerosols are, they can be divided into two categories, viz.

1 Atmospheric aerosols: aerosols dispersed in the whole atmosphere. In fact, the whole atmosphere is an aerosol system, mainly formed by natural processes, but the part formed by artificial processes has become an increasingly serious problem;

2 Industrial aerosols: aerosols formed in a factory or in a scientific laboratory. The scale of it is smaller than that of atmospheric aerosols. The scales of aerosols in the narrow gap of a fibre filter or an aerosol sampler are even smaller. The industrial aerosols are usually waste material produced in industrial production. However, sometimes people manufacture particular aerosols to increase production efficiency, like various atomizer techniques and fluidised beds techniques.

Obviously, the object of aerosol research is much more complicated than that of cloud research. It is impossible to expect that there is a unified classification which is acceptable by all researchers for the aerosol systems like that for the cloud systems. In fact, cloud systems are themselves some particular aerosol systems. It is thus clear that this is a very complicated problem. On the other hand, it is proper to classify aerosol according to the particle size from the view-point of the mechanics of aerosols. This is because of that the mechanical and other physical properties of the aerosol particles all depend on the particle size . Table 2 of Chapter 1 presents part relations between them.

Table 2 of chapte 1 shows that large nucleus and part of Aitken nucleus form a transition region of the mechanical and optical properties. The variations of the particle properties are simply due to the variations of the values of a series of the dynamic similarity -parameters. Table 3 of Chapter 1 gives the value of the dynamic similarity -parameters calculated for four typical particles ($a = 0.01, 0.1, 1, 10\mu$m) with density of 2gm/cm^3.

According to the NACA standard atmospheric data, the mean free path of the air molecule is of order $7.37 \times 10^{-2}\mu$m under one atmosphere pressure with temperature being 15^0C[3]. Thus, the value of the Knudsen number (Kn) is larger than unity for the aerosol particle with $a = 0.01\mu$m, so the gas medium can no longer be treated as a continuous medium. There will not be a disturbing flow field induced by the moving particles. Thus the resistance is proportional to the total number of the air molecules which strike the surfaces of the particles, and to the surface area of the particle, that is, to a^2. Similarly, the mass transfer and the heat transfer between the particle and the medium are also proportional to the total number of the air molecules which hit the surface of particles, and also to a^2. Conversely, when the Knudsen number is smaller than unity, e.g. when a is larger than 1μm, the medium can be regarded as a continuous medium again. There will be a disturbing flow field around the moving particles. Then, the resistance is the result of the force exerted by the flow field on the particles. Table 3 of Chapter 1 shows that the flow field around the moving particles is a low-Reynolds-number flow where the Reynolds number is defined as Re and the flow is described by the linear Stokes equation . From this we may obtain

Table 2. The aerosol particle size and its physical properties

particle radius a (μm)	10^{-3}	10^{-2}	10^{-1}	1	10
category	Aitken nucleus		large nucleus	giant nucleus	
resistance of medium	$\propto a^2$		$\propto af_1$	$\propto a$	
condensation (evapouration) rate	$\propto a^2$		$\propto af_2$	$\propto a$	
cooling (heating) rate	$\propto a^3$		$\propto af_3$	$\propto a$	
transportation of particle concentration	Brownian diffusion is dominant		transition region	gravitational sedimentation is dominant	
light scattering	$\propto a^6$		transition region	$\propto a^2$	

the Stokes resistance which is proportional to a. The molecular diffusivity of vapour D_v and the thermal diffusivity D_H are of the same order as the kinetic viscosity ν. Then, the Schimdt number and the Prandtl number are of the order of unity. The similarity-parameter P_{ev} (Péclet number for vapour diffusion) of the convective diffusion equation for the vapour and P_{eH} (Péclet number for heat conductivity) are of the same order as the Reynolds number, and thus are both smaller than unity. The convection terms in these two convection diffusion equations can be neglected. They then reduce to pure diffusion or pure heat conduction equation. As well known, the mass transfer rate (either condensation rate or evaporation rate) and the heating or cooling rate obtained from these parabolic equations are both proportional to a. The results are thus similar to the Stokes resistance. As the radius varies in a range from 0.01 μm to 0.1 μm, the gas medium can still be regarded as a continuous medium. However, it is necessary to make some modifications which are functions of Knudsen number, like f_1, f_2 and f_3 in Table 2 of Chapter 1. For the resistance modification, Cuningham (1910) gave an empirical formula for f_1, viz.[3]

$$f_1 = \frac{1}{1 + \alpha Kn},\qquad (1.1.1)$$

where α is a dimensionless constant and equals to 0.86. On the other hand, if the particle is very large as the radius a approaches to 100μm, Re is no longer smaller than unity. The problem becomes a high-Reynolds-number problem. For instance,

Table 3. Some dynamic similarity -parameters for four typical aerosol particles

a (μm)	D_0 (cm/s)	V_s (cm/s)	Kn	Re
0.01	1.56×10^{-4}	3.1×10^{-5}	7.37	4.2×10^{-10}
0.1	2.35×10^{-6}	4.3×10^{-4}	7.37×10^{-1}	5.8×10^{-8}
1	1.29×10^{-7}	2.7×10^{-2}	7.37×10^{-2}	1.1×10^{-5}
10	1.19×10^{-8}	2.5	7.37×10^{-3}	1.0×10^{-2}

a (μm)	P_{ev}	P_{eH}	P_{eB}
0.01	1.3×10^{-10}	1.5×10^{-10}	2.0×10^{-7}
0.1	1.7×10^{-8}	2.1×10^{-8}	1.8×10^{-3}
1	1.1×10^{-5}	1.3×10^{-5}	2.1×10
10	1.0×10^{-2}	1.2×10^{-2}	2.1×10^{5}

for a cloud droplet with radius $a = 30\mu$m, its Reynolds number equals to 0.41. For a large cloud droplet with $a = 100\mu$m, its Reynolds number equals to 9.4. Experiments have shown that it is necessary to make some modification if $Re > 0.4$ (that is if the radius of a cloud droplet is large than 30 μm). Obviously, the modifications must be a dimensionless function of Re. The above characters are not suitable to the problem of transportation of particle concentration. The values of Brownian diffusiity D_0 given in Table 3 of Chapter 1 show that they are inversely proportional to radius a, and cannot be a constant like D_v, D_H and ν. Thus, although when the particle radius is smaller than 0.1 μm, the Péclet number for the particle transportation P_{eB} is also smaller than unity like P_{ev}, P_{eH} and Re, and the Brownian diffusion term is dominant. Yet the Péclet number P_{eB} is larger than unity when the particle radius a is larger than 1μm, then the gravitational convection term is dominant, as the Brownian diffusion term is negligible. Thus the referred large nucleus in the could physics (including part of large Aitken nucleus) becomes a transition region for the dynamic feature of aerosol particles. Dividing aerosol particles into three categories (Aitken nucleus , large nucleus and giant nucleus) thus becomes a better classification. Of course, the could physics classification is not exactly consistent with the variations of the dynamic feature of the aerosol particles. This is because the classification in could physics does not differentiate the three particle categories according to their dynamic feature. On the other hand, Table 2 of Chapter 1 also indicates that there are some difficulties when we divide them into the categories exactly according to the dynamic feature of aerosol particles. The transition regions of various dynamic feature do not exactly coincide with each other. For such a complicated system like the aerosol system, we should not place undue hopes on the problem of their exact classification.

2. The Size Distribution of Aerosol Particles

In laboratories one can prepare monodisperse aerosol system to serve some particular purposes. In nature, there are cases in which the aerosols are monodisperse. For example, the pollens of trifoliate grasses are nearly monodisperse. The radii of their pollens are in the range from 24.8μm to 26.9μm. However, such an instance is very rare. Most of the aerosol dispersions formed either by natural processes or artificial processes are polydisperse. Hence, the research on the size distribution of aerosol particles has practical and immediate significance.

We have already been aware that all the properties of aerosol particles including their dynamic feature, optical feature and electrical feature depend on the particle size . Hence, it is necessary to investigate the size distribution of the particles in order to correctly estimate, analyze and explain the experimental results about the aerosol dispersion.

Besides, the size distribution of aerosol particles is a result of its micro-dynamic processes, and is a significant characteristic quantity of the micro-process of aerosols. Thus, to investigate the size distribution of aerosols is also an important method to understand the micro-processes of the aerosol dispersion.

There are two methods to express the size distribution of the aerosol particles.

2.1. Differential Distribution (Radius Distribution)

We denote the particle number with radius between a and $a + \Delta a$ by $n(a)\Delta a$. The dimension of $n(a)$ is then $[L^{-4}]$, or $[l^{-1}L^{-3}]$ where l expresses the dimension of particle radius and L expresses the dimension of geometry length of the space. The relation between the normalized differential spectrum $f(a)$ and $n(a)$ is

$$f(a) = \frac{n(a)}{n_0}, \tag{1.2.1}$$

where n_0 denotes the total particle number per unit volume, and

$$n_0 = \int_0^\infty n(a)\mathrm{d}a, \qquad \int_0^\infty f(a)\mathrm{d}a = 1. \tag{1.2.2}$$

Besides the number density, sometimes we also use the concept of volume concentration ϕ which is the volume occupied by particles per unit space volume, then $\phi = \int_0^\infty \frac{4}{3}\pi a^3 n(a)\mathrm{d}a$, (assuming the particle has the spherical shape). In addition, we have the mass concentration $g_o = \int_0^\infty \frac{4}{3}\pi a^3 \rho n(a)\mathrm{d}a$, where ρ is the particle density (assuming the aerosol has a single chemical composition). The relations between the differential spectrum of particle volume concentration $\phi_1(a)$, of particle mass concentration $g(a)$ and the differential spectrum of particle number, $n(a)$ are given by

$$\phi_1(a) = \frac{4}{3}\pi a^3 n(a), \qquad g(a) = \frac{4}{3}\pi a^3 \rho n(a). \tag{1.2.3}$$

In some problems, it is more convenient to discuss the distribution versus the particle volume v (where v denotes the volume of one particle), or the distribution versus particle mass m (where m denotes the mass of one particle). The corresponding differential spectrum are denoted by $n_1(v)$ and $n_2(m)$ respectively, which imply that there are $n_1(v)\Delta v$ or $n_2(m)\Delta m$ particles with particle volume between v and $v + \Delta v$ or with particle mass between m and $m + \Delta m$. It is evident that the relation between them and the distribution in particle radius $n(a)$ should be

$$n_1(v) = \frac{1}{4\pi a^2}n(a), \qquad n_2(m) = \frac{1}{4\pi a^2 \rho}n(a). \qquad (1.2.4)$$

2.2. Integral Distribution (Cumulative Distribution)

It is usually to use integral distribution N, which is defined as

$$N = \int_a^\infty n(a)\mathrm{d}a. \qquad (1.2.5)$$

We sometimes call it the remainder distribution. To take solid particle as an example, when we use a sieve to sieve the particles, the remainders of the particles on the sieve is just N. Another integral distribution N_1 is defined as

$$N_1 = \int_0^a n(a)\mathrm{d}a. \qquad (1.2.6)$$

It is sometimes called scavenged distribution. The total number of the solid particles under the sieve is N_1, and it is clear that by the definitions of N, N_1 and n_0, we have a relation

$$N + N_1 = n_0, \qquad (1.2.7)$$

and

$$n(a) = -\frac{\mathrm{d}N}{\mathrm{d}a}. \qquad (1.2.8)$$

Because of the large disperse range of the atmospheric aerosols it is often inconvenient to use the radius distribution in linear scale. This leads to use of the log-radius distribution $n^*(a)$, where $n^*(a)$ is given by

$$n^*(a) = -\frac{\mathrm{d}N}{\mathrm{d}(\log a)}. \qquad (1.2.9)$$

Here, $\log a$ is the decimal logarithms of a. From Eq. (1.2.9) and Eq. (1.2.8) we may obtain a relation between $n^*(a)$ and $n(a)$, viz.

$$n(a) = \frac{1}{\ln 10}\frac{n^*(a)}{a}, \qquad (1.2.10)$$

where $\ln 10$ is the Naperian logarithms of 10. Hence, if $n^*(a)$ is a negative power of a, $n(a)$ is also a negative power of a, and is one order smaller than $n^*(a)$.

In 1958, Junge discovered that the differential spectrum in the range of large and giant nucleus for the atmospheric aerosols $n^*(a)$ is proportional to a^{-3} (i.e. $n(a)$ is proportional to a^{-4}). The a^{-4} law for $n(a)$ has been proved to be correct basically by the observations made at later time. Of course, individual observations may have powers different from -4, as a negative power between -3 and -5. However, average speaking, the general trends do coincide with a^{-4} law discovered by Junge. Incidentally, because of the a^{-3} law of $n^*(a)$, then $\phi_1^*(a)$ is a constant and is independent of a. Although the particle volume is proportional to a^3, the $n^*(a)$ is inversely proportional to a^3, yet in the same log-radii interval, their total particle volume should be the same.

Besides the a^{-4} spectrum presented by Junge, there are four-parameter spectra presented by some other authors, i.e.

$$n(a) = Aa^\beta \exp(-Ba^\gamma)$$

where A, B, β and γ are constants. Sometimes, one may set $\gamma = 1$, then the spectra become the three-parameter spectra. Of course, the more parameters the spectrum has, the better the spectrum approximates the observations. However, from the theoretical viewpoint, it is better to keep fewer parameters. This is because the universality of the empirical formula is wider if contained parameters are fewer, and it is more valuable for understanding of the main physical mechanism of the empirical spectrum formation. By this point, Junge's spectrum is the best. In fact, there emerged a lot of theoretical works on the formation of Junge's spectrum, after Junge presented his a^{-4} spectrum. We will discuss these works later in Chapter 7.

In his book "The mechanics of aerosols" Fuchs introduced some empirical formulae of the particle size distribution for the industrial aerosols, bioaerosol and cloud droplets. For a detailed description of these spectra, the reader is referred to his book[2] and we will not repeat here. At this point, it is necessary to stress that those formulae are all for average spectra. The average spectra are of course very valuable if we statistically analyze some physical phenomena which are related to the aerosols. For instance, it is very useful to use the average spectrum of atmospheric aerosols to discuss the transportation of atmospheric radiation. On the other hand, the average spectra have some drawbacks. The Best's spectrum or the Levin's log-normal distribution for the cloud droplet are both average spectra and have only one maximum. However, many observations have shown existence of the second maximum in cloud droplet spectra since 60's. This is an important character of convective clouds. It reflects some significant growing mechanism of cloud droplets. However, the second maximum only exists in individual spectra. It will be smoothed out when we average the individual spectra due to the stochastic feature of the position of the second maximum. This is the limitation of the average spectra.

3. Shape and Structure of Aerosol Particles

It is the surface tension of liquid droplets that makes the droplets into the spherical shape, that is, into a mechanically stable form. The surface tension of the liquid droplet always makes the interface to contract to minimum. Thus the shape of the droplet can remains spherical. When two droplets coagulated, at the beginning the shape of the new droplet becomes a dumbbell which is obviously non-spherical. However, soon the surface tension contracts the new droplet into a new spherical sphere. Only for the large drop, say $a > 140\mu$m, can the liquid drop gradually deviate from the spherical shape, and becomes an oblate sphere. Finally the water drop may be broken down when its radius reaches 3mm. However, all of these are beyond the scope of this book, and we regard the liquid droplet as a spherical particle throughout this book.

The shape and structure of the solid particles are quite a complicated problem. For the particles which are formed naturally, only the particles which are frozen from liquid droplets have spherical shape. Besides this all the others are non-spherically shaped.

The solid particle formed by sublimation of supersaturated gas on a sublimation nucleus has the shape of crystal e.g. ice crystal, snow crystal, etc. The solid particles formed either by natural weathering or by mechanical smash all have irregular shape, and roughly speaking they can be divided into three classes, viz.

1 Isotropic in length—three dimensional irregular body;
2 Scale in two dimensions is larger than the rest dimension—plate-like body;
3 Scale in one dimension is large than the other two dimensions—rod-like or needle-like body.

There are various aggregates of solid particles formed by coagulation in an unstable system. An aggregate may at most have several millions particles flocculated together. They may in general be divided into two classes: chain-like aggregate and body-like aggregate.

It is impossible to discuss the particle size distribution exactly when the particles have irregular shape. We only assume them to be of spherical shape, and study the particle distribution versus their equivalent radius. Two equivalent radii are often used. The equivalent radius in particle volume a_v is the radius of an assumed sphere whose volume is equal to that of the particle. The equivalent radius in settling speed a_s is the radius of an assumed sphere whose settling speed is equal to that of the particle. The former is obtained by measuring the weight of the particle, while the latter by measuring the settling speed of the particle. a_v is equal to a_s only for spherical particles. Otherwise, $a_v \neq a_s$. It is thus evident that the size distribution of non-spherical particles is more complicated than that of spherical particles. It has in itself some indefiniteness.

· Besides the particle size , the particle shape is another factor which determines the dynamic feature of the particle. Table 2 shows that within the range of viscous flow the resistance to the particle is proportional to the particle radius a. The proportionality depends on the particle shape . For the rigid spherical particle, the proportionality is $6\pi\mu V$ where μ is the viscosity of the gas medium and V is the velocity of the moving particle. This is the well known Stokes law, which was obtained by Stokes in last century, and will be reintroduced in some details in Chapter 3. For the non-spherical particle, the situation is more complicated. Among the various bodies, only the spherical body is isotropic. All the others are anisotropic. Thus, in addition to the particle size , the resistance exerted on the non-spherical particle in motion depends also on the moving direction. Then the proportionality is $6\pi\mu V\kappa'$ where κ' is a dimensionless constant depending on the moving direction. For the case of elongated rotation ellipsoid, the dimensionless constant κ'_c is given by the following expression when it is moving along its polar axis, namely [2]

$$\kappa'_c = \frac{4}{3}\left(\beta^2 - 1\right) / \left\{\frac{2\beta^2 - 1}{\sqrt{\beta^2 - 1}}\ln\left(\beta + \sqrt{\beta^2 - 1}\right) - \beta\right\}, \qquad (1.3.1)$$

where β is the ratio of the major axis to the minor axis of the ellipsoid. On the other hand the dimensionless constant κ'_a is given by Eq. (1.3.2) when it is moving in the direction perpendicular to the polar axis, viz.[2]

$$\kappa'_a = \frac{8}{3}\left(\beta^2 - 1\right) / \left\{\frac{2\beta^2 - 1}{\sqrt{\beta^2 - 1}}\ln\left(\beta + \sqrt{\beta^2 - 1}\right) + \beta\right\}. \qquad (1.3.2)$$

Table 4 of Chapter 1 gives some κ'_c and κ'_a for various β[2], where κ' is defined as

$$\kappa' = \frac{1}{3}\kappa'_c + \frac{2}{3}\kappa'_a$$

Table 4. The values of κ for various β

β	κ'_c	κ'_a	κ'
2	1.20	1.38	1.32
3	1.40	1.73	1.62
4	1.60	2.06	1.91
6	1.97	2.68	2.44
8	2.31	3.26	2.94
10	2.65	3.81	3.42
20	4.16	6.38	5.64

It is noted that both κ'_c and κ'_a approach unity as β approaches unity and their differences decrease. Conversely, both κ'_c and κ'_a monotonically increase as β increase, and their differences also increase.

For a slow motion where the value of Reynolds number is small (say, $Re < 0.1$), the inertia moment of rotation of the particle is zero. the orientation of the particle thus remains unchanged during its motion. Only when $Re > 10$, the particle chooses the orientation along which the resistance is maximum when it is in motion. Within the range of low-Reynolds-number flow , if the angle between the particle axis and its direction of motion is θ at the initial instant, then during the whole process of its motion thereafter the resistance will be described by[2]

$$F = -6\pi\mu V a(\kappa_c' \cos^2\theta + \kappa_a' \sin^2\theta). \qquad (1.3.3)$$

When the size of the particle is so small that the Brownian motion becomes significant, the particle will rotate stochastically due to the rotational Brownian motion . After averaging all the possible directions chosen by the particle, Eq. (1.3.3) reduces to

$$F = -6\pi\mu V a\kappa' = -6\pi\mu V a\left(\frac{1}{3}\kappa_c' + \frac{2}{3}\kappa_a'\right). \qquad (1.3.4)$$

where κ' is the average directional factor of the particle making stochastic Brownian rotation. Table 4 of Chapter 1 also gives its value in the fourth column.

To obtain the analytical solution for the disturbing flow field of a non-spherical object is a very difficult problem. The only exact solution for the elongated rotating ellipsoid was obtained by Lamb, and for most cases we can only give some qualitative analysis (see Chapter 3). Therefore, the dynamic analysis in this book has to be restricted to the case of spherical particle only.

CHAPTER 2
THE MOTIONS OF VISCOUS FLUIDS

1. Introduction

It has been made quite clear from the previous chapter that the dynamics of viscous flows is the theoretical basis of the mechanics of aerosols. Hence we will introduce preliminarily the dynamics of viscous fluids before concerning the mechanics of aerosols. In his famous work "An introduction to fluid dynamics", Batchelor (1967) gave penetrating and incisive analyses on the physics of the motions of viscous fluids , the essence of viscous stress and especially the motion of a uniform incompressible viscous fluid . These are very important for the present subject. Therefore, Batchelor's book (1967)[4] will be the basic reference book of this chapter and the next chapter.

The viscous stress is internal friction which the viscous fluid may offer to resist attempts to deform them. The viscous stress occurs only when the motion of the viscous fluid is non-uniform and tends to smooth out the non-uniformity . However the resistance cannot prevent the deformation from occurring, or, equivalently, that the viscous stress vanishes as the rate of deformation tends to zero. From the view point of the molecular kinematics, the viscous stress is one kind of transport phenomena —transport of momentum of molecules when the distribution of the fluid velocity is non-uniform. It is evident that, if the fluid velocities on the two sides of a surface element are different, any random molecular motions will result in the establishment of a net flux of momentum of molecules, viz. a tangential component of stress , and that the sign of the flux, as well as of the stress will be such as to eliminate the difference of the velocities between the two sides. Transport of momentum thus constitutes internal friction and a fluid exhibiting internal friction is said to be viscous. And most common fluids, like air and water in particular, are viscous fluids .

We proceed now to consider a quantitative relation between the net transfer of momentum and the non-uniformity of the velocity distribution. For simplifying the discussion we first consider the cases in which the relevant intensity is a scalar quantity, which we shall denote by C.

Direct attempts to calculate the net flux vector f of the quantity C from considerations of the involved molecular processes are almost out of question for liquids, and achieve only limited success in the case of gas. Some hypothesis is needed. The hypothesis was initially based on measurements of the flux vector in particular physical contexts and used only to those contexts, but has since been recognized as having more general significance.

The first part of the hypothesis is that the flux vector depends only on the local properties of the medium and the local values of C and ∇C. The idea here is simply that the transport across a surface element is determined by molecular motions, and that over this region C can be approximated by a linear function of position, provided some condition of the type

$$\left|\frac{\partial C}{\partial x}\right| / \left|\frac{\partial^2 C}{\partial x^2}\right| \gg \text{mean free path of the molecules } \lambda_m$$

is satisfied. The second part of the hypothesis is that, for sufficiently small values of the magnitude $|\nabla C|$ the flux vector varies linearly with the components of ∇C. The flux vector is known to vanish with $|\nabla C|$, so that the hypotheses may be expressed as

$$f_i = k_{ij}\frac{\partial C}{\partial x_j}. \tag{2.1.1}$$

Both f_i and $\partial C/\partial x_j$ are vectors, and the requirement that Eq. (2.1.1) be valid for all choices of the coordinate system shows that the transport coefficient k_{ij} must be a second-order tensor. In many materials the molecular structure is statistically isotropic, in that case k_{ij} must have a form in which any directional distinction is absent. All sets of orthogonal axes must then be principal axes of the coefficient k_{ij}, which is possible only if

$$k_{ij} = -k\delta_{ij}. \tag{2.1.2}$$

The scalar coefficient k is positive, since the quantity C is transported down the gradient of intensity. Then, the expression for the flux vector is here

$$\mathbf{f} = -k \nabla C. \tag{2.1.3}$$

We now turn to the problem of finding the quantitative relation between the net flux of momentum (viscous stress) and the non-uniformity of the velocity distribution in a viscous fluid.

It is well known that the stress tensor σ_{ij} of a viscous fluid can be regarded as the sum of an isotropic part, $-p\delta_{ij}$, where the pressure p at a point in a moving fluid is the mean normal stress with sign reversed, i.e. $p = \frac{-1}{3}\sigma_{ii}$, and a remaining non-isotropic part, d_{ij}, where d_{ij} is the viscous stress tensor , then, we have

$$\sigma_{ij} = -p\delta_{ij} + d_{ij}. \tag{2.1.4}$$

The flux of momentum across a material surface element which is represented by the viscous stress is assumed to depend only on the instantaneous distribution of fluid velocity in the neighbourhood of the element, or, more precisely, on the departure from uniformity of that distribution. The local velocity gradient is thus the parameter of the flow field with most relevance to the viscous stress.

On the same hypothesis as that introduced in the above paragraph, we may assume that d_{ij} is approximately a linear function of the various components of the velocity gradient for sufficiently small magnitudes of these components. Thus the hypothesis is expressed as

$$d_{ij} = A_{ijkl}\frac{\partial u_k}{\partial x_l},$$

and then

$$d_{ij} = A_{ijkl}e_{kl} - \frac{1}{2}A_{ijkl}\epsilon_{klm}\omega_m, \qquad (2.1.5)$$

according to the decomposition principle of flow fields, where e_{kl} is the rate of strain tensor and ω is the vorticity , i.e.

$$e_{kl} = \frac{1}{2}\left(\frac{\partial u_k}{\partial x_l} + \frac{\partial u_l}{\partial x_k}\right), \qquad \omega = \nabla \times \mathbf{u}. \qquad (2.1.6)$$

Both d_{ij} and $\partial u_i/\partial x_j$ are second-order tensor, therefore A_{ijkl} must be a fourth-order tensor, and is necessarily symmetrical in the indices i and j like d_{ij}. It also takes a simple form when the molecular structure of the fluid is statistically isotropic. All gases have isotropic structure, and so do simple liquids, like water etc. Then A_{ijkl} is an isotropic tensor for these cases.

From the Cartesian tensor analysis we know that the basic isotropic tensor is the Kronecker delta tensor, and all isotropic tensor of even order can be written as the sum of products of delta tensor. Thus

$$A_{ijkl} = \mu\delta_{ik}\delta_{jl} + \mu'\delta_{il}\delta_{jk} + \mu''\delta_{ij}\delta_{kl}, \qquad (2.1.7)$$

where μ, μ', μ'' are scalar coefficients, and since A_{ijkl} is sysmmetrical in i and j, then $\mu' = \mu$. It will be observed that A_{ijkl} is symmetrical in indices k and l also, and that as a consequence the term containing ω drops out of Eq. (2.1.5), giving

$$d_{ij} = 2\mu e_{ij} + \mu''\Delta\delta_{ij}, \qquad (2.1.8)$$

where Δ denotes the rate of expansion, i.e.

$$\Delta = e_{kk} = \nabla \cdot \mathbf{u} = \frac{\partial u_k}{\partial x_k}. \qquad (2.1.9)$$

Finally we recall that by definition d_{ij} makes zero contribution to the mean normal stress , whence

$$d_{ii} = (2\mu + 3\mu'')\Delta = 0$$

for arbitrary values of Δ, implying that

$$2\mu + 3\mu'' = 0.$$

On choosing μ as the one independent scalar constant, we obtain for the viscous stress tensor the expression

$$d_{ij} = 2\mu\left(e_{ij} - \frac{1}{3}\Delta\delta_{ij}\right). \qquad (2.1.10)$$

This expression was obtained by Saint–Venant (1843) and Stokes (1845) in essentially the above way, after having been derived by Navier (1822) and Poisson (1829) from specific assumptions concerning the molecular mechanism of internal friction.

The significance of the parameter μ can be seen from the form taken by the relation Eq. (2.1.10) in the special case of a simple shearing motion. With $\partial u_1/\partial x_2$ as the only one non-zero velocity derivative, all components of d_{ij} are zero except the tangential stresses

$$d_{12} = d_{21} = \mu \partial u_1/\partial x_2. \tag{2.1.11}$$

Thus μ is a constant of proportionality between the rate of shear and the resisting tangential force termed the viscosity of the fluid.

The first part of the hypothesis that in the neighbourhood of the surface element u can be approximated by a linear function of position implies that the Knudsen number (Kn) must be smaller than unity. This is true for the aerosol particles with radii larger than 1μm. On the other hand when $Kn \geq 1$, the value of the viscosity μ needs correction. This is the Cuningham correction for μ which has already mentioned in Chapter 1. When $Kn \gg 1$, for example for the small Aitken nucleus with radius smaller than $10^{-2}\mu$m, the medium of air cannot be treated as a continual medium again. Under these circumstances it is needed to use the molecular kinematical theory.

The second part of the hypothesis implies that the linear relation between the rate of strain and the viscous stress is accurate only for small magnitudes of the velocity gradient. However, experiments have shown that this linear relation may hold over a remarkably wide range of values of the rate of strain. For water and air, the linear law appears to be accurate under all except possibly the most extreme conditions, such as with a shock wave. This phenomenon becomes understandable when the molecular mechanism of internal friction is considered. The bulk relative motion of the fluid can cause only a small change in the statistical properties of the molecular motion when the characteristic time of the bulk motion is long compared with the one of the molecular motion (which in the case of a gas would be given by the average time between molecular collisions). For air at normal temperature and pressure the average time between collisions is about 10^{-10}sec; it is evident that common practical values of the bulk rate of strain are indeed 'small' in the above sense.

Fluids for which the linear law Eq. (2.1.10) between the viscous stress and the rate of strain tensor does hold accurately are usually said to be 'Newtonian fluid s', in recognition of the fact that the simple relation Eq. (2.1.11) for a simple shearing motion was first proposed by Newton (1687). Of course, the medium of air for aerosol particles is a Newtonian fluid .

Substituting Eq. (2.1.4) and Eq. (2.1.10) into equation of motion for a Newtonian fluid gives

$$\rho \frac{Du_i}{Dt} = \rho F_i - \frac{\partial p}{\partial x_i} + \frac{\partial}{\partial x_j}\left\{2\mu\left(e_{ij} - \frac{1}{3}\Delta\delta_{ij}\right)\right\}. \tag{2.1.12}$$

This is the Navier–Stokes equation of motion. Here

$$\frac{D}{Dt} = \frac{\partial}{\partial t} + \mathbf{u}\cdot\nabla. \tag{2.1.13}$$

The operator D/Dt is meaningful only when applies to a field variable (that is, a function of $\mathbf{x}$ and t), and is said to give a time derivative following the motion of a given fluid element. F_i in Eq. (2.1.12) is the resultant vector of the volume forces, per unit mass of fluid. The viscosity μ in Eq. (2.1.12) has no relation with pressure p. As is well known from the kinetic theory of gases, the viscosity μ is equal to $\alpha \rho v_a \lambda_m$, where the coefficient α is approximately $\frac{2}{3}$ for air at ordinary temperature and pressure, ρ is the density of air, v_a is the speed of sound (which is a good representative for the rms molecular velocity), λ_m is the mean free path. Although the density is proportional to pressure, however, the mean free path is proportional inversely to pressure. Thus the dependence of μ on p is cancelled. On the other hand, the viscosity of air increases with temperature.

Table 1 of Chapter 2 gives this relation. The unit of μ is poise in recognition of the important contribution to the dynamics of viscous flows of Poiseuille, 1 poise=1 gm/cm.sec. Table 1 of Chapter 2 shows that the variations of μ with temperature are small.

Table 1. The variation of viscosity μ and kinematic viscosity ν of air

$T(^0C)$	$\rho(\text{gm}/\text{cm}^3)$	$\mu(\text{gm}/\text{cm.sec})$	$\nu(\text{cm}^2/\text{sec})$
-50	1.582×10^{-3}	1.45×10^{-4}	0.092
0	1.293	1.71	0.132
10	1.247	1.76	0.141
15	1.225	1.78	0.145
20	1.205	1.81	0.150
30	1.165	1.86	0.160
40	1.127	1.90	0.169

The viscosity μ of air is actually a constant for most problems of the mechanics of aerosols. The Navier–Stokes Eq. (2.1.12) reduces to

$$\rho \frac{Du_i}{Dt} = \rho F_i - \frac{\partial p}{\partial x_i} + \mu \left(\frac{\partial^2 u_i}{\partial x_j \partial x_j} + \frac{1}{3} \frac{\partial \Delta}{\partial x_i} \right). \qquad (2.1.14)$$

Usually the characteristic velocity of the flow field in the area of the mechanics of aerosols is much smaller than the sound speed. The mass conservation equation reduces here to $\Delta = \nabla \cdot \mathbf{u} = 0$. The medium of air for aerosol particles is an incompressible fluid . Then Eq. (2.1.14) becomes

$$\rho \frac{D\mathbf{u}}{Dt} = \rho \mathbf{F} - \nabla p + \mu \nabla^2 \mathbf{u}. \qquad (2.1.15)$$

The body force $\mathbf{F}$ represents the action of the earth's gravitational field in the space of aerosols, and we shall put $\mathbf{F} = \mathbf{g}$ with $\mathbf{g}$ assumed to be uniform over the fluid volume.

Under these conditions, the equation of motion for a viscous fluid Eq. (2.1.15) becomes

$$\frac{\mathbf{Du}}{\mathbf{Dt}} = -\frac{1}{\rho}\nabla p + \mathbf{g} + \nu\nabla^2\mathbf{u}, \tag{2.1.16}$$

where $\nu(= \frac{\mu}{\rho})$ is the kinematic viscosity , and is effectively a diffusivity for the velocity $\mathbf{u}$. Its unit is Stokes in recognition of the important contribution to the dynamics of viscous flows of Stokes, and 1Stokes = 1cm^2/sec. The factor ρ in μ now is cancelled, then ν is inversely proportional to p, and increases with temperature too. The last column in Table 5 shows the variation of ν with temperature at the pressure of one atmosphere. Obviously, ν is actually a constant for most problems of the mechanics of aerosols.

It may be seen from Eq. (2.1.16) that when ρ is uniform (for most problems of the mechanics of aerosols ρ is indeed uniform) the force per unit volume due to gravity is balanced exactly by a pressure equal to $\rho\mathbf{g}\cdot\mathbf{x}$. This suggests that the pressure p should be written as

$$p = p_0 + \rho\mathbf{g}\cdot\mathbf{x} + P, \tag{2.1.17}$$

where p_0 is a constant and $p_0 + \rho\mathbf{g}\cdot\mathbf{x}$ is the pressure which, in a moving fluid, gives rise to a pressure gradient in balance with the force of gravity . P is the remaining part of the pressure and evidently arises from the effect of the motion of the fluid, being determined by the equation

$$\frac{\mathbf{Du}}{\mathbf{Dt}} = -\frac{1}{\rho}\nabla P + \nu\nabla^2\mathbf{u}. \tag{2.1.18}$$

There is no such P term in general uses, and it is referred here as the modified pressure. Gravity no longer appears in the equation of motion when the modified pressure P is introduced, and provided it is also absent from the boundary conditions (note that in most problems of the mechanics of aerosols, gravity does not appear in boundary conditions) we may infer that gravity has no effect on the velocity distribution in the fluid. In subsequent discussions of motions of a uniform fluid for which the boundary conditions involve only the velocity (the mechanics of aerosols is always of such a boundary conditions), we shall introduce the modified pressure and write the equation of motion in the form Eq. (2.1.18) in the later context . However, following the general practice, this modified pressure will be denoted by the same symbol p. The symbol P is thus no longer used.

In this and the next chapter the effect of the viscous stress will be examined. In order to reveal this effect clearly and to get physical insight, we shall assume that the fluid is an incompressible fluid ; as already seen, this is a valid approximation for aerosol dynamics , the principal restriction being that the speed of the fluid should be small compared to the sound speed everywhere. It will also be assumed that no other effects which could cause the density of a material element to vary significantly are present. Under these conditions the equation describing the balance of internal energy and the thermodynamic equation of state are irrelevant, being replaced by the property of invariance of the density of an element of fluid, i.e.

$$\mathrm{D}\rho/\mathrm{D}t = 0.$$

The mass conservation equation reduces to

$$\nabla \cdot \mathbf{u} = 0. \tag{2.1.19}$$

Fortunately, the conditions of the flow field induced by the aerosol particles satisfy all the above restrictions, Eq. (2.1.9) will be thus used in the mechanics of aerosols.

We are now turning to the problem of finding the conditions on the velocity and stress at a material boundary. There are general two transition relations at a surface for each quantity, one representing continuity of the appropriate intensity across the surface, and one representing continuity of the normal component of the flux vector . Fluid momentum is one such transportable quantity, the associated intensity and flux vector being velocity and stress respectively.

The first of the above two transition relations is simply that the tangential component of velocity is continuous across a material boundary separating a fluid and another medium. The justification for this condition can be regarded as lying in the fact that any discontinuity in velocity across a material surface would lead almost immediately to a very large stress at the surface whose direction is to tend to eliminate the relative velocity of the two masses; the condition of continuity of the velocity is thus not an exact law, but a statement of what may be expected to happen, approximately in normal circumstances. The effectiveness of the viscous stress which smooths out the discontinuity in fluid velocity depends on the magnitude of the viscosity . Clearly there will be some special circumstances in which the viscous stress is relatively weak, and in which a steep velocity gradient promoted by some other causes is able to persist; and in such cases it may be convenient to speak of a 'discontinuity' in velocity.

The case where a boundary separates a fluid mass and a solid mass is particularly important in practice, and is the usual case with the dynamics of rigid aerosol particles. Continuity of the tangential component of velocity across the boundary is here referred to as the no-slip condition . The validity of the no-slip condition at a fluid–solid interface was debated for some years in last century, there being some doubt about whether molecular interaction at such an interface leads to momentum transfer of the same nature as that at the fluid–fluid surface; but the absence of slip at a rigid wall is now amply confirmed by direct observations and by the correctness of its many consequences under normal conditions. The one important exception is the flow of a gas at such low density that the mean velocity of the molecules varies appreciably over a distance of one mean free path. It seems that here there can be a non-zero jump in velocity and also in temperature at a rigid wall, since the number of collisions of the molecules in a volume element is not large enough for even an approximate equilibrium to be established. When the separation between two solid aerosol particles is as small as a mean free path, the non-zero jump in velocity may occur at the rigid boundary also, and the no-slip condition is replaced by the Maxwell molecular slip condition . We will discuss this situation later in Chapter 5 and 6.

The second of the two transition relations is that the difference between the stresses on two sides of the boundary is a normal force which is the surfsce tension , as represented by the following expression:

$$\sigma''_{ij}n_j - \sigma'_{ij}n_j = -\gamma \left(\frac{1}{R_1} + \frac{1}{R_2} \right) n_i, \qquad (2.1.20)$$

where σ''_{ij} is a stress tensor on the side of the interface to which the unit normal vector **n** points, and σ'_{ij} is the stress tensor on the other side of the interface, R_1 and R_2 are radii of curvature of the interface in any two orthogonal planes containing **n**, being regarded as positive when the corresponding centre of curvature lies on the side of the interface to which **n** points, and γ is the surface tension, for an interface separating air and pure water at 15^0C. $\gamma = 73.5$ dyn/cm.

When making the above relation Eq. (2.1.20) more explicit by using the expressions Eq. (2.1.4) and Eq. (2.1.10) for the stress tensor σ''_{ij} and σ'_{ij} it is convenient to take separately the components of the surface force normal to the boundary (direction **n**) and tangential to it (direction **t**,). For the tangential component we have

$$\mu'' e''_{ij} t_i n_j = \mu' e'_{ij} t_i n_j, \qquad (2.1.21)$$

and, for the normal component,

$$p'' - 2\mu'' \left(e''_{ij} n_i n_j - \frac{1}{3}\Delta'' \right) = p' - 2\mu' \left(e'_{ij} n_i n_j - \frac{1}{3}\Delta' \right) + \gamma(R_1^{-1} + R_2^{-1}) \qquad (2.1.22)$$

at each point of a boundary between two fluids.

It is worthwhile to contrast the forms of the two transition relations at a material boundary—continuity of velocity and continuity of stress in the two extreme cases in which the medium on one side of the boundary is either wholly rigid or of negligible density and viscosity . At a fluid–solid interface, both normal and tangential components of velocity are continuous across the boundary, so that if the velocity of the rigid side is given we then have an applicable boundary condition for the distribution of velocity in the fluid. However, the stress in a rigid side is unknown and no applicable boundary condition for the stress distribution in the fluid is available.

The other extreme case may be typified as a liquid–gas interface, the density and viscosity of a gas being much smaller than those of a liquid under normal conditions. It is evident from the form of the Navier–Stokes equation Eq. (2.1.15) that the magnitude of pressure variation in a fluid diminishes with ρ and μ, so that, provided the velocities and velocity derivatives in the gas and liquid are of comparable magnitude, the pressure variations in the gas are much smaller than those in the liquid; and the frictional stresses are likewise smaller in the gas. As an approximation the stress everywhere in the gas may be taken as $-p_0\delta_{ij}$, where p_0 is the uniform gas pressure. Equating the jump of stress across the interface to the normal force due to surfsce tension therefore yields the following approximate boundary conditions for the flow in the liquid:

$$e_{ij} t_i n_j = 0, \qquad (2.1.23)$$

$$p - 2\mu \left(e_{ij} n_i n_j - \frac{1}{3}\Delta \right) = p_0 - \gamma(R_1^{-1} + R_2^{-1}) \qquad (2.1.24)$$

at each point of the interface; and it will be possible to put $\Delta = 0$ in view of the effective incompressibility of liquid. The relation Eq. (2.1.23) and Eq. (2.1.24) are appropriate to what is called a free surface of the liquid. The condition of continuity of velocity across the interface is not useful here because the velocity distribution in the gas is not of interest and may remain unknown.

Further insight into the effect of surface forces on the motion of a fluid may be obtained by considering the energy balance for the fluid of volume τ contained within a material surface S.

We now calculate the internal energy balance for a mass of homogeneous fluid. The rate at which work is being done on the fluid in the material volume τ is the sum of a contribution

$$\int u_i F_i \rho d\tau$$

from the resultant body force, and a contribution

$$\int u_i \sigma_{ij} n_j dS = \int \frac{\partial(u_i \sigma_{ij})}{\partial x_j} d\tau,$$

from the surface forces exerted at the boundary by the surrounding matter. Thus the total rate of work per unit mass of fluid, acting on the material element, is

$$u_i F_i + \frac{u_i}{\rho}\frac{\partial \sigma_{ij}}{\partial x_j} + \frac{\sigma_{ij}}{\rho}\frac{\partial u_i}{\partial x_j} = u_i \frac{\mathrm{D} u_i}{\mathrm{D} t} + \frac{\sigma_{ij}}{\rho}\frac{\partial u_i}{\partial x_j}$$

in terms of the equation of motion. It will be seen that the first of the two terms arising from the rate of work done by surface forces on an element of the fluid, viz. $\rho^{-1} u_i \partial \sigma_{ij}/\partial x_j$, is associated with the small difference between the stresses on opposite sides of the element and contributes (together with the rate of work done by the body forces) to the gain of kinetic energy of bulk motion of the element, and the second, viz. $\rho^{-1} \sigma_{ij} \partial u_i/\partial x_j$, is associated with the small difference between the velocities on opposite sides of the element and represents the work done for deforming the element without changing its velocity. This work done for deforming the element is manifested wholly as an increase in the internal energy of the fluid. Thus, the rate of change of internal energy per unit mass of the material element of fluid due to the surface forces is

$$\left(\frac{\mathrm{D} E}{\mathrm{D} t}\right)_{\sigma_{ij}} = \frac{\sigma_{ij}}{\rho}\frac{\partial u_i}{\partial x_j} = \frac{\sigma_{ij} e_{ij}}{\rho} = -\frac{p\Delta}{\rho} + \frac{2\mu}{\rho}\left(e_{ij} e_{ij} - \frac{1}{3}\Delta^2\right). \qquad (2.1.25)$$

which is divided into two separate contributions to the work done for deforming the element made by the isotropic or pressure part of the stress and by the viscous part (the deviatoric part) of the stress. The first term on the right-hand-side of Eq. (2.1.25) represents rate of change of compression energy, and is zero for an incompressible fluid. The second term is non-negative, showing that any shearing motion in the fluid is inevitably accompanied by a unidirectional transfer of energy from the kinetic energy to internal energy . And for an incompressible fluid we may write

$$\Phi = \frac{2\mu}{\rho} e_{ij} e_{ij} \qquad (2.1.26)$$

for the rate of dissipation of mechanical energy, per unit mass of fluid, due to viscosity

In the above introductory exposition of the dynamics of viscous fluid, we have described the equations governing an incompressible viscous fluid with suitable boundary conditions. We are now prepared for an analysis of the motion of fluid. However, a set of governing equations is too complicated for a direct mathematical approach. Basic difficulty lies in the non-linearity of the governing equations of fluid in motion. The feature of the Navier–Stokes equation Eq. (2.1.18) which causes most analytical difficulties is the nonlinearity in the velocity u arising in the expression Eq. (2.1.13) for the acceleration of a fluid element in the Eulerian specification of the flow field. Progress in theoretical fluid dynamics has been made hitherto mainly by approximations, and the viscous flow approximation which is the theoretical basis of the mechanics of aerosols is one of the most successful approximations. In the viscous flow approximation, the nonlinear term of velocity u is either exactly equal to or approximately to zero. The nonlinear Navier–Stokes equation is thus reduced to linear equation. In these circumstances, the flow field possesses additivity. And for some cases exact analytical solution for the velocity distribution and the pressure distribution can be obtained. Some of the solutions are very important for the mechanics of aerosols. From the next section to Chapter 3 we will discuss them.

2. Two Exact Viscous Flow Fields

We first consider the viscous flow in which the nonlinear fluid inertia force $\mathbf{u} \cdot \nabla \mathbf{u}$ is exactly equal to zero. The simplest of such cases is the unidirectional flow s in which the velocity vector has the same direction everywhere, and is thus independent of the distance in the flow direction. The convective rate of change of velocity then vanishes identically and the acceleration of a fluid element is equal to $\partial \mathbf{u}/\partial t$, with only one non-zero component.

We put the Cartesian coordinate system such that the x-axis is in the direction of u. Then the velocity components v, w equal zero , whence the y- and z-components of the equation of motion Eq. (2.1.18) is

$$\frac{\partial p}{\partial y} = 0, \qquad \frac{\partial p}{\partial z} = 0, \qquad (2.2.1)$$

and

$$p = p(x, t), \qquad (2.2.2)$$

where p stands for the modified pressure as explained in section 1of this chapter. The x-component of the equation of motion is

$$\rho \frac{\partial u}{\partial t} = -\frac{\partial p}{\partial x} + \mu \left(\frac{\partial^2 u}{\partial y^2} + \frac{\partial^2 u}{\partial z^2} \right), \qquad (2.2.3)$$

and since neither the first nor the last term depends on x, we can write

$$\frac{\partial p}{\partial x} = -G(t). \qquad (2.2.4)$$

When G is positive, the pressure gradient represents a body-force in the direction of the positive x-axis.

In the cases of steady unidirectional flow to be considered in this section, $\partial u/\partial t = 0$, $-G$ is a constant pressure gradient, and Eq. (2.2.3) becomes

$$\frac{\partial^2 u}{\partial y^2} + \frac{\partial^2 u}{\partial z^2} = -\frac{G}{\mu}. \tag{2.2.5}$$

The fluid density ρ does not appear in Eq. (2.2.5), since the total acceleration of every fluid element is zero.

It is now necessary to solve Eq. (2.2.5), subject to the boundary conditions, which in general will prescribe the value of the pressure gradient $-G$ and the value of u at certain values of y and z.

2.1. Poiseuille Flow

The case of flow in a long tube with round section under a steady pressure with difference imposed at the two ends of the tube was studied by Hagen in 1839 and by Poiseuille in 1840. In a cylindrical coordinate system whose origin coincides with the centre of the round section, the equation of motion Eq. (2.2.5) becomes

$$\frac{1}{r}\frac{d}{dr}\left(r\frac{du}{dr}\right) = -\frac{G}{\mu}, \tag{2.2.6}$$

since the flow field is axially symmetric about the axis of the long tube. The corresponding solution of Eq. (2.2.6) is

$$u = \frac{G}{4\mu}\left(-r^2 + A\log\frac{r}{r_0} + B\right). \tag{2.2.7}$$

There is a singularity at r=0 unless A=0, so we just choose this value for the integral constant A. According to the no-slip condition at the tube boundary $r = a$, we know that

$$u = 0 \qquad \text{at} \qquad r = a, \tag{2.2.8}$$

hence we find

$$u = \frac{G}{4\mu}(a^2 - r^2), \tag{2.2.9}$$

A quantity of practical importance is the flux volume passing any section of the tube, its value is

$$Q = \int_0^a u2\pi r dr = \frac{\pi G a^4}{8\mu} = \frac{\pi a^4(p_0 - p_1)}{8\mu l}, \tag{2.2.10}$$

where p_0 and p_1 are the pressures at the inlet and outlet of a portion of the tube with length l. By experiments with water Hagen and Poiseuille established that the flux varies by the first power of the pressure drop down the tube and by the fourth power of the tube radius. The accuracy that Q/a^4 is constant observed in experiments is a strong evidence for the assumption that 'no-slipping' exists at the tube wall and for the hypothesis that the viscous stress varies linearly with rate of strain. The relation Eq. (2.2.10) thus provides a method of measuring the viscosity of the fluid.

We see also from Eq. (2.1.26) that the rate at which mechanical energy per unit mass of fluid is dissipated by viscosity , is here

$$\Phi = \frac{\mu}{\rho} \left(\frac{du}{dr} \right)^2 = \frac{G^2 r^2}{4 \rho \mu}. \tag{2.2.11}$$

Thus the total instantaneous rate of dissipation in the fluid which fills the length l of the round tube is

$$\int_0^a \Phi \rho 2\pi l r \, dr = \frac{\pi l G^2 a^4}{8\mu} = lQG. \tag{2.2.12}$$

It is noted that the dissipation of mechanical energy is not proportional to viscosity μ, but is inversely proportional to μ. The result is contrary to expectation. At first sight, it would be proportional to μ, but at the same time it is proportional to the square of the velocity gradient which is inversely proportional to μ. This leads to a result in which Φ is inversely proportional to μ.

2.2. Plane Poiseuille Flow and Couette Flow

The governing equation is here

$$\frac{d^2 u}{dy^2} = -\frac{G}{\mu}, \tag{2.2.13}$$

and without losing of generality we may suppose that the flow takes place between the planes $y = 0$ and $y = d$. If the two bounding planes are rigid and move in the x-direction with speed 0 and U respectively, the no-slip condition s at the two planes are then

$$u = 0 \qquad at \qquad y = 0, \tag{2.2.14}$$

$$u = U \qquad at \qquad y = d. \tag{2.2.15}$$

The solution of Eq. (2.2.13) with boundary conditions Eq. (2.2.14) and Eq. (2.2.15) is

$$u = \frac{G}{2\mu} y(d - y) + \frac{Uy}{d}. \tag{2.2.16}$$

When the two rigid planes are not in relative motion, i.e. $U = 0$, the flow reduces to two-dimensional Poiseuille flow, and the velocity profile is parabolic. When the applied pressure gradient G is zero, we obtain then Couette flow—a simple shearing motion with a linear profile, namely each thin layer of fluid is in steady motion under the action of equal and opposite frictional forces on its two sides.

At the end of this section we will give a remark on the stability of the steady unidirectional flow . Most steady unidirectional flow s are found in practice to be unstable under certain conditions. The simple case of Poiseuille flow was in fact the first flow field to be used for the purpose of systematically investigating the phenomenon of hydrodynamic instability. Reynolds experimentally established in 1883 that, provided the flux of fluid along the tube was sufficiently small, the flow field Eq. (2.2.9) characterized by a parabolic distribution was realized, and accidental disturbances were obliterated; at higher speeds, however, the flow showed intermittent oscillations and ultimately became permanently unsteady and wildly irregular (the phenomenon of turbulence). Although the conditions under which the various steady unidirectional flow s are stable, and consequently capable of being set up in practice, are not known with precision in all cases, it is generally true that these flows are stable for small enough values of the dimensionless number UL/ν, where U and L are the velocity and the length representing the motion and (lateral) extent of the flow . Reynolds' estimate of the critical value of this parameter for the case of Poiseuille flow (with L equal to the tube diameter and U equal to the mean velocity over the cross-section) was about 6400, implying that he was able to observe steady unidirectional flow of water at 20^0C for values of UL less than about 64cm^2/sec. Thus here, as in other cases, the conditions under which stable steady unidirectional flow occurs are within the range of conditions encountered in the nature and in the laboratory. In the analyses of the filtering efficiency of fibre filters in the mechanics of aerosols, the flow in the capillary made up by the gap between straight fibres of the filter is usually assumed to be Poiseuille flow under small aperture conditions, i.e. $h/R < 1$, h being the radius of the gap between fibres, and R being the radius of fibres (Fuchs 1964[2]). From the view point of stability of Poiseuille flows, it is permissible to make such an assumption.

3. Setting up of Viscous Flows

As shown in section 2 of this Chapter, the pressure gradient in unidirectional flow is in the direction of the streamlines and is a function of t only. In nearly all the cases of unidirectional flow to be considered in this section the fluid motion is due entirely to some kind of unsteady motion of the boundaries, the pressures far upstream and far downstream remain equal throughout the motion. Thus $G = 0$ for such cases, and Eq. (2.2.3) reduces to

$$\frac{\partial u}{\partial t} = \nu \left(\frac{\partial^2 u}{\partial y^2} + \frac{\partial^2 u}{\partial z^2} \right). \qquad (2.3.1)$$

This 'diffusion' equation is exactly the same as that governing the two-dimensional distribution of temperature in a stationary medium with thermal diffusivity ν, and we can make use of a number of mathematical results which are already obtained in this connection. The solution at any time t is obtained in terms of the distribution u $(y, z, 0)$ at the initial instant $t=0$ as

$$u(y,z,t) = \frac{1}{4\pi\nu t} \int \int_{-\infty}^{\infty} u(y',z',0)\exp\left\{ -\frac{(y-y')^2}{4\nu t} - \frac{(z-z')^2}{4\nu t} \right\} \mathrm{d}y'\mathrm{d}z'. \qquad (2.3.2)$$

3.1. Setting up of Couette Flow

Suppose now that the fluid is bound by two rigid boundaries at $y=0$ and $y=d$ and is initially at rest, and then the fluid motion is due to that the lower plate is suddenly set to the steady velocity U along its own plane, meanwhile the upper plate being held stationary. The governing equation of fluid motion is

$$\frac{\partial u}{\partial t} = \nu \frac{\partial^2 u}{\partial y^2}, \qquad (2.3.3)$$

with the initial condition

$$u(y,0) = 0 \qquad \text{for} \quad 0 < y \le d, \qquad (2.3.4)$$

and the boundary conditions

$$u(0,t) = U, \ u(d,t) = 0 \qquad \text{for} \quad t > 0. \qquad (2.3.5)$$

The appropriate solution of Eq. (2.3.3) may conveniently be found by introducing a new variable

$$w(y,t) = U(1 - \frac{y}{d}) - u, \qquad (2.3.6)$$

which is the difference between the Couette flow (the steady asymptotic form of u) and u. The equation of motion Eq. (2.3.3) with the initial condition Eq. (2.3.4) and the boundary condition Eq. (2.3.5) becomes

$$\frac{\partial w}{\partial t} = \nu \frac{\partial^2 w}{\partial y^2}, \qquad (2.3.7)$$

$$w(y,0) = U\left(1 - \frac{y}{d}\right), \qquad (2.3.8)$$

$$w(0,t) = w(d,t) = 0. \qquad (2.3.9)$$

The boundary conditions now are transformed to the homogeneous boundary conditions at $y=0$ and $y=d$. A particular solution for w which satisfies these two boundary conditions is found to be

$$w_n = A_n \exp\left(-n^2 \pi^2 \frac{\nu t}{d^2}\right) \sin \frac{n\pi y}{d},$$

where n is an integer. We now try to satisfy the condition on w at $t=0$ by using the whole set of the solutions, that is, we seek values of the integral constants A_n such that

$$\sum_{n=1}^{\infty} A_n \sin \frac{n\pi y}{d} = w(y,0) = U\left(1 - \frac{y}{d}\right).$$

This requires

$$A_n = \frac{2}{d} \int_0^d U\left(1 - \frac{y}{d}\right) \sin \frac{n\pi y}{d} dy = \frac{2U}{\pi n}. \qquad (2.3.10)$$

Thus the velocity distribution is given by

$$u(y,t) = U\left(1 - \frac{y}{d}\right) - \frac{2U}{\pi}\sum_{n=1}^{\infty}\frac{1}{n}\exp\left(-n^2\pi^2\frac{\nu t}{d^2}\right)\sin\frac{n\pi y}{d}; \qquad (2.3.11)$$

the form of this Fourier series reflects the discontinuity in u with respect to y at $y=0$ when $t=0$.

As was to be expected, the velocity tends asymptotically to Couette flow which is a steady flow between two rigid planes in relative motion (see section 2of this chapter). The rapidity increases with n, and the first term ($n=1$) survives longer than others. As soon as this first term dominates the series, the declination from the Couette flow (the asymptotic steady state) decays approximately exponentially, with a 'half-life' equal to $d^2/(\pi^2\nu)$.

3.2. Setting up of Poiseuille Flow in a Pipe

Second, we consider a case in which the flow is due, not to a moving boundary, but to a pressure gradient. The fluid contained in a long pipe of round cross-section is initially at rest, then is set to motion by a difference between the suddenly imposed pressures at the two ends of the pipe and maintained by external means. This pressure difference produces immediately a uniform axial pressure gradient, $-G$ say, throughout the fluid, and so the equation to be satisfied by the axial velocity u is

$$\frac{\partial u}{\partial t} = \frac{G}{\rho} + \nu\left(\frac{\partial^2 u}{\partial r^2} + \frac{1}{r}\frac{\partial u}{\partial r}\right). \qquad (2.3.12)$$

where G is a constant. The boundary and initial conditions are

$$u(a,t) = 0 \qquad \text{for all } t, \qquad (2.3.13)$$

$$u(r,0) = 0 \qquad \text{for} \quad 0 \le r \le a. \qquad (2.3.14)$$

Eq. (2.3.12) is different from Eq. (2.3.1) by an inhomogeneous term G/ρ. The asymptotic velocity distribution is in parabolic form of Poiseuille flow Eq. (2.2.9). Then, Eq. (2.3.12) can be made homogeneous by using the declination of the velocity from its steady asymptotic form Poiseuille flow as a dependent variable w. With the new variable w given as

$$w(r,t) = \frac{G}{4\mu}(a^2 - r^2) - u, \qquad (2.3.15)$$

the equation to be solved becomes

$$\frac{\partial w}{\partial t} = \nu\left(\frac{\partial^2 w}{\partial r^2} + \frac{1}{r}\frac{\partial w}{\partial r}\right), \qquad (2.3.16)$$

with

$$w(a,t) = 0 \qquad \text{for all } t, \qquad (2.3.17)$$

$$w(r,o) = \frac{G}{4\mu}(a^2 - r^2) \qquad \text{for } 0 \leq r \leq a. \tag{2.3.18}$$

A particular solution of this equation which satisfies the boundary condition at $r = a$ is

$$J_0\left(\lambda_n \frac{r}{a}\right) \exp\left(-\lambda_n^2 \frac{\nu t}{a^2}\right),$$

where J_0 is the Bessel function of the first kind of order zero and λ_n is the positive roots of $J_0(\lambda) = 0$. An appropriate superposition of the terms of this complete solution set can also meet the condition at $t=0$. Thus w is given by the Fourier–Bessel series

$$w(r,t) = \frac{G}{4\mu} \sum_{n=1}^{\infty} A_n J_0\left(\lambda_n \frac{r}{a}\right) \exp\left(-\lambda_n^2 \frac{\nu t}{a^2}\right),$$

where the coefficients A_n are taken to satisfy

$$a^2 - r^2 = \sum_{n=1}^{\infty} A_n J_0\left(\lambda_n \frac{r}{a}\right), \tag{2.3.19}$$

that is,

$$A_n = \frac{2a^2}{J_1^2(\lambda_n)} \int_0^1 x(1 - x^2) J_0(\lambda_n x) dx$$

$$= \frac{8a^2}{\lambda_n^3 J_1(\lambda_n)} \tag{2.3.20}$$

The velocity distribution is then given as

$$u(r,t) = \frac{G}{4\mu}(a^2 - r^2) - \frac{2Ga^2}{\mu} \sum_{n=1}^{\infty} \frac{J_0(\lambda_n \frac{r}{a})}{\lambda_n^3 J_1(\lambda_n)} \exp\left(-\lambda_n^2 \frac{\nu t}{a^2}\right). \tag{2.3.21}$$

Calculations of $u(r,t)$ have shown that u approaches its asymptotic Poiseuille distribution $\frac{G}{4\mu}(a^2 - r^2)$ very quickly. The difference between u and its limit form is small as long as $t\nu/a^2 = 0.75$. Hence the smaller the radius a of the pipe is, the quicker u approaches the Poiseuille flow. If $a = 10\mu$m, then the time needed to approach Poiseuille flow is only 5μsec. Therefore, in analyzing the filtering efficiency of the fibre filter in the mechanics of aerosols, we have every reason to assume the flow field in the gap between the fibres of the filter to be Poiseuille flow.

4. Similarity in the Mechanics of Viscous Fluids and Aerosol Dynamics

The motion of an incompressible viscous fluid which is effectively of uniform density ρ is governed by the equations

$$\rho\left(\frac{\partial u_i}{\partial t} + u_j \frac{\partial u_i}{\partial x_j}\right) = -\frac{\partial p}{\partial x_i} + \mu \frac{\partial^2 u_i}{\partial x_j \partial x_j}, \tag{2.4.1}$$

$$\frac{\partial u_i}{\partial x_i} = 0, \tag{2.4.2}$$

where p is the modified pressure . We propose to consider the effect of the values of the parameters ρ and μ on the flow. To see this it is useful to write these equations in terms of dimensionless variables, so that the effect of changing the values of ρ and μ is dissociated from the effect of mere changes of units. No parameters with the dimensions of length and velocity occur in the above equations, so that we must look to the boundary and initial conditions with dimensional quantities, namely substitute dimensionless variables for x and u.

Let us suppose that the specification of the boundary and initial conditions for a particular flow involves some representative length L (which might be the maximum diameter of an interior boundary, or the distance between enclosing boundaries, for instance, the diameter of a moving spherical particle, or the distance of the gap between two rigid spheres close to each other) and some representative velocity U (which for example, might be the steady speed of a rigid boundary, the steady speed of a moving rigid spheres, or the steady relative velocity of two rigid spherical particles in relative motion), in such a way that these conditions can be expressed in the dimensionless forms

$$u' = \text{given function of } t' \text{ at some given } x',$$

$$u' = \text{given function of } x' \text{ at some given } t',$$

where

$$u' = \frac{u}{U}, \quad t' = \frac{tU}{L}, \quad x' = \frac{x}{L}$$

Then with these new variables, and

$$p' = \frac{p - p_0}{\rho U^2},$$

where p_0 is some representative value of the (modified) pressure in the fluid, the governing equations become

$$\frac{\partial u_i'}{\partial t'} + u_j' \frac{\partial u_i'}{\partial x_j'} = -\frac{\partial p'}{\partial x_i'} + \frac{1}{Re} \frac{\partial^2 u_i'}{\partial x_j' \partial x_j'}, \tag{2.4.3}$$

$$\frac{\partial u_i'}{\partial x_i'} = 0, \tag{2.4.4}$$

where $Re = \frac{\rho L U}{\mu}$, and is the Reynolds number already defined in Chapter 1.

The equations now contain explicitly only the dimensionless parameter Re, and the solution for the dependent variables u' and p' that satisfies boundary conditions can depend only on

1 the independent variables x' and t',
2 the Reynolds number Re, and

3 the dimensionless ratios specifying the boundary and initial conditions, all these
ratios can be described as specifying the 'geometry' of the boundary and initial
conditions.

This substitution to dimensionless variables which is a seemingly superficial step,
is very revealing. It shows firstly that, once the solution for a particular flow field is
known and is expressed in dimensionless form, a triply-infinite family of solutions may
be obtained from it by choosing values of ρ, L, U and μ in such a way that the value
of Re keeps unchanged. All those flows that satisfy the same boundary and initial
conditions when expressed in dimensionless form, and for which the corresponding
values of ρ, L, U and μ differ without the values of the combination $\rho LU/\mu$ being
different, are described by one and the same non-dimensional solution; and all such
flows are said to be dynamical similar, since the magnitudes of the various terms in
the equation of motion representing the forces (viscous, pressure, and inertia) acting
at a given non-dimensional position and instant in the fluid are in the same ratio in
all such flows.

This principle of dynamical similarity is widely used as means for obtaining information about an unknown flow field from 'model test'. By that , one can achieve
such information from specially designed experiments which are carried out under
practical physical conditions instead of solving the unknown flow field theoretically.
For instance, the scientists who are working in the area of the mechanics of aerosols
often wish to predict the velocity with which small solid or liquid particles will separate from an aerosol system, and, at the beginning, they need to know the terminal
velocity of an isolated small particle of known size and density and of simplified
shap—spherical— falling through air. Direct measurement on the fall of a single
particle is difficult, because the very small size of aerosol particles makes observation
awkward. Dynamical similarity can now be used to argue that the flow about the
falling spherical particle is the same, when expressed non-dimensionally in terms of
the velocity U and diameter L of the sphere, as that about a much larger sphere
moving at such a speed and through such a fluid that $\rho LU/\mu$ has the same value as
in the first case. The value of μ/ρ for lubricating oil is about 31 times that for air,
and that for glycerine it is about 53 times; thus a dynamically similar flow field can
be obtained in one of these liquids with larger spheres which are easier to be observed
in experiments, and the retarding or 'drag ' force D on the sphere due to the fluid
can be observed for a number of values of L and U. The relation

$$D = -\int m_i \sigma_{ij} n_j \mathrm{d}A$$

$$= -\rho U^2 L^2 \int m_i \left\{ -p'\delta_{ij} + \frac{1}{Re}\left(\frac{\partial u_i'}{\partial x_j'} + \frac{\partial u_j'}{\partial x_i'}\right) \right\} n_j \mathrm{d}A, \qquad (2.4.5)$$

where the unit vector $\mathbf{m}$ specifies the direction of the sphere's motion, the integration
is over the surface area A of the sphere, and $\mathrm{d}A' = \mathrm{d}A/L^2$, shows that the dimensionless 'drag coefficient' $D/\rho U^2 L^2$ is the same for all the dynamically similar flow
fields corresponding to the given value of Re, and the model test provides values of

this drag coefficient for Re covering the value range which is appropriate to a falling aerosol particle. The terminal velocity of the aerosol particle can then be calculated from its known size and density.

The Eq. (2.4.3) and Eq. (2.4.4) in dimensionless variables reveal that, for a given geometry of the boundary and initial conditions, there is no more than a single-infinite family of different solutions in the dimensionless form, different members of the family corresponding to different values of Re. In other words, for a given geometry of the boundary and initial conditions, the effect on a flow field of various ρ, L, U or μ can be described uniquely by a change of Re. The fact that Re is the parameter determining a flow field for boundaries with a given form was first recognized by Stokes (1851), although the later work by Reynolds (1883) on the onset of turbulence in flow through tubes led it being termed the Reynolds number .

The relation Eq. (2.4.5) for the drag on a moving object may thus be expressed in a general form

$$\frac{D}{\rho U^2 L^2} = \text{function of } Re \text{ alone,} \qquad (2.4.6)$$

which is valid for the flow fields with geometrically similar boundary and initial conditions. All other dimensionless parameters of the flow are likewise functions of Re. A practical problem in the dynamics of an incompressible uniform viscous fluid thus frequently reduces to theoretically or experimentally determine the form of the relevant unknown function of Re over a certain range of values of Re.

The magnitude of the Reynolds number Re may be regarded as an estimate of the relative importance of the non-viscous and viscous forces acting on a unit volume of the fluid. The right-hand-side of the equation of motion Eq. (2.4.1) contains the pressure force $-\partial p/\partial x_i$ and the viscous force $\mu \partial^2 u_i/\partial x_j \partial x_j$, and the sum of the two equals minus of the so-called inertia force $(-Du_i/Dt)$. These three forces together are in equilibrium, and the balance between them can be expressed by the ratio of any two. Since the pressure force usually plays a passive role, being set up in the fluid as a consequence of motions of a rigid boundary or of the existence of frictional stresses (although this is not so in cases of flow due to an applied pressure gradient, such as Poiseuille flow), it is customary to characterize the flow by the ratio of the magnitudes of the inertia and viscous forces. At any point in the fluid this ratio is

$$\frac{|\rho Du_i/Dt|}{|\mu \partial^2 u_i/\partial x_j \partial x_j|} = Re \frac{|Du_i'/Dt'|}{|\partial^2 u_i'/\partial x_j' \partial x_j'|}.$$

Thus if each of $|Du_i'/Dt'|$ and $|\partial^2 u_i'/\partial x_j' \partial x_j'|$ is of order unity, which is likely , if the flow field is a simple one and L and U are truly representative parameters (although there are usually special places in the fluid where these dimensionless quantities are very small), Re measures the relative magnitude of the inertia and viscous forces. For a given geometry of the boundary and initial conditions, changes in Re correspond to changes in the relative magnitude of inertia and viscous forces—although again the argument is not rigorous , because Du_i'/Dt' and $\partial^2 u_i'/\partial x_j' \partial x_j'$ are themselves dependent on Re and we must assume that they remain at order of unity. In particular, the effect

of making $Re \ll 1$ is to make the inertia force much smaller than the viscous force, so
that pressure and viscous forces are dominant in the flow fluid, conversely the effect
of making $Re \gg 1$ is to make the inertia force much greater than the viscous force,
so that inertia and pressure forces are dominant. At values of Re of order unity, all
the three forces presumably are important in the equation of motion. It is necessary
to indicate at this point that this feature of the Reynolds number (and also other
similarity-parameter which will be discussed later in this section) has very important
significance. In theoretical analyses of the fluid mechanics this feature has become a
very powerful tool to simplify the governing equation of the fluid motion, being the
main basis of various perturbative approximations. A complete set of perturbation
method in the fluid mechanics has since been developed, and thus pushes the modern
study on the fluid mechanics forward. This has exceeded which Stokes or Reynolds
might have expected when they proposed the characteristic number.

   None of the flow fields investigated hitherto in this chapter illustrates these general
remarks very well, because they are all specially simple in one way or another. In
the case of steady unidirectional flow , the inertia force is everywhere identically zero,
so that there is no possibility of affecting the force balance by changing any of the
boundary parameters; nor is there in the cases of unsteady unidirectional motion
caused by moving boundaries, because the pressure force is zero everywhere. If a flow
is generated from rest due to a steady motion of a planar boundary with a velocity
U, the velocity distribution (see Eq. (2.3.16)) can be written as

$$\frac{u(y,t)}{U} = 1 - \frac{y}{d} - \frac{2}{\pi} \sum_{n=1}^{\infty} \frac{1}{n} \exp\left(-\pi^2 n^2 \frac{tU}{d} \frac{1}{Re}\right) \sin\left(\pi n \frac{y}{d}\right),$$

where $Re = \rho dU / \mu$. This has the general form expected in the case in which one
length and one velocity are needed to specify the boundary conditions. As already
remarked, the pressure forces are zero everywhere, so that the inertia and viscous
forces are of equal magnitude everywhere, no matter what the value of Re may be. In
these circumstances, a change of Re is entirely equivalent in its effect to the change
of the time scale, larger values of Re correspond to a slower approach to the ultimate
steady state—Couette flows.

   There are many other dynamical similarity -parameters in fluid mechanics. We
will here introduce some of them which are of special importance in the mechanics of
aerosols.

   For the problem of mass transfer to or from an aerosol particle the process of it
can be described by the conservative convective diffusion equation

$$\frac{\partial C}{\partial t} + u_j \frac{\partial C}{\partial x_j} = D_m \frac{\partial^2 C}{\partial x_j \partial x_j}, \qquad\qquad (2.4.7)$$

where C denotes the concentration of the diffusible quantity in fluid (e.g. vapor
concentration), D_m the molecular diffusivity (e.g. vapor diffusivity D_v) and u_i is
the flow field around the aerosol particle. There are no representative length L and
representative velocity U in Eq. (2.4.7). However we can determine them from the

boundary and initial conditions too. For the falling spherical particles under gravity through air which is otherwise at rest , the diameter L of the particle may be the representative length of the flow field and the terminal velocity of it may be the representative velocity. The governing Eq. (2.4.7) reduces to its dimensionless form with these representative parameters

$$\frac{\partial C'}{\partial t} + u'_j \frac{\partial C'}{\partial x'_j} = \frac{1}{Pe} \frac{\partial^2 C'}{\partial x'_j \partial x'_j},$$ (2.4.8)

where

$$\mathbf{x}' = \frac{\mathbf{x}}{L}, \ t' = \frac{tU}{L}, \ \mathbf{u}' = \frac{\mathbf{u}}{U} \ \text{and} \ C' = \frac{C - C_0}{C_1 - C_0},$$

In the expression C_0 is the steady vapor concentration far away from the particle, C_1 is the saturated vapor concentration at the surface of the particle. The dimensionless parameter Pe in Eq. (2.4.8) is the Péclet number in the process of mass transfer and is defined as

$$Pe = \frac{UL}{D_m}.$$ (2.4.9)

The process is a condensation process when $C_0 \gg C_1$ (or evaporation if $C_0 \ll C_1$). D_m thus becomes D_v—the vapor molecular diffusivity—and the Péclet number is P_{ev} which has already been discussed in Chapter 1. Because of the dependence of $\mathbf{u}$ on Re, the solution of C' will be determined both by Pe and by Re. Then we have

$$C' = f(\mathbf{x}', t; Pe, Re).$$ (2.4.10)

Besides the Péclet number and the Reynolds number there is another dimensionless parameter which is termed as the Schmidt number (see Chapter 1) defined as

$$Sc = \nu / D_m.$$ (2.4.11)

Of course, these three dimensionless parameters are not independent of each other. The relation among Pe, Re and Sc is read as

$$Pe = Sc \cdot Re.$$ (2.4.12)

According to the dynamical similarity principle, as discussed in the above paragraphs of this section, for a given geometry of boundary and initial conditions and for a flow field with a given Reynolds number the dimensionless mass transfer rate Nusselt number (Nu) is determined by the Péclet number only, i.e.

$$Nu = \frac{Q}{Q_0} = f(Pe),$$ (2.4.13)

where Q is the dimensional mass transfer rate, viz.

$$Q = -\int_A D_m \nabla C \cdot \mathbf{n} dS,$$ (2.4.14a)

where **n** is the unit vector outward normal to the surface of the particle, and Q_0 is a characteristic value of Q, which is usually taken as the zero-Péclet number mass transfer rate, i.e.

$$Q_0 = 4\pi a D_v (C_1 - C_0). \qquad (2.4.14b)$$

where a is the radius of the falling aerosol particle.

It is similar to the above exposition in which we have argued about the dynamical significance of the Reynolds number . In general we may think that the magnitudes of $|DC'/Dt'|$ and $|\partial^2 C'/\partial x'_j \partial x'_j|$ are both of order unity. Thus, the Péclet number is the ratio of the convective term to the diffusive term in Eq. (2.4.7). As a result, the changes of the values of the Péclet number correspond to the changes in the relative magnitude of convective term and diffusive term—although again the argument is not rigorous, because DC'/Dt' and $\partial^2 C'/\partial x'_j \partial x'_j$ are themselves dependent on Pe and we must assume that they remain of order unity. Fortunately experiments have shown the validity of the assumption in general. Therefore, the diffusive term in Eq. (2.4.7) is dominant when $Pe \ll 1$ and as the first order of approximation the convective term can be neglected. This is the case for which we have already seen in Chapter 1. On the other hand, the convective term is dominant if $Pe \gg 1$ and the diffusive term can be neglected in general as the first order of approximation . Although Eq. (2.4.8) is a linear equation, the difficulty in solving this kind of equations is very similar to the nonlinear equation, since the coefficient **u** is a four-dimensional function of time-space . Thus, the dynamical significance of the Péclet number makes itself a powerful tool in solving equation Eq. (2.4.8). We can use the perturbation method developed in the fluid mechanics to deal with the problem of mass transfer too. And the Péclet number plays a very important role in solving the convective-diffusion equation just like the Reynolds number 's role in solving the Navier–Stokes equation. It is needed to point out that the magnitude of Sc is of order unity in aerosol systems. For example, $Sc = 0.6$ for the case of vapour diffusion through air. Thus, the order of the magnitude of P_{ev} is always the same as the order of the magnitude of Re in aerosol just as indicated in Table 3of Chapter 1. However, it is not so for certain cases of hydrosol systems. The diffusivity of some molecule or ion in hydrosol systems is of order of as small as $10^{-5} \text{cm}^2/\text{sec}$, even $10^{-6} \text{cm}^2/\text{sec}$ for some large molecule. Hence, the order of the magnitude of Sc can be as large as $10^3 - 10^4$ for certain hydrosol systems. Their Péclet number can be still larger than unity even for the Reynolds number being as small as 10^{-2}. This constitutes a special problem of diffusive processes at low Reynolds number but high Péclet number in hydrosols.

Table 3 of Chapter 1 also gives the Péclet number P_{eH} of heating or cooling processes, as well as the Péclet number P_{eB} of the particle transfer processes. The definitions of these Péclet number s are similar to the Péclet number of mass transfer

Eq. (2.4.7) can be changed into a convective-conduction equation by changing concentration C into temperature T, and the diffusivity D_m into heat conductivity D_H. The Péclet number Pe defined as Eq. (2.4.9) becomes P_{eH} if we replace D_m in Eq. (2.4.9) by D_H. It is also given in Table 3 of Chapter 1. The dimensionless heat transfer rate is stilled termed as the Nusselt number (Nu) which is defined similarly by equation Eq. (2.4.13) where Q is the dimensional heat transfer rate and Q_0 is usually the zero-Péclet number heat transfer rate, $Q_0 = 4\pi a D_H(T_1 - T_0)$ for a spherical particle. For a given geometry of the boundary and initial conditions, and a given flow field with Reynolds number prescribed, Nu is determined by P_{eH} only. The third dimensionless parameter is named as the Prandtl number (Pr) which is defined by ν/D_H. The relation between P_{eH}, Pr and Re is

$$P_{eH} = Pr \cdot Re. \qquad (2.4.15)$$

For the case of heat transfer in aerosol system, $Pr = 0.7$. Therefore, the magnitudes of P_{eH} shown in Table 3 of Chapter 1 are similar to the Reynolds number , which are also small. However, some exceptions exist in hydrosol systems. For example, $Pr = 10$ for the case of the medium being water, $Pr = 200$ for the case of the medium being glycerine. In these circumstances, P_{eH} can be larger than unity, while the Reynolds number can be as small as 0.1, even 0.01.

Finally, when D_m in Eq. (2.4.9) is changed into D_0 which is the Brownian diffusivity of aerosol particles, then Eq. (2.4.9) gives the Péclect number P_{eB} in the process of aerosol particle transfer. Generally speaking, it is the ratio of the convective term to the Brownian diffusion term. This is similar to Re, P_{ev} and P_{eH}. What is different from the above cases lies in that the Brownian diffusivity is no longer a constant but is inversely proportional to the radius a of the aerosol particles. Thus the Schmidt number is no longer a constant too, but is proportional to the aerosol particle radius a. When a increases from 0.01μm to 10μm, Sc increases form 10^3 to 10^7. The feature of aerosol particle transfer as indicated in Table 2 of Chapter 1 is then different from the processes of mass transfer and heat transfer . Despite the smallness of the Reynolds number of aerosol particles, the Péclet number P_{eB} can be large than unity for the case of giant nuclei transfer in which process the convective term dominates the Brownian diffusive term. In this range, a special process is formed which is characterized by low Reynolds number and high Péclet number . Of course for the case of Aitken nuclei the process is still at low Reynolds number and low Péclet number .

In the mechanics of aerosols some times we should consider the Stokes number . It is the ratio of the inertia force of an aerosol particle to the resistance force when the particle is in motion. It is termed Stokes number because the resistance force is usually described by Stokes resistance law. For the case of a spherical aerosol particle with radius a and density ρ, the Stokes number is given by

$$Stk = \frac{2\rho a^2}{9\mu}\frac{U}{L}, \qquad (2.4.16)$$

where μ is the viscosity of air, U and L are the representative magnitudes of the velocity and the length of the moving particles respectively. When $Stk \ll 1$, the inertia force of aerosol particles can be neglected as the lowest order approximation, then the applied force is always balanced by the resistance force exerted by the medium. This is just the case for the aerosol particles with radius $a < 10\mu$m. Therefore we will neglect the inertia force of aerosol particles in subsequent chapters.

In Chapter 1 we considered another dimensionless parameter, the Knudsen number (Kn) defined as the ratio of the mean free path of the air molecules to the representative length of the flow field. In the dynamics of aerosol particles the definition of it is λ_m/a where a is the radius of the particles. The dynamical significance of Kn has been discussed in Chapter 1, and we will not repeat again.

5. Viscous Force Dominated Flows

As already remarked, the presence of the nonlinear term $\mathbf{u} \cdot \nabla \mathbf{u}$ in the expression for the acceleration makes solution of the equation of motion very difficult for general but the simplest flow fields. However, flow fields in which $\mathbf{u} \cdot \nabla \mathbf{u}$ can be exactly zero everywhere are rare both in nature and in engineering practice. It happens that in some circumstances of practical interest of aerosol, since the nonlinear term, although not identically zero, is small, and may be neglected as an appropriate approximation. When there is no periodic force exerted on the boundary (e.g. the surface of an aerosol particle), the representative magnitude of time can be accounted by the representative velocity U and representative length L. Then the unsteady inertia force $\partial \mathbf{u}/\partial t$ is of the same order as the nonlinear inertia force of air and can be neglected as an approximation when $Re \ll 1$. This is just the case as indicated in Table 3 of Chapter 1. When there is a periodic force exerted on the surface of the aerosol particle (e.g. aerosol particles moving in a sound field), it is necessary to consider the Strouhal number (Str) which is the ratio of the unsteady inertia force $\partial \mathbf{u}/\partial t$ to the nonlinear inertia force $\mathbf{u} \cdot \nabla \mathbf{u}$, viz.

$$Str = \frac{nL}{U},$$

where n is the frequency of the periodic force. Obviously, only under the conditions in which $Str \cdot Re \sim O(1)$, can $\partial \mathbf{u}/\partial t$ not be neglected. This implies that the frequency n must be as high as

$$n \sim \frac{\nu}{L^2},$$

then $\partial u_i/\partial t$ is of the same order as $\nu \partial^2 u_i/\partial x_j \partial x_j$. For the case of an aerosol particle with radius a being equal to 1μm the unsteady inertia force $\partial u_i/\partial t$ can always be neglected when $n \ll 10^7 \mathrm{sec}^{-1}$. Thus, the whole inertia force $D\mathbf{u}/Dt$ can be neglected in the mechanics of aerosols and the Navier–Stokes equation Eq. (2.1.18) reduces to the Stokes equation

$$\nabla p = \mu \nabla^2 \mathbf{u}, \tag{2.5.1}$$

and

$$\nabla \cdot \mathbf{u} = 0. \tag{2.5.2}$$

If the boundary conditions involve $\mathbf{u}$ alone (this is the case for the mechanics of aerosols), the problem is to find the appropriate solution of

$$\nabla^2 \omega = 0, \quad \nabla \cdot \omega = 0, \tag{2.5.3}$$

where

$$\omega = \nabla \times \mathbf{u}, \tag{2., 5.4}$$

and ω is the vorticity of the flow field, the pressure then can be found from Eq. (2.5.1). On the other hand, if the boundary conditions are given in terms of p alone, the equation to be solved is

$$\nabla^2 p = 0. \tag{2.5.5}$$

Then the velocity **u** can be obtained from Eq. (2.5.1) and Eq. (2.5.2). In either events, the distributions of p and **u** do not depend on μ, and μ determines only the relative magnitudes of **u** and p (or of **u** and a relative pressure $p - p_0$, as more precise). The harmonics thus play a very important role in low-Reynolds-number fluid mechanics. Note that Eq. (2.5.1) is only formally the same as the governing equation in a steady unidirectional flow . However an essential difference exits between these two flows. For the former case, the nonlinear inertia force is approximately equal to zero, whereas it is exactly equal to zero for the later case. Nevertheless we can make use of the solution of the steady unidiectional flow to approximate the solution of Eq. (2.5.1). Here there are some examples.

5.1. Quasi-Poiseuille Flow

In the first example the nonlinear term in the equation of motion is small everywhere because, for essentially geometrical reasons, the velocity **u** varies only slowly along the streamlines. In the case of flow in a cylindrical tube due to an applied pressure difference at two distant ends, $\mathbf{u} \cdot \nabla \mathbf{u} \neq 0$, but since the viscous force is non-zero for the cylindrical tube selecting a sufficiently slow rate of variation of the cross-section we can always make the ratio of $|\mathbf{u} \cdot \nabla \mathbf{u}|$ to the viscous force negligibly small.

Consider first the simple case of steady flow along a round tube whose radius a varies slowly with distance x along the centre-line. A constant difference between the pressures at two distant ends is maintained, and the resulting axial pressure gradient $-G$ will also vary slowly with respect to x. In the neighbourhood of any spot x, within several tube radii upstream and downstream, the tube radius and the axial pressure gradient are approximately uniform with values $a(x)$ and $-G(x)$, and the approximate expression for the axial velocity obtained from the Poiseuille flow Eq. (2.2.9) is

$$u(x,r) = \frac{G}{4\mu}(a^2 - r^2), \tag{2.5.6}$$

where r denotes radial distance from the centre-line. Then if Q is the constant volume flux along the tube

$$Q = \frac{\pi a^4 G}{8\mu}, \tag{2.5.7}$$

Eq. (2.5.6) can be written as

$$u(x,r) = U\left(1 - \frac{r^2}{a^2}\right), \qquad U = \frac{2Q}{\pi a^2}. \tag{2.5.8}$$

The streamlines are not exactly unidirectional but are inclined to the axis at a small angle whose magnitude is of order $da/dx, = \alpha(x)$, so that in addition to the axial component of velocity u there is a radial component v of order αu. It follows from Eq. (2.5.7) that

$$p_1 - p_2 = \frac{8\mu Q}{\pi} \int_{x_1}^{x_2} a^{-4} dx, \qquad (2.5.9)$$

so that if the pressures p_1, p_2 at the two ends $x = x_1, x = x_2$ are given, and the tube geometry is known, Q can be calculated and thence $G(x)$.

The expression Eq. (2.5.8) is evidently a valid approximation for sufficiently small value of α, and we can find a specific condition for its validity that a representative value of the magnitude of each of the terms $\rho u \partial u/\partial x$, $\rho v \partial u/\partial r$ is $\alpha \rho U^2/a$. On the other hand, for the viscous forces $\mu \bigtriangledown^2 u$ we have the representative magnitude $\mu U^2/a$, showing that the solution Eq. (2.5.8) is consistent with that at the situation where the inertia force is neglected, provided

$$\alpha \frac{\rho a U}{\mu} \ll 1. \qquad (2.5.10)$$

Therefore, the nonlinear term can be neglected even for large Reynolds number case if $\alpha \ll 1$. In fact, for the long straight tube of round cross-sections, $\alpha = 0$, and the Reynolds number can be large provided the flow is stable (of course the flow will be unstable and becomes turbulence if Re is larger than a certain critical value).

In Poiseuille flow the Reynolds number has no dynamical significance, the inertia force is zero everywhere, however, the Re cannot be equal to zero. On the other hand, when $Re \ll 1$ (e.g. for the case of flow in the gap between the fibres of an aerosol filter, the velocity of the flow is very slow, the gap is very small, then $Re \ll 1$), the relation Eq. (2.5.10) can be satisfied even for α being quite large. Thus, the inertia force is negligible and the flow can be approximated by the Poiseuille flow. So far, we have seen from the analyses of steady unidirectional flow, unsteady unidirectional flow and flow in a tube with varying cross-section that the Poiseuille flow is a sufficiently good approximation in the analyses of the filtering efficiency of a fibre filter of aerosol particles.

5.2. Lubrication Theory

It is a matter of common experience that two solid objects can easily slide one over another when there is a thin layer of fluid between them and under certain conditions a high pressure may be set up inside the fluid layer. For instance, a sheet of paper dropped onto a smooth floor will often 'float' on a film of air between it and the floor and will be able to glide horizontally for some distance before coming to rest. The fact that high pressure exists in the fluid layer is widely used in engineering practice. In some cases the fluid layer may be used to support a load, and is then called a lubrication bearing. The lubrication theory is an important one in the dynamics of viscous flows, and later we will see that it plays a very important role in the process of coagulation between aerosol particles.

The essence of the phenomenon is that, as the thickness of the fluid layer between the two solid boundaries is so small, the rate of strain and the viscous stress in the fluid layer are very large, this large stress then can be used, by a suitable choice of the configuration of the fluid layer (see below), to develop a large pressure. To see how this is done, we shall consider a simple case of one solid object with a plane surface gliding steadily over another solid object, the surface of the gliding block is of a finite length l in the direction of motion and of a great width so that the motion may be regarded as two dimensional; this case was first analyzed by Reynolds in 1886. It is indicated the necessary condition for producing a large pressure is that the two plane surfaces are slightly inclined to one another. The origin of the rectangular coordinate system will be set at the entrance of the surface of the lower block as fixed relative to the upper solid body; the lower solid surface then moves in its own plane, at speed U, and the whole flow field is steady with respect to these axes.

The thickness d of the fluid layer is small everywhere compared to l, and is d_1 at $x=0$ where the pressure is p_0, $d = d_2$ at $x = l$ where the pressure is p_l. Provided $d_1 > d_2$, one can have $d_2 = d_1 - \alpha l$ and $d = d_1 - \alpha x$ where α is the inclination angle between these two plane surfaces. We then examine the possibility that the velocity distribution near any cross section of the layer where the thickness and pressure gradient have certain values, is approximately the same as in a uniform layer ($\alpha = 0$) with the thickness and the pressure gradient having the same values everywhere. Provided the fluid velocity is of the same order of magnitude as U, which is true as will be seen later, the argument leading to Eq. (2.5.10) applies here. Thus the condition for the two-dimensional Poiseuille and Couettee flow approximation to be valid is

$$\alpha \frac{\rho d U}{\mu} \ll 1. \tag{2.5.11}$$

This requirement is usually satisfied under practical conditions of lubrication . We can therefore proceed to make use of the solutions found for uniform channels in the manner described above.

In the neighbourhood of any spot, where the thickness of the fluid layer is d and the pressure gradient is $-G$, in accordance with Eq. (2.2.16) we have

$$u = \frac{G}{2\mu} y(d - y) + U \left(\frac{d - y}{d} \right). \tag{2.5.12}$$

The volume flux, per unit width of the fluid layer, is

$$Q = \int_0^d u \, dy = \frac{G d^3}{12\mu} + \frac{1}{2} U d, \tag{2.5.13}$$

and Q must be independent of x. This requires the pressure gradient to vary with d according to the relation

$$\frac{dp}{dx} = -G = 6\mu \left(\frac{U}{d^2} - \frac{2Q}{d^3} \right). \tag{2.5.14}$$

Integration of Eq. (2.5.14) gives

$$p - p_0 = \frac{6\mu}{\alpha} \left\{ U \left(\frac{1}{d} - \frac{1}{d_1} \right) - Q \left(\frac{1}{d^2} - \frac{1}{d_1^2} \right) \right\}. \qquad (2.5.15)$$

Now the sliding block may be supposed to be completely immersed in the fluid, with narrow passages for the fluid on one side of the block only, so that the pressures at the two ends are approximately the same, i.e. $p_l = p_o$. This condition enables us to determine Q from Eq. (2.5.15):

$$Q = U \frac{d_1 d_2}{d_1 + d_2}, \qquad (2.5.16)$$

and then the expression for the pressure becomes

$$p - p_0 = \frac{6\mu U}{\alpha} \frac{(d_1 - d)(d - d_2)}{d^2 (d_1 + d_2)}. \qquad (2.5.17)$$

The volume flux and the pressure distribution in the lubrication layer can now be calculated when the sliding velocity U and the inclination angle of the sliding block are known. The pressure increment $p - p_0$ is one-signed throughout the whole layer, and is positive when $d_1 > d_2$. Thus a lubrication layer will generate a positive pressure, and will be able to support a load along the normal of the layer surface, only when the layer is so arranged that the relative motion of the two surfaces tends to drag fluid (by viscous stresses) from the wider to the narrower end of the layer. The pressure increment has a single maximum in the layer at

$$d_{max} = \frac{2 d_1 d_2}{d_1 + d_2},$$

and the maximum increment $p_{max} - p_0$ is given by

$$p_{max} - p_0 = \frac{3\mu U}{2\alpha} \frac{(d_1 - d_2)^2}{d_1 d_2 (d_1 + d_2)} = \frac{3\mu l U}{2} \frac{(d_1 - d_2)}{d_1 d_2 (d_1 + d_2)}. \qquad (2.5.18)$$

From this, we see that $(p_{max} - p_0) \to \infty$ as $d_1 \to 0$ (assuming $(d_1 - d_2)/d_1$ to be of order unity), showing that very high pressures can be set up in very thin films. Note that for all viscous flows due to a moving boundary the contraction configuration of the lubrication layer is an essential condition for generating a positive pressure. If $d_1 = d_2 = d$, then $p = p_0 = $ constant. No positive pressure will be generated and the flow reduces to the Couette flow. The pressure distribution will be uniform no matter how thin the thickness of the fluid layer is. Another example is an important case of a lubrication layer between a rotating circular shaft in a circular bearing of slightly larger radius. The pressure will be uniformly distributed if the shaft and the bearing are concentric and the width of the gap between them is the same everywhere. The shaft and the bearing must be eccentric so that a positive pressure can be generated where the gap is at minimum, and a negative pressure can be generated where the gap is at maximum. The lubrication layer between the shaft and the bearing will thus be able to support a load along the centre-line. This is the 'bearing–principle' which has widely applications in engineering practice. Finally, it is necessary to indicate that the contraction configuration of lubrication layer is no longer an essential condition for the positive pressure to be generated when the flow is due to an applied pressure difference (not due to a moving boundary). Consider the Poiseuille flow in a long straight tube with round cross-section with length l and radius a. The pressures applied at two distant ends are p_0 and p_l respectively. The total pressure difference $\pi a^2(p_o - p_l)$ must be balanced by the total viscous friction exerted at the wall when the motion is a steady flow. Thus,

$$2\pi a l \tau = \pi a^2(p_o - p_l),$$

where τ is the viscous friction at the wall of the tube. Then we have

$$\frac{p_o - p_l}{\tau} = \frac{2l}{a}.$$

Obviously, $p_o - p_l \to \infty$ as $a \to 0$, whereas the cross-section remains constant, and there is not any contraction which occurs along the direction of motion. Therefore, for the case of viscous flow due to an applied pressure difference , the fact that the transverse scale of the viscous flow approaches zero is the only essential condition for generating a very large positive pressure, the contraction of the flow is no longer needed.

CHAPTER 3
SOME LOW-REYNOLDS-NUMBER FLOWS

1. Dynamics of Isolated Particles

When a body with representative linear dimension d is in a steady translational motion, with speed U, through a fluid which is undisturbed, d and $U(=|U|)$ are a representative length and velocity for the flow field as a whole. The inertia forces on the fluid are therefore to be of order $\rho U^2/d$ and the viscous forces to be of order $\mu U/d^2$. The ratio of these two forces is estimated as $\rho dU/\mu, = Re$, so that when $Re \ll 1$ the inertia forces may be negligible. Because of the boundary conditions being given in terms of u alone, we may use the vorticity equations Eq. (2.5.3) and Eq. (2.5.4) and rewrite Eq. (2.5.1) as

$$\nabla\left(\frac{p-p_0}{\mu}\right) = \nabla^2 u = -\nabla \times \omega, \qquad (3.1.1)$$

where p_0 is the uniform pressure far from the body and the right-hand relation exists only for the case of incompressible fluid . We choose a co-ordinate system relative to which the fluid is stationary at infinity. The boundary conditions for a rigid body moving with velocity U are then

$$u = U \quad \text{at the body surface,} \qquad (3.1.2a)$$

$$u \to 0 \text{ and } p - p_0 \to 0 \text{ as } |x| \to \infty. \qquad (3.1.2b)$$

We shall make use here of the fact that equations Eq. (3.1.1) and Eq. (2.5.2),and the boundary conditions Eq. (3.1.2), are linear and homogeneous in $u, (p-p_0)/\mu$ and U. The expressions of u and $(p-p_0)/\mu$ must therefore be linear and homogeneous in U. Several examples which are important for the mechanics of aerosols are given below.

1.1. A Rigid Spherical Particle

The flow field induced by a rigid sphere in a translational motion was first solved by Stokes (1851). This case is important in a variety of physical contexts, such as the settling of solid substances in liquid or gas, and the fall of mist droplets in air. The quantity of the greatest practical interest is the drag force exerted by the fluid on the sphere, since from this the terminal velocity for the free fall under gravity can be calculated.

We choose the origin of the co-ordinate system to be at the instantaneous position of the centre of the sphere with radius a. The distributions of u and $(p-p_0)/\mu$ must be symmetrical about the axis through the centre of the sphere and parallel to U, and the vector u lies in a plane through that axis. The differential operators in Eq. (3.1.1) and Eq. (2.5.2) are independent of the choice of co-ordinate system, so that $(p-p_0)/\mu$ and u depend only on the vector x but not any other combination of components of x. The parameters U and a complete the list of quantities on which $(p-p_0)/\mu$ and u can depend (although if the object had been of any shape other than spherical, the vectors specifying orientation of the object and the scalar shape parameters would have had to be included). It follows that $(p-p_0)/\mu$ must be of the form $\mathbf{U}\cdot\mathbf{x}F$, where a^2F is a dimensionless function of $\mathbf{x}\cdot\mathbf{x}/a^2(=r^2/a^2)$ alone. Since $p-p_0$ satisfies the Laplace's equation, and vanishes at infinity, it can be represented as a series of spherical solid harmonics functions with negative powers in r; and the only term of the series which is compatible with this form is r^{-2} (the 'dipole' term). Thus

$$\frac{p-p_0}{\mu} = \frac{C\mathbf{U}\cdot\mathbf{x}}{r^3}, \tag{3.1.3}$$

and

$$\boldsymbol{\omega} = \frac{C\mathbf{U}\times\mathbf{x}}{r^3}, \tag{3.1.4}$$

where C is a constant.

The velocity corresponding to this vorticity distribution is found in terms of the stream function ψ. With a spherical polar co-ordinate system (and $\theta = 0$ in the direction of U) the azimuthal component of $\boldsymbol{\omega}$, ω_ϕ, is defined as

$$\omega_\phi = \frac{1}{r}\frac{\partial(ru_\theta)}{\partial r} - \frac{1}{r}\frac{\partial u_r}{\partial \theta}, \tag{3.1.5}$$

where

$$u_r = \frac{1}{r^2\sin\theta}\frac{\partial\psi}{\partial\theta}, u_\theta = \frac{-1}{r\sin\theta}\frac{\partial\psi}{\partial r}. \tag{3.1.6}$$

Thus we find from Eq. (3.1.4) that

$$\frac{\partial^2\psi}{\partial r^2} + \frac{\sin\theta}{r^2}\frac{\partial}{\partial\theta}\left(\frac{1}{\sin\theta}\frac{\partial\psi}{\partial\theta}\right) = -\frac{CU\sin^2\theta}{r}, \tag{3.1.7}$$

where $U = |\mathbf{U}|$.

From the inner boundary condition of Eq. (3.1.2) we set that

$$u_r = U\cos\theta \quad \text{at} \ r = a, \tag{3.1.8a}$$

$$u_\theta = -U\sin\theta \quad \text{at} \ r = a. \tag{3.1.8b}$$

According to Eq. (3.1.8) and Eq. (3.1.7), the solution for ψ must be proportional to $\sin^2\theta$. We therefore put

$$\psi = U\sin^2\theta f(r), \tag{3.1.9}$$

which may be seen to be equivalent to a velocity vector of the form

$$\mathbf{u} = \mathbf{U}\left(\frac{1}{r}\frac{df}{dr}\right) + \mathbf{x}\frac{\mathbf{x} \cdot \mathbf{U}}{r^2}\left(\frac{2f}{r^2} - \frac{1}{r}\frac{df}{dr}\right). \tag{3.1.10}$$

The equation for the unknown function f is

$$\frac{d^2 f}{dr^2} - \frac{2f}{r^2} = -\frac{C}{r}, \tag{3.1.11}$$

and its the general solution is

$$f(r) = \frac{1}{2}Cr + Lr^{-1} + Mr^2. \tag{3.1.12}$$

The terms containing the new constants L and M represent an irrotational motion.

Now the outer boundary condition of Eq. (3.1.2) demands that $f/r^2 \to 0$ as $r \to \infty$; and the inner boundary condition Eq. (3.1.8) requires $f(a) = \frac{1}{2}a^2$. Hence

$$M = 0, \quad L = \frac{1}{2}a^3 - \frac{1}{2}Ca^2 \text{ and } C = \frac{3}{2}a. \tag{3.1.13}$$

The stream function is thus

$$\psi = Ur^2\sin^2\theta\left(\frac{3}{4}\frac{a}{r} - \frac{1}{4}\frac{a^3}{r^3}\right). \tag{3.1.14}$$

Then

$$\mathbf{u} = \mathbf{U}\left(\frac{3}{4}\frac{a}{r} + \frac{1}{4}\frac{a^3}{r^3}\right) + \mathbf{x}\frac{\mathbf{x} \cdot \mathbf{U}}{r^2}\left(\frac{3}{4}\frac{a}{r} - \frac{3}{4}\frac{a^3}{r^3}\right). \tag{3.1.15}$$

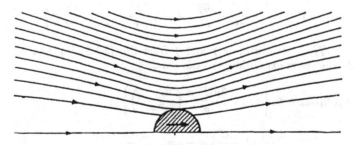

Fig. 1. Streamlines for low-Reynolds-number flow induced by a moving rigid sphere

A sketch of the streamline s is shown in Figure 1 of Chapter 3. The streamline s are symmetrical about a plane perpendicular to U and through the centre of the sphere ($\theta = \frac{\pi}{2}$), as is implied by the linearity of u in U; reversing the direction of U merely leads to a change of the sign of u every where. It will also be noticed that the disturbance due to the sphere extends to a considerable distance from the sphere. The asymptotic behavior of the velocity at large r is r^{-1}. It decays very slowly. This is an important feature of viscous flows . We shall see later in Chapters 5 and 7 that this feature can produce a great impact on the dynamics of aerosol or hydrosol particles. The slow r^{-1} decay of the disturbance makes two moving spheres having a long-distance hydrodynamic interaction . The long distance hydrodynamic interaction can lead to some very difficult problems to be solved . For example, it can make the integral not absolutely convergent when we calculate the mean fall speed of particles suspended in air or liquid.

These features of the solution are consequences of neglecting of the inertia term in the Navier–Stokes equation . The equation for the vorticity , viz. $\nabla^2 \boldsymbol{\omega} = 0$, shows that the flow represented by Eq. (3.1.14) is effectively due solely to the steady molecular diffusion of vorticity to infinity in all directions. The sphere being a source of vorticity is a consequence of the no-slip condition . The term $\partial \omega / \partial t$ which is present in the full equation for $\boldsymbol{\omega}$, and which represents the effect of the continual change of the position of the sphere with respect to the axes, has been neglected here, and the molecular diffusion spreads the vorticity as far ahead of the sphere as behind it; it is as if the sphere were stationary and acted purely as a source of vorticity . The vorticity distribution shows the decrease as r^{-2} to be expected for the diffusion of each component of $\boldsymbol{\omega}$ from a stationary steady dipole source.

We now try to find the force exerted by the fluid on the sphere and evaluate the stress tensor at $r = a$. The i-th component of the force per unit area exerted on the sphere at a position denoted by $\mathbf{x} = a\mathbf{n}$ is

$$n_j(\sigma_{ij})_{r=a} = n_j \left\{ -p\delta_{ij} + \mu \left(\frac{\partial u_i}{\partial x_j} + \frac{\partial u_j}{\partial x_i} \right) \right\}_{x=a},$$

and for a velocity of the form Eq. (3.1.10) this may be found as

$$\left\{ -pn_i + \mu n_i \mathbf{U} \cdot \mathbf{n} \left(-\frac{f''}{r} + \frac{6f'}{r^2} - \frac{10f}{r^3} \right) + \mu U_i \left(\frac{f''}{r} - \frac{2f'}{r^2} - \frac{10f}{r^3} \right) \right\}_{r=a},$$

where f' denotes df/dr. Substituting p and f from Eq. (3.1.3) Eq. (3.1.12) and Eq. (3.1.13) into the above expression, one can obtain

$$n_j(\sigma_{ij})_{r=a} = n_i \left\{ -p_0 + \frac{3\mu \mathbf{U} \cdot \mathbf{n}}{a} \left(\frac{2C}{a} - 3 \right) \right\}$$

$$+ \frac{3\mu U_i}{a} \left(1 - \frac{C}{a} \right) = -p_0 n_i - \frac{3\mu U_i}{2a}. \tag{3.1.16}$$

The first term on the right-hand-side of Eq. (3.1.16) is simply the same uniform normal stress as in the fluid at infinity, and makes no contribution to the total force on the sphere, which is a retarding or drag force parallel to $\mathbf{U}$ and its magnitude is

$$D = 6\pi\mu aU. \tag{3.1.17}$$

The expression Eq. (3.1.17) is usually known as the Stokes' law for the resistance to a moving sphere at small Reynolds number . We have mentioned it many times in the previous chapters.

Parts of the Stokes' resistance come from the pressure of the flow field. From Eq. (3.1.3) we see that

$$p = p_0 + \mu C \frac{\mathbf{U} \cdot \mathbf{x}}{r^3}.$$ (3.1.18)

The contribution of it to the resistance should then be

$$(-n_j p \delta_{ij})_{r=a} = p_0 n_i - \left(\mu C \frac{\mathbf{U} \cdot \mathbf{x}}{r^3} n_i \right)_{r=a}$$

The resultant force of the first term on the right-hand-side of the above expression is zero. The second term on the right-hand-side is a radial component. The component of it along the direction of $\mathbf{U}$ is

$$-\frac{3\mu}{2a} U \cos^2 \theta.$$

Integrating the above expression on the whole surface of the sphere we find the contribution of the pressure to the resistance D_p

$$D_p = -\frac{3\mu}{2a} U \int_0^{\pi} \cos^2 \theta (2\pi a^2) \sin \theta \mathrm{d}\theta = -2\pi \mu a U = \frac{1}{3} D.$$ (3.1.19)

Thus, the contribution of the pressure to the resistance is one third of it. Obviously the rest two third comes from the viscous stress . The value of it is $4\pi \mu a U$.

It is common practice to express the resistance exerted by the fluid in terms of a dimensionless coefficient obtained by dividing the resistance by $\frac{1}{2}\rho U^2$ and by the area of the body projected on to a plane normal to $\mathbf{U}$; thus the drag coefficient is here

$$C_D = \frac{D}{\frac{1}{2}\rho U^2 \pi a^2} = \frac{24}{Re} \quad \left(\text{where} \quad Re = \frac{2aU\rho}{\mu} \right).$$ (3.1.20)

Experimental results have shown that the above relation is right if $Re < 0.4$. The observed data will deviate from the Stokes' drag coefficient $(24/Re)$ when $Re > 0.4$. The departure becomes large when $Re > 1$, and the values of drag coefficient predicted by Stokes' law are always smaller than those obtained by experiments if $Re > 0.4$. This is easy to understand from the Helmholtz's minimum dissipation theorem. The rate of dissipation in the Stokes's flow is less than that in any other solenoidal velocity field (in which the inertia forces of the fluid cannot be neglected again) with the same value of the velocity at all points of the boundary within the concerned region. This implies that the resistance of the viscous flow is minimal.

It is now a simple matter to calculate the terminal velocity $\mathbf{V}_t$ of a sphere while falling freely under gravity through air which is otherwise at rest. When the Stokes number of the sphere is zero the terminal velocity is the actual velocity of the sphere anywhere. Taking into account the buoyancy force exerted on the sphere, we find for the terminal velocity $\mathbf{V}_s$ of the sphere of density ρ_p

$$6\pi\mu a\mathbf{V}_s = \frac{4}{3}\pi a^3(\rho_p - \rho)\mathbf{g}$$

that is,

$$\mathbf{V}_s = \frac{2}{9}\frac{a^2\mathbf{g}}{\nu}\left(\frac{\rho_p}{\rho} - 1\right). \qquad (3.1.21)$$

The corresponding values of the Reynolds number for a sphere falling with its terminal velocity is

$$Re = \frac{4}{9}\frac{a^3 g}{\nu^2}\left(\frac{\rho_p}{\rho} - 1\right). \qquad (3.1.22)$$

For a water droplet (assumed to be rigid) falling through air at 20^0C and a pressure of one atmosphere we have $\rho = 1.2053 \times 10^{-3}$gm/cm^3, $\rho_p = 1$gm/cm^3, $\nu = 0.15$cm^2/sec, making the Reynolds number $Re = 1.60 \times 10^7 a^3$, a being in centimeters. Hence, for all droplets with radius smaller than 30 μm , the Stokes's law is applicable. If the density of the rigid particle is 2gm/cm^3, we then have $Re = 3.2 \times 10^7 a^3$, a being also in centimeters. For all such particles with radius smaller than 25 μm , the Stokes's law is also applicable. The density of the particles shown in Table 3 of Chapter 1 is 2gm/cm^3, the values of their Reynolds number is much smaller than 0.4. Therefore, we can calculate their terminal velocity $\mathbf{V}_s$ according to Eq. (3.1.21).

1.2. A Spherical Droplet

When a droplet of a fluid which is different from the medium is in translational motion, it can cause a flow inside the liquid droplet. This phenomenon was first investigated by Hadamard (1911). It is of interest to see if this internal circulation affects the drag significantly. We shall suppose that the two fluids are immiscible and the surface tension at the interface is sufficiently strong to keep the 'droplet' approximately spherical against any deforming effect of viscous forces. The condition for this is that γ/a (where γ is the coefficient of surface tension) should be large compared to the normal stress due to the motion, of order $\mu U/a$, that is,

$$\gamma \gg \mu U. \qquad (3.1.23)$$

It will also be assumed that the Reynolds number of the motion within the droplet is small compared to unity, like that of the motion outside the droplet.

For the case of water droplets suspended in air at 20^0C, $\gamma = 72.8$dyn/cm, for the case of oil droplets suspended in water at 20^0C, $\gamma = 20$dyn/cm. Therefore, the assumption of spherical liquid drop is sufficiently valid even for the droplets with radius as large as several millimeters. However, we have pointed out in Chapter 1 that the deformation can occur much earlier than the above estimation. For instance, the water droplet deforms a little, into a flattened sphere, when its radius is about

140μm. The deformation is proportional to its size when its radius is within the range from 500μm to 2250μm. All the raindrops can break up when the radius of them is larger than 3000μm. The deformation and break up of water drops are very complicated problems. Many other factors such as the inertia of the fluid , internal circulation, turbulence play important roles in these processes when the drop is large, and its Reynolds number is larger than unity. However, it is beyond the scope of the book. All the values of the Reynolds number of the aerosol particles indicated in Table 1 and Table 3 of Chapter 1 are much smaller than unity. Thus the form of the droplets remains spherical in the following discussions.

The argument used to determine the velocity and pressure distributions for the case of a rigid sphere can be modified without difficulty. The motions both inside and outside the sphere are axisymmetric and satisfy the equations Eq. (3.1.1) and Eq. (2.5.2) (although with different values of the viscosity). u and $p - p_0$ must vanish at infinity, as discussed before, and $\bar{u}$ and $\bar{p} - \bar{p}_0$ (where the overbar indicates a quantity relating to the internal fluid and its motion) are finite everywhere within the sphere. Because of that there is no break up, the normal components of the velocities must be continuous across the surface of the sphere, that is, that

$$\mathbf{n} \cdot \mathbf{u} = \mathbf{n} \cdot \bar{\mathbf{u}} = \mathbf{n} \cdot \mathbf{U} \quad \text{at } r = a. \tag{3.1.24}$$

Replacing the no-slip condition at the surface of a rigid sphere there are certain dynamical matching conditions. No relative motion of the two fluids can occur at the interface. Thus the tangential components of the velocity of the two fluids must be continuous across the surface of the sphere, i.e.

$$\mathbf{n} \times \mathbf{u} = \mathbf{n} \times \bar{\mathbf{u}} \quad \text{at } r = a. \tag{3.1.25}$$

Finally, the tangential stress exerted at the interface by the external fluid must be equal and opposite to that exerted by the internal fluid, then we have

$$\epsilon_{mki} n_k n_j (\sigma_{ij} - \bar{\sigma}_{ij}) = 0 \quad \text{at } r = a. \tag{3.1.26}$$

The equations and boundary conditions are linear and homogeneous in u, $p - p_0, \bar{u}, \bar{p} - \bar{p}_0$ and U, so that the relations Eq. (3.1.3) to Eq. (3.1.13) still stand except that the constant C is no longer equal to $\frac{3}{2}a$, and must be determined by the internal motion. $\bar{p}$ satisfies the Laplace's equation, like p, and the appropriate solution, in analogue to Eq. (3.1.3), is

$$(\bar{p} - \bar{p}_0)/\mu = \overline{C}\mathbf{U} \cdot \mathbf{x},$$

where $\overline{C}$ is a constant. And the internal vorticity is

$$\bar{\omega} = -\frac{1}{2}\overline{C}\mathbf{U} \times \mathbf{x}.$$

The equation for the stream function $\bar{\psi}$, similar to Eq. (3.1.7), is then

$$\frac{\partial^2 \bar{\psi}}{\partial r^2} + \frac{\sin\theta}{r^2} \frac{\partial}{\partial \theta} \left(\frac{1}{\sin\theta} \frac{\partial \bar{\psi}}{\partial \theta} \right) = \frac{1}{2}\overline{C}Ur^2\sin^2\theta. \tag{3.1.27}$$

From the boundary condition Eq. (3.1.24) and Eq. (3.1.27), $\overline{\psi}$ should also be proportional to $\sin^2\theta$, thus we set

$$\overline{\psi} = U\sin^2\theta \overline{f}(r). \qquad (3.1.28)$$

Substituting Eq. (3.1.28) into Eq. (3.1.27) we obtain

$$\overline{f}'' - \frac{2\overline{f}}{r^2} = \frac{1}{2}\overline{C}r^2. \qquad (3.1.29)$$

A particular integral of Eq. (3.1.29) is $\frac{1}{20}\overline{C}r^4$, and the complementary function is $\overline{L}r^{-1} + \overline{M}r^2$. Because $\overline{u}$ and $\overline{p} - \overline{p}_0$ are finite everywhere, $\overline{L}$ must be zero, hence

$$\overline{f} = \frac{1}{20}\overline{C}r^4 + \overline{M}r^2. \qquad (3.1.30)$$

According to the boundary condition Eq. (3.1.24) the relation between $\overline{M}$ and $\overline{C}$ should be

$$\overline{M} = \frac{1}{2} - \frac{1}{20}\overline{C}a^2. \qquad (3.1.31)$$

The velocity within the sphere is thus

$$\overline{\mathbf{u}} = \mathbf{U} - \frac{1}{10}\overline{C}\{\mathbf{U}(a^2 - 2r^2) + \mathbf{xU}\cdot\mathbf{x}\}. \qquad (3.1.32)$$

To determine C and $\overline{C}$ from the dynamical matching conditions, from Eq. (3.1.25) we have

$$C - \frac{1}{2}a = \frac{1}{10}\overline{C}a^3 + a. \qquad (3.1.33)$$

Only the term containing U_i in the general expressixon Eq. (3.1.16) for the stress across the interface contributes to the tangential component and matching this tangential component requires

$$\frac{3\mu}{a^2}(a - C) = \frac{3}{10}\overline{\mu}a\overline{C}. \qquad (3.1.34)$$

Hence equations Eq. (3.1.33) and Eq. (3.1.34) give

$$C = \frac{1}{2}a\frac{2\mu + 3\overline{\mu}}{\mu + \overline{\mu}}, \quad \overline{C} = -\frac{5}{a^2}\frac{\mu}{\mu + \overline{\mu}}. \qquad (3.1.35)$$

The resultant force exerted on the interface by the external fluid is now obtained by integrating the force per unit area Eq. (3.1.16) over the surface of the sphere. Then we have

$$D = 4\pi\mu aU\frac{\mu + \frac{3}{2}\overline{\mu}}{\mu + \overline{\mu}}. \qquad (3.1.36)$$

For a rigid sphere, $\overline{\mu}/\mu \to \infty$, and Eq. (3.1.36) reduces to Eq. (3.1.17). The case of a spherical gas bubble moving through liquid corresponds to the other extreme, $\overline{\mu}/\mu = 0$. Then $D = 4\pi\mu aU$. The drag exerted on a spherical gas bubble is smaller than that exerted on a rigid sphere by $2\pi\mu aU$. However, observation of the terminal speed of very small gas bubbles suggests that the drag is often close to the value $6\pi\mu aU$ other than the expected value $4\pi\mu aU$; this is because any surface-active impurities present in the liquid are likely to form a mesh of large molecules at the bubble surface and to cause the interface to act partially like a rigid surface (Levich 1962)[5]. The same is true for the case of cloud and fog droplets suspended in air. There are a lot of surface-active impurities in air. Therefore, cloud and fog droplets can be regarded as rigid spheres, if the values of their Reynolds number are smaller than unity.

1.3. Particles of Arbitrary Shape

Although it is difficult to work out the details of the flow of moving objects with a shape which is not spherical, at small Reynolds number, (the only analytical solution which has been obtained by Lamb (1932) and was mentioned in Chapter 1 is that for the case of a rigid ellipsoid body), some general results are available in textbooks.

At first the inertia force can be neglected in a viscous flow , the governing forces are then viscous stresses . From the Stokes equation we see that the magnitude of the viscous forces per unit volume is of order $\mu U/d^2$, where d is the representative length of the body. Then the magnitude of the total volume of the body is of order d^3. Hence, the magnitude of the resultant force exerted on the surface of the body is of order μUd. Explicit knowledge can be obtained when we consider only the situations of the far field.

Arguments like those used at the beginning of this section show that, for an object of arbitrary shape in translational motion with velocity **U**, both **u** and $(p - p_0)/\mu$ are linear and homogeneous in **U**. Furthermore, a change of size of the body without change of its shape simply changes the length scale of the whole flow field, so that for a body of given shape $\mathbf{u}/U$ and $(p - p_0)d/\mu U$ are dimensionless functions of $\mathbf{x}/d$.

Both the tangential and excess normal stress es in the fluid are linear in **U**, so that the resultant force exerted by the object is given by an integral

$$F_i = - \int \sigma_{ij} n_j \mathrm{d}A, \qquad (3.1.37)$$

taken over the body surface, and is proportional to μUd. The Eq. (3.1.1) governing a flow with negligible inertia forces is equivalent to

$$\frac{\partial \sigma_{ij}}{\partial x_j} = 0.$$

It follows from application of the divergence theorem that the integral in Eq. (3.1.37) has the same value for any surface in the fluid enclosing the object and in particular for a sphere of large radius centred at the origin. Hence

$$F_i = - \int_{r \to \infty} \lim(r\sigma_{ij}x_j)\mathrm{d}\Omega(\mathbf{x}), \qquad (3.1.38)$$

where $d\Omega(\mathbf{x})$ is an infinitesimal element of solid angle at the direction of **x**. This relation shows that in the case of a flow induced by to a moving object which exerts a finite force on the fluid, $p - p_0$ and the rate-of-strain tensor must both decrease at least as rapidly as r^{-2} for $r \to \infty$.

We know also that $p - p_0$ is a harmonic function and can be represented as a series of spherical solid harmonics of negative powers in r. The first non-zero term of this series is power r^{-2}, so that

$$\frac{p - p_0}{\mu} \sim \frac{P_{ij}U_j dx_i}{r^3} \tag{3.1.39}$$

is the asymptotic form as $r \to \infty$, where P_{ij} is a numerical tensor coefficient depending only on the object shape. The vorticity ω also satisfies the Laplace's equation, and may be written as a similar series. Terms of the same power in the series of $(p - p_0)/\mu$ and of ω are related by the governing Eq. (3.1.1), and it may be seen that if the leading term of the series of $(p - p_0)/\mu$ is $\alpha \cdot \nabla r^{-1}$, that of ω is $\alpha \times \nabla r^{-1}$. Consequently we have

$$\omega_l \sim \epsilon_{ikl}\frac{P_{ij}U_j dx_k}{r^3}, \tag{3.1.40}$$

as $r \to \infty$. Finally we may obtain the asymptotic form for the velocity, which is determined by Eq. (3.1.40) and the requirement that u is solenoidal. We find

$$u_k \sim \frac{1}{2}P_{ij}U_j \left(\frac{d}{r}\delta_{ik} + \frac{d}{r^3}x_i x_k \right), \tag{3.1.41}$$

as $r \to \infty$.

It is now possible to relate the coefficient P_{ij} to the force F by evaluating the stress at a spherical surface of large radius (see Eq. (3.1.38)). It leads to a result

$$F_i = 4\pi\mu d P_{ij}U_j. \tag{3.1.42}$$

It appears that the single numerical tensor P_{ij} is sufficient for a specification of the total force on the fluid and the asymptotic expressions of the pressure and velocity, when a body of given shape moves with translational velocity U, and in the case of a spherical body of radius $\frac{1}{2}d$ composed of fluid with viscosity $\bar{\mu}$, we know from the preceding calculations that

$$P_{ij} = \frac{1}{2}\delta_{ij}\frac{\mu + \frac{3}{2}\bar{\mu}}{\mu + \bar{\mu}},$$

and $P_{ij} = \frac{3}{4}\delta_{ij}$ when the body is a rigid sphere.

The flow at large distances from the object is axisymmetric about the direction of the vector $P_{ij}U_j$. Consequently we may represent the flow in this region in terms of a stream function . With spherical polar coordinates (r, θ, φ) and the axis $\theta = 0$ in the direction of the vector $P_{ij}U_j$—which is also the direction of the force F—we find from Eq. (3.1.9) and Eq. (3.1.41), Eq. (3.1.42), Eq. (3.1.43) that

$$\psi = \frac{1}{2}dr P_{ij}U_j \sin^2\theta, \tag{3.1.43}$$

and

$$\psi = \frac{F}{8\pi\mu}r\sin^2\theta, \tag{3.1.44}$$

in this region, where F is the magnitude **F**. When an object of arbitrary shape moves through fluid at small Reynolds number , the distant flow field depends only on the resultant force exerted on the fluid and is not affected by the continual change of position of the object.

These general results have a convenient form for application to the case of small particles, either solid or fluid, falling freely under gravity . If the volume τ and density $\bar{\rho}$ of the particles are known, the distributions of velocity and pressure far from the particles are immediately obtained from the above formulae by setting

$$F = (\bar{\rho} - \rho)\tau g;$$

details of the shape of the particles are irrelevant, and it presumably also does not matter whether the particles continually turn over and change their orientations relative to the direction of gravity or whether they move on a path which is not vertical.

The flow field represented by Eq. (3.1.44) is later (see Chapter 5) referred to as being due to the existence of a 'Stokeslet' at the origin.

2. The Oseen Approximation

Whitehead (1889) attempted to improve the Stokes's approximation for the solution of a flow field induced by a moving sphere by iteration method. Substituting the first approximation Eq. (3.1.14) into the inertia term of the full Navier–Stokes equation yields the iteration equation. A particular integral that satisfies the surface condition can be easily found. However, the velocity does not behave properly at infinity, and no complementary function can be added to correct it. In the next order approximation the velocity would become infinite at infinity.

The nonexistence of a second order approximation to the Stokes's solution for unbound uniform flow past a sphere is known as the Whitehead's paradox. Whitehead himself regarded it as an indication that discontinuities must arise in the flow field associated with the formation of a dead-wake. However, this explanation is now known to be incorrect.

Just as the d'Alembert's paradox was resolved by Prandtl's discovery (1905) that a flow with high Reynolds number is a singular perturbation problem, so the Whitehead's paradox were shown by Oseen (1910) to arise from the singular nature of flow with low Reynolds number . The region of nonuniformity is a thin layer near the surface of the object at high Reynolds number , whereas the region is vicinity of the infinity for low Reynolds number . The source of the difficulty can be understood by examining the relative magnitude of the terms neglected in the Stokes's approximation.

According to the solution Eq. (3.1.15), an estimate of the magnitude of the viscous force $\mu \nabla^2 \mathbf{u}$ is $\mu a U / r^3$. If the sphere velocity is exactly steady, and the rate of change of $\mathbf{u}$ at a fixed point is due simply to changing its position relative to the point concerned, the operator $\partial/\partial t$ is equivalent to $-\mathbf{U} \cdot \nabla \mathbf{u}$ and the inertia force is

$$\rho \left(\frac{\partial \mathbf{u}}{\partial t} + \mathbf{u} \cdot \nabla \mathbf{u} \right) = \rho(-\mathbf{U} \cdot \nabla \mathbf{u} + \mathbf{u} \cdot \nabla \mathbf{u}). \tag{3.2.1}$$

In terms of Eq. (3.1.15), the first term on the right-hand-side is estimated as $\rho U^2 a / r^2$, whereas the second term as $\rho U^2 a^2 / r^3$. These two terms are of the same order near the sphere, but the first is dominant at places far from the sphere. Thus the ratio of the order of magnitude of the neglected inertia forces to that of the retained viscous forces is

$$\rho \frac{U^2 a}{r^2} \bigg/ \frac{\mu U a}{r^3} = \frac{\rho a U}{\mu} \frac{r}{a} = \frac{1}{2} Re \frac{r}{a}. \tag{3.2.2}$$

At positions near the sphere the solution is indeed self-consistent when $Re \ll 1$, but it seems that the inertia force corresponding to the solution becomes comparable to the viscous force at distances from the sphere of order a/Re. The solution Eq. (3.1.15) is evidently not valid at the large distances from the sphere. This is the singular feature of the Stokes flow discovered by Oseen in 1910. Oseen also showed how it is possible to improve the equation and thereby to remove the inconsistency. Oseen's improvement applies to the cases in which the object is moving with a steady velocity **U** and the flow is steady relative the object, in this situation the local inertia force is as given in Eq. (3.2.1). Since the first one of the two terms on the right-hand-side of Eq. (3.2.1) becomes dominant at large r, and is responsible for inertia force being comparable to the viscous force at sufficiently large r, Oseen suggested that among the two contributions to the inertia force, only the first term retained in the equation of motion. The second term is again neglected upon the assumption that $Re \ll 1$; provided $|u|$ falls off at least as rapidly as r^{-1} as r increases, this second term remains small relative to the viscous force no matter r is large or small. Near the sphere the two terms in Eq. (3.2.1) are of the same order and will both be small compared to the viscous force, provided $Re \ll 1$, so that in this region the suggested equation is equivalent to Eq. (3.1.1).

The Oseen equation for a flow due to a moving object at small Reynolds number is therefore

$$\rho \frac{\partial \mathbf{u}}{\partial t} = -\rho \mathbf{U} \cdot \nabla \mathbf{u} = -\nabla p + \mu \nabla^2 \mathbf{u}. \tag{3.2.3}$$

To describe the flow field one also needs the continuity Eq. (2.5.2) and the boundary condition Eq. (3.1.2) along with the Oseen equation .

The solution of these new equations for the case of a moving sphere is not known in a closed form, but an approximate solution has been found by Lamb (1911). In terms of the stream function , this approximate solution is

$$\psi = U a^2 \left[-\frac{1}{4} \frac{a}{r} \sin^2\theta + 3(1 - \cos\theta) \frac{1 - \exp\{-1/4 Re(1 + \cos\theta) r/a\}}{Re} \right], \tag{3.2.4}$$

at the instant when the centre of the sphere coincides with the origin, where $Re = 2aU\rho/\mu$ as before. This expression is readily seen to satisfy the Eq. (3.2.3) exactly, and it also makes $\mathbf{u} \to 0$ as $r \to \infty$. Near the sphere, where r/a is of order unity and $Re r/a \ll 1$, it becomes

$$\psi = U a^2 \sin^2\theta \left\{ -\frac{1}{4} \frac{a}{r} + \frac{3}{4} \frac{r}{a} + O\left(Re \frac{r}{a}\right) \right\}, \tag{3.2.5}$$

and therefore coincides with the Stokes's solution Eq. (3.1.14) with a relative error of order Re. This is just the degree of approximation to which Eq. (3.2.3) represents the equation of motion, so that Eq. (3.2.4) is as an accurate solution of Eq. (3.2.3) as is expected.

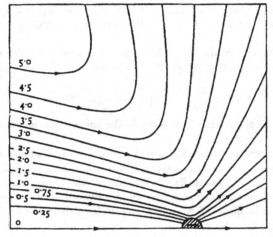

Fig. 2. Streamlines for the outer part of the Oseen flow induced by a moving rigid sphere

Figure 2 of Chapter 3 shows the streamline s corresponding to Eq. (3.2.4) with neglecting of the first term within the square brackets which is significant only near the sphere when $Re \ll 1$. The qualitative differences between the Oseen and Stokes solutions in the outer part of the flow field are evident. The streamline s are no longer symmetrical about the plane $\theta = \frac{1}{2}\pi$, as was expected from the fact that the governing equation does not remain satisfied after a change of the signs of u and U. Far from the sphere the flow tends to become radial, as if is radiated from a source of fluid at the sphere, except within a 'wake' directly behind the sphere. Analytically, we see from Eq. (3.2.4) that when $Rer/a \gg 1$ the flow has different forms according to whether $1 + \cos\theta$ is small compared to unity. At positions where $1 + \cos\theta$ is not small, the stream function becomes

$$\psi \sim Ua^2 \frac{3}{Re}(1 - \cos\theta). \qquad (3.2.6)$$

It is well known that the stream function of a point source of fluid volume, which is situated at the origin, is given by

$$\psi = \frac{m}{4\pi}(1 - \cos\theta), \qquad (3.2.7)$$

where m is termed the 'strength' of the source, and is equal to the total outward flux of the fluid volume across any closed surface which encloses the origin. The associated irrotational velocity field is given by

$$\mathbf{u} = \frac{m\mathbf{x}}{4\pi r^3}. \tag{3.2.8}$$

Comparing Eq. (3.2.7) with Eq. (3.2.6), one can see that Eq. (3.2.6) describes the outward radial flow from a source at the origin emitting $12\pi a^2 U/Re$ units of volume per second.

On the other hand, within the wake, where $1 + \cos\theta$ is of the same order of magnitude as $4a/rRe$ (that is, where $\pi - \theta$ is small and of order $(8a/rRe)^{1/2}$), we have

$$\psi \sim Ua^2 \frac{6}{Re} \left[1 - \exp\left\{ -\frac{Re}{8} \frac{r}{a} (\pi - \theta)^2 \right\} \right], \tag{3.2.9}$$

which describes a compensating flow towards the sphere, the inflow velocity being $3Ua/2r$ on the axis $\theta = \pi$.

Far from the sphere, the vorticity is zero in the outward radial flow from a source at the origin, so is only confined to the wake, which is bound by a parabola of revolution on which $(\pi - \theta)^2 r/a$ is of order Re^{-1}. Whereas in the Stokes approximation the vorticity diffuses out in all directions from an effectively stationary sphere, here the motion of the sphere is allowed for, as may be seen from the equation for ω obtained from Eq. (3.2.3) and Eq. (2.5.2).

$$\frac{\partial \omega}{\partial t} = -\mathbf{U} \cdot \nabla\omega = \nu \nabla^2 \omega. \tag{3.2.10}$$

This equation for each component of ω is of the same form as that satisfied by temperature in a stationary conducting medium through which a steady source of heat is moving with a steady velocity $\mathbf{U}$. The vorticity generated by the sphere is left behind as the sphere moves on, in a wake which becomes narrower as Re increases.

We may mow confirm that the solution Eq. (3.2.4) is self-consistent while the Stokes's solution is not; that is, we show that the neglected term $\rho\mathbf{u} \cdot \nabla\mathbf{u}$, evaluated by means of Eq. (3.2.4) is small compared to any term retained in the equation of motion, when $Re \ll 1$. In the region near the sphere, where r/a is of order unity, Eq. (3.2.4) reduces to the Stokes's solution (with an error of order Re), for which $\rho|\mathbf{u} \cdot \nabla\mathbf{u}|$ is already known to be small compared to $\mu|\nabla^2\mathbf{u}|$, the ratio of these terms being of order Re. Far from the sphere, in the region where Rer/a is of order unity, and Eq. (3.2.4) first differs significantly from the Stokes solution, the magnitude of $\mathbf{u}$ as given by Eq. (3.2.4) is of order Ua/r, or URe; hence the ratio of the neglected term $\rho|\mathbf{u} \cdot \nabla\mathbf{u}|$ to the retained term $\rho|\mathbf{U} \cdot \nabla\mathbf{u}|$ is of order Re and is again small. Still further from the sphere, where $r/a \gg Re^{-1}$, $|\mathbf{u}|$ is even smaller in comparison with U.

It appears then that the approximate form of the equation of motion suggested by Oseen has a solution such that the approximation is self-consistent over the whole of the flow field when $Re \ll 1$. Near the sphere this solution has the same form as the Stokes's solution—and so leads to the same expression, $6\pi\mu aU$, for the resistance experienced by the sphere—with a relative error of order Re. Since Eq. (3.2.4) is evidently a good approximation to the solution of the complete equations of motion which is valid for $Re \ll 1$ over the whole flow field, therefore Eq. (3.2.4) can be naturally considered as the first order approximation of a successive approximation to a solution of a practical flow field equation. And the second approximation to the drag coefficient has been found by Kaplun and Lagerstorm 1957; Proudman and Pearson 1957 to be

$$C_D = \frac{24}{Re}\left(1 + \frac{3}{16}Re\right). \tag{3.2.11}$$

In comparison with experiments, it is noted that the above formula agrees with the measured drag for a slightly larger range of the Reynolds number better than the Stokes's law.

3. Flow due to Longitudinal Relative Motion of Two Spheres

We have expounded the theories of an isolated particle moving in unbound space at low Reynolds number in the preceding sections. These theories are important in the mechanics of aerosols. However, the isolated particle theory is far away from the needs of further development of the mechanics of aerosols. Thus it is necessary to develop a theory of two-sphere motion, three-sphere motion, even multi-particle motion for a more complete understanding the essences of the mechanics of aerosols. Owing to the difficulty of the problems, we can only investigate the two-sphere case at the moment. This is an important development in low-Reynolds-number hydrodynamics since 1960s. Two simple examples will be discussed in this section, i.e. in the neighbourhood of a moving sphere, there is another sphere or a rigid plane. This is another extreme in contrast with the case of isolated particle. The lubrication theory play an important role in this case and an exact asymptotic solution can be obtained. The results are significant in many practical problems of aerosol mechanics.

3.1. A Circular Disk Moving Towards a Rigid Plane

A circular disk of radius a is parallel to and at a small distance d $(d \ll a)$ from a rigid plane. The space between them is occupied by fluid. The pressure p_0 at the edge of the disk is the atmospheric pressure. The disk is in motion with a steady speed U in the direction normal to the plane. We now try to find the resistance.

We choose a cylindrical coordinates (r, θ, z) such that the plane $z = 0$ coincides with the rigid plane, and the z-axis directs to the disk and passes through the centre of the disk. The motion is axisymmetric about the z-axis so that the azimuthal angle is irrelevant. The mass-conservation equation for the flow field in the gap between the disk and the rigid plane is thus

$$\frac{1}{r}\frac{\partial(ru)}{\partial r} + \frac{\partial w}{\partial z} = 0, \tag{3.3.1}$$

where u is the r-component of the velocity and w is the z-component of the velocity of the flow field in the gap. The representative length scale along z-direction is

d. The representative length scale along r-direction is a. Therefore, from the mass-conservation Eq. (3.3.1) we know that w can be neglected compared to u since $d \ll a$. Thus, the equation of z-component of the motion reduces to

$$\frac{\partial p}{\partial z} = 0. \tag{3.3.2}$$

The distribution of the pressure in the gap will be uniform along z-direction. It depends only on r. The equation of the r-component of the motion is

$$\rho \left[u \frac{\partial u}{\partial r} + w \frac{\partial u}{\partial z} \right] = -\frac{\partial p}{\partial r} + \mu \left[\frac{\partial^2 u}{\partial z^2} + \frac{1}{r} \frac{\partial}{\partial r} \left(r \frac{\partial u}{\partial r} \right) - \frac{u}{r^2} \right]. \tag{3.3.3}$$

The inertia force can be neglected, if the Roynolds number Re is small where

$$Re = \frac{\rho d U}{\mu}.$$

The first term in the bracket on the right-hand-side of Eq. (3.3.3) is dominant since $d \ll a$. Thus Eq. (3.3.3) reduces to

$$\frac{\partial p}{\partial r} = \mu \frac{\partial^2 u}{\partial z^2} = -G. \tag{3.3.4}$$

with boundary conditions

$$u = 0 \qquad \text{at} \quad z = 0 \quad \text{and} \quad z = d. \tag{3.3.5}$$

Obviously, equations Eq. (3.3.4) and Eq. (3.3.5) are the equations in lubrication film at low Reynolds number , and the solution for u is given as

$$u = \frac{G}{2\mu} z(d - z). \tag{3.3.6}$$

The volume flux Q_r at r is

$$Q_r = \int_0^d 2\pi r u \, dz = \frac{\pi r d^3 G}{6\mu} = -\pi r^2 U. \tag{3.3.7}$$

This gives the pressure distribution from the integral Eq. (3.3.7), i.e.

$$p(r) = p_0 + \frac{3\mu U}{d^3}(r^2 - a^2). \tag{3.3.8}$$

The total resistance F to the motion of the disk can be obtained by integrating Eq. (3.3.8)

$$F = -\int_0^a (p - p_0) 2\pi r \, dr = \frac{-34\pi \mu a^4}{2d^3} U = -16\mu a U f. \tag{3.3.9}$$

The resistance on an isolated disk falling under gravity in an unbound space is $16\mu aU$. This can be obtained from the Lamb's flat ellipsoid resistance formula by putting the ratio of long axis to short axis $\beta \to \infty$ (see Chapter 1). Comparing with Eq. (3.3.9), we see that there is a correction factor f,

$$f = \frac{3\pi}{32}\frac{1}{\epsilon^3}, \qquad (3.3.10)$$

where $\epsilon = d/a$. It is indicated from Eq. (3.3.10) that f will tend to infinite as $\epsilon \to 0$. The resistance on an isolated disk is a constant, whereas the resistance on a disk moving in a bound space is not a constant, which may tend to infinity as the gap between the disk and the rigid plane tends to zero. This is one of the important differences between objects moving in an unbound space and moving in a bound space. We have already pointed out that the lubrication resistance tends to infinity as the gap tends to zero is a common phenomenon in low-Reynolds-number flow s. When Reynolds number is small, the disk cannot reach the rigid plane during a finite time interval.

3.2. A Rigid Sphere Moving Towards Another Rigid Sphere

A rigid sphere A of radius a is in steady motion with speed U relative to a stationary sphere B of radius b. We still choose the cylindrical coordinates (r, θ, z) such that the z-axis coincides with the line of centres and the direction of z-axis is from the centre of sphere B to the centre of sphere A. The origin of the coordinates is at the intersection of z-axis and the surface of sphere B. In this coordinate system, the centre of sphere A is $(0, 0, a + d)$, and that of sphere B is at $(0, 0, -b)$, where d is the minimum distance between surfaces of the two spheres and

$$d \ll a \quad \text{and} \quad b.$$

When the Reynolds number Re is small where

$$Re = \frac{\rho dU}{\mu} \ll 1,$$

the flow field in the gap between the two spheres is still an axisymmetric lubricating flow field and is governed by the equations Eq. (3.3.2), Eq. (3.3.4) and Eq. (3.3.1) with the boundary conditions

$$u = 0 \quad \left(\text{at} \quad z = -\frac{r^2}{2b} \quad \text{and} \quad z = d + \frac{r^2}{2a}\right). \qquad (3.3.11)$$

The r-component of velocity u can be obtained by integrating Eq. (3.3.4) with boundary condition Eq. (3.3.11), i.e.

$$u = \frac{G}{2\mu}\left(z + \frac{r^2}{2b}\right)\left(d + \frac{r^2}{2a^2} - z\right). \qquad (3.3.12)$$

The volume flux Q_r at r is

$$Q_r = \int_{-\frac{r^2}{2b}}^{d+\frac{r^2}{2a}} 2\pi r u \, dz = \frac{\pi r G}{6\mu}\left(d + \frac{r^2}{2a} + \frac{r^2}{2b}\right)^3 = -\pi r^2 U. \qquad (3.3.13)$$

Integrating Eq. (3.3.13) yields the pressure distribution

$$p - p_0 = \frac{-3\mu}{(a^{-1}+b^{-1})}\frac{1}{[d+\frac{a^{-1}+b^{-1}}{2}r^2]^2}U. \qquad (3.3.14)$$

Because of the resistance of the lubricating flow film being infinity as the film thickness tends to zero, the total resistance F on sphere A from the whole flow field can be obtained by integrating Eq. (3.3.14),

$$F = \int_0^\infty 2\pi r(p - p_0)\,dr$$

$$= \frac{-6\pi\mu U}{(a^{-1}+b^{-1})^2}\frac{1}{d}$$

$$= -6\pi\mu a U f. \qquad (3.3.15)$$

Comparing Eq. (3.3.15) with Eq. (3.1.17), it is evident that the total resistance on a sphere moving in a bound space (towards another stationary sphere) is different from that on a sphere moving in an unbound space (isolated sphere) by a factor f where

$$f = \frac{\lambda^2}{(1+\lambda)^2}\epsilon^{-1}. \qquad (3.3.16)$$

Here λ is the size ratio and ϵ is the dimensionless gap, i.e.

$$\lambda = \frac{b}{a}, \quad \epsilon = \frac{d}{a}. \qquad (3.3.17)$$

The total resistance on an isolated moving sphere at low-Reynolds-number is a finite constant, whereas that on a sphere which is nearly contacting another stationary sphere is not a constant, and will tend to infinity as $\epsilon \to 0$. The moving sphere cannot contact the stationary sphere within a finite time interval. We will see later that this feature can significantly affect the processes of sedimentation and coagulation of aerosol particles. Failure in understanding the resistance of the lubricating film between two particles tending to infinity leads underestimation of the key effect of van der Waals attractive potential on coagulation in Fuchs' book[2].

When the gap between surfaces of the two spheres is not so small, the effect of the flow field outside the gap cannot be neglected again and must be taken into account. This is a singular perturbation problem, since the resistance inside the gap tends to infinity (inner region) as $\epsilon \to 0$, whereas that outside the gap remains finite (outer region). Using the method of matched asymptotic expansions we may obtain the asymptotic expansion for the factor f, viz.

$$f = C_1\epsilon^{-1} + C_2\ln\epsilon^{-1} + C_3 + C_4\epsilon\ln\epsilon^{-1} + C_5\epsilon + O(\epsilon^2\ln\epsilon^{-1}), \qquad (3.3.18)$$

where C_1 through C_5 are known functions of λ. We will discuss this method in some detail in next section.

4. Flow due to Transverse Relative Motion of Two Spheres

The discussion is similar to the previous section and begins from the sphere–plane problem.

4.1. *A Rigid Sphere Moving Parallel to A Rigid Nearby Plane*

It is supposed that the fluid motion is generated by a rigid sphere of radius a, moving with uniform velocity $(\mathcal{U}, 0, 0)$ referred to a system of Cartesian coordinates (ax, ay, az) in which the plane is $z = 0$ and, instantaneously, the coordinates of the sphere centre are $(0, 0, a(1 + \varepsilon))$. Then the minimum clearance between the sphere and the plane is εa. It is further supposed that the fluid is incompressible, has a constant density and viscosity , and the Reynolds number Re

$$Re = \frac{\mathcal{U} a \rho}{\mu}$$

is sufficiently small to permit neglecting the inertia terms in the Navier–Stokes equation s. The equations governing the fluid motion are therefore the Stokes equations Eq. (3.1.1) and Eq. (2.5.2). The boundary conditions are

$$\mathbf{u} \to 0 \qquad \text{as} \quad (x, y, z) \to \infty, \tag{3.4.1}$$

and

$$\mathbf{u} = 0 \qquad \text{at} \quad z = 0. \tag{3.4.2}$$

Further at any point on the rigid sphere the components (u, v, w) of **u** in cylindrical coordinates $(ar, 0, az)$, if one draws a line through the center of the sphere and normal to the plane, the intersecting point on the plane can be chosen as the origin of the coordinate system, are given by

$$u = \mathcal{U}\cos\theta, \quad v = -\mathcal{U}\sin\theta, \quad w = 0. \tag{3.4.3}$$

The method for solving this problem is to divide the flow field into an inner region in which r, z are both small, the velocity gradients and pressure are large and flow properties are similar to those in lubrication problems, and an outer region comprising the rest of the flow field in which essentially ε may be set equal to zero, the velocity gradients and pressures are moderate and the flow properties are obtained by means of the Fourier–Bessel transforms. We first consider the inner region and consider Stokes equations in cylindrical coordinates system, i.e.

$$\frac{1}{\rho} \frac{\partial p}{\partial r'} = \nu \left(\nabla'^2 u - \frac{u}{r'^2} - \frac{2}{r'^2} \frac{\partial v}{\partial \theta} \right), \tag{3.4.4}$$

$$\frac{1}{\rho r'}\frac{\partial p}{\partial \theta} = \nu \left(\nabla'^2 v + \frac{2}{r'^2}\frac{\partial u}{\partial \theta} - \frac{v}{r'^2} \right), \tag{3.4.5}$$

$$\frac{1}{\rho}\frac{\partial p}{\partial z'} = \nu \nabla'^2 w, \tag{3.4.6}$$

and

$$\frac{1}{r'}\frac{\partial (r'u)}{\partial r'} + \frac{1}{r'}\frac{\partial v}{\partial \theta} + \frac{\partial w}{\partial z'} = 0, \tag{3.4.7}$$

where $r' = ar$, $z' = az$ and $\nabla'^2 = a^{-2}\nabla^2$.

In the inner region we write (u, v, w) as

$$\frac{u}{U} = U\cos\theta, \quad \frac{v}{U} = V\sin\theta, \quad \frac{w}{U} = W\cos\theta, \tag{3.4.8}$$

and the pressure p as

$$\frac{ap}{\mu U} = P\cos\theta. \tag{3.4.9}$$

Substituting Eq. (3.4.8) and Eq. (3.4.9) in Eq. (3.4.4)—Eq. (3.4.7) yields

$$\frac{\partial P}{\partial r} = L_0^2 U - \frac{2(U + V)}{r^2}, \tag{3.4.10}$$

$$\frac{-P}{r} = L_0^2 V - \frac{2(U + V)}{r^2}, \tag{3.4.11}$$

$$\frac{\partial P}{\partial z} = L_1^2 W, \tag{3.4.12}$$

and

$$\frac{\partial U}{\partial r} + \frac{U + V}{r} + \frac{\partial W}{\partial z} = 0, \tag{3.4.13}$$

where the operator L_0^2 is defined by $L_0^2 = \frac{\partial^2}{\partial r^2} + \frac{1}{r}\frac{\partial}{\partial r} + \frac{\partial^2}{\partial z^2}$ and the operator L_m^2 is defined by $L_m^2 = L_0^2 - m^2/r^2$. The clearance $a\delta$ between the rigid sphere and the rigid plane, expressed as a function of r, is given by

$$\delta = 1 + \varepsilon - (1 - r^2)^{1/2}. \tag{3.4.14}$$

Now δ and therefore z are of order $O(\varepsilon)$, which together with Eq. (3.4.14) one may suggest to introduce new variables, R, Z, defined as

$$r = \varepsilon^{1/2} R, \quad z = \varepsilon Z. \tag{3.4.15}$$

Thus Eq. (3.4.14) now changes to a new form as

$$\delta = \varepsilon H + \frac{1}{8}\varepsilon^2 R^4 + O(\varepsilon^3), \tag{3.4.16}$$

where $H = 1 + \frac{1}{2}R^2$.

The boundary conditions determine that U and V are of order $O(1)$. Consequently Eq. (3.4.13) results that $W = O(\varepsilon^{1/2})$ and Eq. (3.4.11) results that $P = O(\varepsilon^{-\frac{3}{2}})$, which suggests that we may expect the solutions of P, U, V and W of the forms

$$P(r, z) = \varepsilon^{-\frac{3}{2}} P_0(R, Z) + \varepsilon^{-\frac{1}{2}} P_1(R, Z) + O(\varepsilon^{\frac{1}{2}}), \qquad (3.4.17)$$

$$U(r, z) = U_0(R, Z) + \varepsilon U_1(R, Z) + O(\varepsilon^2), \qquad (3.4.18)$$

$$V(r, z) = V_0(R, Z) + \varepsilon V_1(R, Z) + O(\varepsilon^2), \qquad (3.4.19)$$

$$W(r, z) = \varepsilon^{\frac{1}{2}} W_0(R, Z) + \varepsilon^{\frac{3}{2}} W_1(R, Z) + O(\varepsilon^{\frac{5}{2}}). \qquad (3.4.20)$$

Substituting Eq. (3.4.14) and Eq. (3.4.17) through Eq. (3.4.20) into Eq. (3.4.10) through Eq. (3.4.13) the following set of equations for U_0, V_0, W_0 and P_0 can be obtained

$$\frac{\partial P_0}{\partial R} = \frac{\partial^2 U_0}{\partial Z^2}, \quad -\frac{P}{R} = \frac{\partial^2 V_0}{\partial Z^2}, \quad \frac{\partial P_0}{\partial Z} = 0, \qquad (3.4.21)$$

$$\frac{\partial U_0}{\partial R} + \frac{U_0 + V_0}{R} + \frac{\partial W_0}{\partial Z} = 0, \qquad (3.4.22)$$

and the following set of equations for U_1, V_1, W_1 and P_1 are

$$\frac{\partial P_1}{\partial R} = \frac{\partial^2 U_1}{\partial Z^2} + \frac{\partial^2 U_0}{\partial R^2} + \frac{1}{R} \frac{\partial U_0}{\partial R} - \frac{2(U_0 + V_0)}{R^2}, \qquad (3.4.23)$$

$$-\frac{P_1}{R} = \frac{\partial^2 V_1}{\partial Z^2} + \frac{\partial^2 V_0}{\partial R^2} + \frac{1}{R} \frac{\partial V_0}{\partial R} - \frac{2(U_0 + V_0)}{R^2}, \qquad (3.4.24)$$

$$\frac{\partial^2 P_1}{\partial Z^2} = \frac{\partial^2 W_1}{\partial Z^2}, \qquad (3.4.25)$$

and

$$\frac{\partial U_1}{\partial R} + \frac{U_1 + V_1}{R} + \frac{\partial W_1}{\partial Z} = 0. \qquad (3.4.26)$$

Similar sets of equations can be written down for U_i, V_i, W_i and $P_i (i > 1)$.

The boundary conditions Eq. (3.4.3) are equivalent to

$$U_i(R, 0) = V_i(R, 0) = W_i(R, 0) = 0 \quad (i \geq 0), \qquad (3.4.27)$$

and

$$\left. \begin{array}{l} U_0(R, \frac{\delta}{\varepsilon}) + \varepsilon U_1(R, \frac{\delta}{\varepsilon}) + \cdots = 1, \\ V_0(R, \frac{\delta}{\varepsilon}) + \varepsilon V_1(R, \frac{\delta}{\varepsilon}) + \cdots = -1, \\ W_0(R, \frac{\delta}{\varepsilon}) + \varepsilon W_1(R, \frac{\delta}{\varepsilon}) + \cdots = 0, \end{array} \right\} \qquad (3.4.28)$$

from which we may deduce that

$$U_0(R, H) = -V_0(R, H) = 1, \quad W_0(R, H) = 0, \qquad (3.4.29)$$

$$U_1(R, H) = -\frac{1}{8} R^4 U_{0Z}(R, H), \qquad (3.4.30)$$

$$V_1(R, H) = -\frac{1}{8}R^4 V_{0Z}(R, H),$$

(3.4.31)

$$W_1(R, H) = -\frac{1}{8}R^4 W_{0Z}(R, H).$$

(3.4.32)

The subscript Z indicates differentiation with respect to Z. Expressions for $U_i(R, H)$, $V_i(R, H)$ and $W_i(R, H)(i > 1)$ can also be written down by equating the coefficients of appropriate higher powers of ε.

Eq. (3.4.21), Eq. (3.4.22), Eq. (3.4.27) with $i=0$ together with Eq. (3.4.29) are of the form which occurs in the classical lubrication theory and they are solved in the usual way in that theory. The appropriate form for this type of Reynolds equations is found to be

$$R^2 P_0'' + (R + 3R^3/H)P_0' - P_0 = -6R^3/H.$$

(3.4.33a)

A special integral of the Reynolds equation is

$$P_0 = 6R/5H^2,$$

(3.4.33b)

and Eq. (3.4.33b) gives rise to the unique solution of the Reynolds equation which decays to zero at $R = \infty$.

From Eq. (3.4.21), Eq. (3.4.22), Eq. (3.4.27) with $i=0$ and Eq. (3.4.33) it can be shown that

$$U_0 = \frac{6 - 9R^2}{10H^3}Z^2 + \frac{2 + 7R^2}{5H^2}Z,$$

(3.4.34)

$$V_0 = -\frac{3}{5H^2}Z^2 - \frac{2}{5H}Z,$$

(3.4.35)

$$W_0 = \frac{8R - 2R^3}{5H^4}Z^3 + \frac{2R^3 - 7R}{5H^3}Z^2.$$

(3.4.36)

The contributions to the Cartesian components of the force exerted by the fluid on the rigid plane at $Z = 0$, which result from the effect of the inner solution are $(\overline{F}_x^i, \overline{F}_y^i, \overline{F}_z^i)$ and are given by

$$\overline{F}_x^i = 6\pi\mu a \mathcal{U}\overline{f}^i, \quad \overline{F}_y^i = \overline{F}_z^i = 0,$$

where

$$\overline{f}^i = \frac{1}{6}\int_0^{R_0} \left\{ \frac{\partial U}{\partial Z} - \frac{\partial V}{\partial Z} \right\}_{z=0} R dR,$$

R_0 being a value of R which is so chosen that the inner solution is valid for $0 \leq R \leq R_0$. The contribution of $\overline{f}_0^i$ to $\overline{f}^i$ in the solution is therefore given by

$$\overline{f}_0^i = \frac{1}{6}\int_0^{R_0} \left\{ \frac{\partial U_0}{\partial Z} - \frac{\partial V_0}{\partial Z} \right\}_{z=0} R dR,$$

(3.4.37)

Using Eq. (3.4.34) and Eq. (3.4.35) one can get

$$\overline{f}_0^i = \frac{8}{15}\ln(1 + \frac{1}{2}R_0) - \frac{2}{5} + \frac{4}{5(2 + R_0^2)}.$$

(3.4.38)

The contributions to the Cartesian components of the force exerted on the rigid sphere due to the inner solution are (F_x^i, F_y^i, F_z^i), and are given by

$$F_x^i = -6\pi\mu a \mathcal{U} f^i, \quad F_y^i = F_z^i = 0,$$

where

$$f^i = \frac{1}{6}\int_0^{\chi_0}\left\{\left[P - \varepsilon^{-\frac{1}{2}}\left(2\frac{\partial U}{\partial R} - \frac{\partial V}{\partial R}\right)\right]\sin\chi \right.$$
$$\left. + \left[\varepsilon^{-1}\left(\frac{\partial U}{\partial Z} - \frac{\partial V}{\partial Z}\right) + \varepsilon^{-\frac{1}{2}}\frac{\partial W}{\partial R}\right]\cos\chi\right\}_{z=\frac{t}{\varepsilon}}\sin\chi d\chi, \qquad (3.4.39)$$

and $\pi - \chi$ is the angle between the radial vector of the sphere and the z-axis, and χ_0 is chosen as the inner solution is valid for $0 \le \chi \le \chi_0$. The contribution f_0^i to f^i from the leading terms of the inner solution is found by using Eq. (3.4.17)–Eq. (3.4.20) and note that $\chi = \varepsilon^{\frac{1}{2}}R + O(\varepsilon^{\frac{3}{2}})$ for small values of χ. The expression of f_0^i is given by

$$f_0^i = \frac{1}{6}\int_0^{R_0}\{RP_0 + \frac{\partial U_0}{\partial Z} - \frac{\partial V_0}{\partial Z}\}_{z=H}RdR = \frac{8}{15}\ln(1 + \frac{1}{2}R_0^2). \qquad (3.4.40)$$

We now turn to the outer region.

The equations governing the flow field which is not in the neighbourhood of the nearest points of the rigid sphere and the rigid plane are again given by Eq. (3.1.1) and Eq. (2.5.2) with the boundary conditions Eq. (3.4.1), Eq. (3.4.2), Eq. (3.4.3). These equations are satisfied when the pressure p and the cylindrical components (u, v, w) of **u** are of the form

$$\left. \begin{array}{rl} \frac{ap}{\mu \mathcal{U}} &= 2Q\cos\theta, \quad \frac{u}{\mathcal{U}} = [rQ + \frac{1}{2}(\chi + \psi)]\cos\theta, \\ \frac{v}{\mathcal{U}} &= \frac{1}{2}[\chi - \psi]\sin\theta, \quad \frac{w}{\mathcal{U}} = [zQ + \varphi]\cos\theta, \end{array}\right\} \qquad (3.4.41)$$

where r, z are cylindrical coordinates which are dimensionless relative to the radius of the sphere, and ψ, χ, ϕ and Q are functions of r, z only, satisfying

$$L_0^2\psi = L_2^2\chi = L_1^2\varphi = L_1^2Q = 0. \qquad (3.4.42)$$

The operator L_0^2 and L_m^2 are the same as before.

Substituting u, v, w from Eq. (3.4.41) into the mass conservation equation, one can be obtain

$$\left[3 + r\frac{\partial}{\partial r} + z\frac{\partial}{\partial z}\right]Q + \frac{1}{2}\left[\frac{\partial\psi}{\partial r} + \left(\frac{\partial}{\partial r} + \frac{2}{r}\right)\chi + 2\frac{\partial\varphi}{\partial z}\right] = 0. \qquad (3.4.43)$$

The boundary conditions on the rigid plane and the rigid sphere require that

$$\varphi + zQ = 0, \quad \chi + rQ = 0, \quad \psi + rQ = 0, \qquad (3.4.44, 45, 46)$$

on the rigid plane and that

$$\varphi + zQ = 0, \quad \chi + rQ = 0, \quad \psi + rQ = 2, \qquad (3.4.47, 48, 49)$$

on the rigid sphere.

A first order approximation to the solution for the flow in the region which is not in the neighbourhood of the nearest points of the sphere and the plane would be of the limit form when $\epsilon \to 0$, namely the sphere is in contact with the plane at $z = 0$. This limit form is found by solving the problem with $\epsilon = 0$ excluding the origin from the region since the boundary conditions Eq. (3.4.46) and Eq. (3.4.49) cannot be satisfied simultaneously there. The fact that the component w of velocity is zero on the z-axis $(0 \leq z \leq \epsilon)$ when $\epsilon \neq 0$ and that the boundary values of w on the sphere surface and the plane are both zero when $\epsilon = 0$ suggests that w remains finite as $\epsilon \to 0$.

In order to facilitate the solution of the problem when $\epsilon = 0$ we transform the coordinates r, z to tangent-sphere coordinates ξ, η by relations

$$r = 2\eta/(\xi^2 + \eta^2), \quad z = 2\xi/(\xi^2 + \eta^2). \qquad (3.4.50)$$

The rigid plane is now at $\xi = 0$, the rigid sphere at $\xi = 1$, the origin $r = z = 0$ is set by $\eta = \infty$ in contrast with the infinity by $\xi = \eta = 0$. It may be shown that a solution of the equation $L_m^2 f = 0$ which is nonsingular at $\xi = \eta = 0$ is given by

$$f = (\xi^2 + \eta^2)^{\frac{1}{2}} \int_0^\infty \{\mathcal{A}(s)\sinh s\xi + \mathcal{B}(s)\cosh s\xi\} J_m(s\eta)ds\}, \qquad (3.4.51)$$

where $\mathcal{A}(s)$ and $\mathcal{B}(s)$ are functions of s which are chosen to guarantee the convergence of the integral.

We shall assume that there is a region of the space occupied by the fluid in which both the inner and outer solution s are valid so that both solutions match in this region. Such an assumption implies that, in this region,

$$\epsilon^{\frac{1}{2}} R \sim \frac{2}{\eta}, \qquad (3.4.52)$$

for large R and η. Furthermore, in this region

$$P \sim \frac{24}{5}\epsilon^{-\frac{3}{2}}R^{-3} \sim \frac{3}{5}\eta^3. \qquad (3.4.53)$$

Consequently, for large values of η,

$$Q \sim \frac{3}{10}\eta^3 \quad \text{and} \quad zQ \sim \frac{3}{5}\xi\eta. \qquad (3.4.54)$$

It is therefore apparent that w cannot be non-singular at the origin unless

$$\varphi \sim -\frac{3}{5}\xi\eta, \qquad (3.4.55)$$

for large values of η, and

$$\chi \sim -\frac{3}{5}\eta^2, \qquad (3.4.56)$$

$$\psi \sim -\frac{3}{5}\eta^2. \tag{3.4.57}$$

In view of Eq. (3.4.42), Eq. (3.4.44) and Eq. (3.4.51) a suitable form for φ is given as

$$\varphi = (\xi^2 + \eta^2)^{\frac{1}{2}} \int_0^\infty A(s)\sinh s\xi J_1(s\eta)ds, \tag{3.4.58}$$

where $A(s) \sim -3/5s^2$ for small s and exponentially deceases for large s.

It is impossible to construct a solution Q of the type Eq. (3.4.51) with $m{=}1$ which makes the integral convergent. Therefore we write

$$Q = Q_0 + \frac{3}{10}Q_1, \tag{3.4.59}$$

where $L_1^2 Q_0 = L_1^2 Q_1 = 0$. A suitable form for Q_1 is found to be

$$Q_1 = \eta(\xi^2 + \eta^2) + (\xi^2 + \eta^2)^{\frac{1}{2}} \int_0^\infty V_1(\xi, s)J_1(s\eta)ds, \tag{3.4.60}$$

where

$$V_1(\xi s) = e^{-s\xi}[\xi^2/s + 3\xi/s^2] - [3e^{-s}\sinh s\xi]/s^3. \tag{3.4.61}$$

The function Q_0 is given by

$$Q_0 = \frac{1}{2}(\xi^2 + \eta^2)^{\frac{1}{2}} \int_0^\infty [B\cosh s\xi + C\sinh s\xi]J_1(s\eta)ds, \tag{3.4.62}$$

where

$$B = SA'' + A' - A/S + 9/5s^3, \tag{3.4.63}$$

$$C = [sA'' - A']K + \frac{9}{5}[e^{-s} - 1]/s^3, \tag{3.4.64}$$

and

$$K = s^{-1} - \coth s. \tag{3.4.65}$$

For the solution χ, we write

$$\chi = \chi_0 + \frac{3}{5}\chi_1, \tag{3.4.66}$$

where $L_2^2 \chi_0 = L_2^2 \chi_1 = 0$. A suitable form for χ_1 is found to be

$$\chi_1 = -\eta^2 + (\xi^2 + \eta^2)^{\frac{1}{2}} \int_0^\infty V_2(\xi, s)J_2(s\eta)ds, \tag{3.4.67}$$

where

$$V_2(\xi, s) = e^{-s\xi}[\xi^2 + 3\xi/s]. \tag{3.4.68}$$

The function χ_0 is given as

$$\chi_0 = (\xi^2 + \eta^2)^{\frac{1}{2}} \int_0^\infty [F\cosh s\xi + G\sinh s\xi]J_2(s\eta)ds, \tag{3.4.69}$$

where

$$F = A - sA' + 9/5s^2, \tag{3.4.70}$$

$$G = [2A - sA']K - 9/5s^2. \tag{3.4.71}$$

To determine a suitable form for the function one can carry out in a similar way, as for Q and χ, we write

$$\psi = \psi_0 + \frac{3}{5}\psi_1. \tag{3.4.72}$$

A suitable form for ψ_1 is found to be

$$\psi_1 = -\eta^2 + (\xi^2 + \eta^2)^{\frac{1}{2}} \int_0^\infty V_0(\xi, s) J_0(s\eta) \mathrm{d}s, \tag{3.4.73}$$

where

$$V_0(\xi, s) - e^{-s\xi}[\xi^2 + \xi/s] + [e^{-s}\mathrm{sinh}s\xi]/s^2. \tag{3.4.74}$$

The form for ψ_0 is found to be

$$\psi_0 = (\xi^2 + \eta^2)^{\frac{1}{2}} \int_0^\infty [D\mathrm{cosh}s\xi + E\mathrm{sinh}s\xi] J_0(s\eta) \mathrm{d}s, \tag{3.4.75}$$

where, in order to satisfy the boundary conditions Eq. (3.4.46) and Eq. (3.4.49),

$$D = sA' + A - 3/5s^2, \tag{3.4.76}$$

and

$$E = sA'K - \frac{3}{5}[e^{-s} - 1]/s^2 + 2(\mathrm{coth}s - 1). \tag{3.4.77}$$

Substituting Q, χ, ψ, φ into the mass-conservation Eq. (3.4.43), we obtain the equation for A as

$$s^3 K'A'' + sA'[s^2 K'' + 3sK' + 2K] - A[s^2 K'' + 4sK' + 2K] + s^2 X'' = 0, \tag{3.4.78}$$

where $K = s^{-1} - \mathrm{coth}s$ and $X = \mathrm{coth}s - 1$.

The solution of Eq. (3.4.78) which we seek for must be $\sim 3/5s^2$ for small values of s and must decay to zero exponentially for large values of s. It has not been possible to obtain such solution in a closed form. It may be easily shown that if A_1 and A_2 are special integrals and complementary function respectively of Eq. (3.4.78) for small values of s, then

$$A_1 = -3/5s^2 + O(1),$$

$$A_2 = s^{\sqrt{10}-2}\{1 - O(s^2)\}.$$

The Eq. (3.4.78) was integrated numerically and was found to be

$$\sim -2s^2 e^{-2s} + O(e^{-2s})$$

for large values of s.

We now calculate the contributions to the force acting on the plane and sphere due to the outer solution .

The leading order contributions to the Cartesian components of the force acting on that part of the rigid plane $\xi = 0$ for which $0 \leq \eta \leq \eta_0$ which result from the outer solution are $(\overline{F}_x^o, \overline{F}_y^o, \overline{F}_z^o)$ and given as

$$\overline{F}_x^o = 6\pi\mu a\mathcal{U}\overline{f}^o, \quad \overline{F}_y^o = \overline{F}_z^o = 0,$$

where

$$\overline{f}^o = \frac{1}{3}\int_0^{\eta_0}\left\{\frac{2}{\eta^2}\frac{\partial Q}{\partial \xi} + \frac{1}{\eta}\frac{\partial \psi}{\partial \xi}\right\}_{\xi=0}\, d\eta. \tag{3.4.79}$$

The contribution $\overline{f}_0^o$ to $\overline{f}^o$ from the solution which we have obtained in above discussions is therefore given as

$$\overline{f}_0^o = \frac{16}{15}\ln\eta_0 - \frac{8}{15} + \frac{1}{3}\int_0^\infty\{s^2A''K + 2(\coth s - 1) - 16e^{-2s}/5s\}ds + p_1(\eta_0), \tag{3.4.80}$$

where $p_1(\eta_0) \to 0$ as $\eta_0 \to \infty$. The total force acting on the rigid plane may now be obtained by combining the contributions of both the inner and outer solution s as given in Eq. (3.4.38) and Eq. (3.4.80), provided that we restrain the concerned r within a region where the inner and outer solution s are both valid , and the region corresponds to that η_0 and R_0 satisfy a relation

$$R_0\eta_0 = 2/\epsilon^{\frac{1}{2}}. \tag{3.4.81}$$

It is assumed here that the inner and outer solution s have an overlapping region in which $1 \ll R \ll \epsilon^{-\frac{1}{2}}$ and $1 \ll \eta \ll \epsilon^{-\frac{1}{2}}$. The first-order approximation, neglecting these terms which approach to zero with ϵ, then gives

$$\overline{F}_x = 6\pi\mu a\mathcal{U}\overline{f}_0, \quad \overline{F}_y = \overline{F}_z = 0,$$

where $\overline{f}_0 = \overline{f}_0^i + \overline{f}_0^o$. Using Eq. (3.4.38), Eq. (3.4.80) and Eq. (3.4.81), then O'Neill and Stewartson (1967) gave [6]

$$\overline{f}_0 = \frac{8}{15}\ln\left(\frac{2}{\epsilon}\right) - \frac{14}{5} + \frac{1}{3}\int_0^\infty\{s^2A''K + 2(\coth s - 1) - 16e^{-2s}/5s\}ds. \tag{3.4.82}$$

The discussion of the force on the rigid sphere follows a similar line. The leading order contributions to the Cartesian components of the force acting on the sphere $\xi = 1$ due to the outer solution are (F_x^o, F_y^o, F_z^o) and are given as

$$F_x^o = -6\pi\mu a\mathcal{U}f^o, \quad F_y^o = F_z^o = 0,$$

where

$$f^o = \frac{1}{3}\int_0^{\eta_0}\left\{r\frac{\partial Q}{\partial \xi} - Q\frac{\partial r}{\partial \xi} + \frac{\partial \psi}{\partial \xi}\right\}_{\xi=1}\frac{\eta\, d\eta}{1+\eta^2}; \tag{3.4.83}$$

and η_0 is defined as before. The leading term f_0^o of f^o is given by

$$f_0^o = \frac{16}{15}\ln\eta_0 - \frac{14}{15} + \frac{1}{3}\int_0^\infty \{s^2 A'' K + 2(\coth s - 1) - 16e^{-2s}/5s\}ds + q_1(\eta_0), \quad (3.4.84)$$

where $q_1(\eta_0) \to 0$ as $\eta_0 \to \infty$.

Combining Eq. (3.4.84) with Eq. (3.4.40) we obtain the total force exerted on the rigid sphere by the fluid to the first order approximation . This is given by O'Neill and Stewartson (1967)[6]

$$F_x = 6\pi\mu a \mathcal{U} f_0, \quad F_y = F_z = 0,$$

where

$$f_0 = \frac{8}{15}\ln(\frac{2}{\epsilon}) - \frac{14}{15} - \frac{1}{3}\int_0^\infty \{s^2 A'' K + 2(\coth s - 1) - 16e^{-2s}/5s\}ds. \quad (3.4.85)$$

4.2. Two Rigid Spheres Moving along the Direction Perpendicular to the Line Connecting Their Centers When the Minimum Clearance between the Two Spheres Approaches to Zero

If we denote the minimum clearance between the moving sphere A of radius a and the stationary sphere B of radius b by ϵa, it is clear that as $\epsilon \to 0$, the flow in the neighbourhood of the nearest points of the spheres becomes strongly sheared with large velocity gradients and pressure whilst in general the flow in the other region of the fluid will remain essentially weakly sheared with velocity gradients and pressure both being moderate. To discuss the properties of the flow in the entire fluid rigorously, it would be necessary to approach the problem in a way in analogy to the above treatment. Such an approach would necessitate to match asymptotic expansion of the solution of the Stokes equation which is valid in the inner strongly sheared flow region, to that in the outer region.

We now let (ar, θ, az) denote cylindrical coordinates with the origin at $r = z = 0$ which sets at the point on sphere B and is nearest to sphere A , while the z-axis is along the line connecting the centres, in which case the equations of the surfaces of sphere A and sphere B will be described by

$$z = 1 + \epsilon - (1 - r^2)^{1/2} \quad \text{(for sphere } A\text{)}, \quad\quad (3.4.86)$$

$$z = -\lambda_1^{-1} - (\lambda_1^{-2} - r^2)^{1/2} \quad \text{(for sphere } B\text{)} \quad\quad (3.4.87)$$

respectively, where $\lambda_1 = a/b$. The forms of Eq. (3.4.86) and Eq. (3.4.87) suggest that within the inner region we can introduce the strained coordinates R, Z defined as

$$r = \epsilon^{\frac{1}{2}}R, \quad z = \epsilon Z, \quad\quad (3.4.88)$$

in which case Eq. (3.4.86) and Eq. (3.4.87) now respectively yield

$$Z = 1 + \frac{1}{2}R^2 + O(\epsilon), \tag{3.4.89}$$

$$Z = -\frac{1}{2}\lambda_1 R^2 + O(\epsilon). \tag{3.4.90}$$

We still have the cylindrical components (u, v, w) of the flow field and pressure p as those given in Eq. (3.4.8) and Eq. (3.4.9). The dimensionless functions U, V, W and P satisfy the same equations as Eq. (3.4.10)–Eq. (3.4.13). The asymptotic expansions of P, U, V, W as $\epsilon \to 0$ are the same as Eq. (3.4.17)–Eq. (3.4.20). Thus the first expansion terms P_0, U_0, V_0 and W_0 satisfy Eq. (3.4.21) and Eq. (3.4.22). The boundary conditions on sphere A is nearly the same as Eq. (3.4.29), the only changed is at H. We will use H_1 instead of H, as $H_1 = 1 + \frac{1}{2}R^2$. In addition, the boundary conditions on sphere B are given as

$$U_0(R, H_2) = V_0(R, H_2) = W_0(R, H_2) = 0, \tag{3.4.91}$$

where $H_2 = -\frac{1}{2}\lambda_1 R^2$. Integrating Eq. (3.4.21) one has

$$U_0 = \frac{1}{2}P_1'Z^2 + A_1 Z + B_1, \quad V_0 = \frac{1}{2}R^{-1}P_0 Z^2 + A_2 Z + B_2, \tag{3.4.92}$$

where A_1, A_2, B_1 and B_2 are functions of R which are determined in terms of the boundary conditions Eq. (3.4.29) and Eq. (3.4.91). After integrating Eq. (3.4.22) with respect to Z from H_2 to H_1 and using Eq. (3.4.92) and Eq. (3.4.91) we obtain the Reynolds equation for the two-sphere case:

$$R^2 P_0'' + [R - 3(1 + \lambda_1)R^3(H_2 - H_1)^{-1}]P_0' - P_0 = 6(1 + \lambda_1)R^3(H_2 - H_1)^{-3}. \tag{3.4.93}$$

A special integral of this equation is

$$P_0 = \frac{6(1 - \lambda_1)R}{5(1 + \lambda_1)(H_1 - H_2)^2}. \tag{3.4.94}$$

When $b \to \infty$, sphere B becomes a plane, thus $\lambda_1 \to 0$ and Eq. (3.4.94) agrees with Eq. (3.4.33). It can be shown that Eq. (3.4.94) is the complete solution of Eq. (3.4.93). From the solution for P_0, we may obtain the solution for U_0, V_0 and W_0 by using equations Eq. (3.4.92) and Eq. (3.4.13). Thus we may calculate the contributions from the inner solution to the Cartesian components of the force exerted on sphere A and sphere B respectively.

Using the method similar to that O'Neill and Stewartson employed (1967)[6] we may also obtain the outer solution for Q, ψ, χ and φ, and then can calculate the contributions from the outer solution to the Cartesian components of the force exerted on sphere A and sphere B respectively.

Matching the inner and the outer solution s at r_0 which is in the overlap region, and combining the contributions from both the inner and outer solutions we may obtain the total force acting on the sphere A and sphere B.

It is indicated that the contribution from the inner region is different from that from the outer region. For example, the leading term $\frac{8}{15}\ln\epsilon^{-1}$ in the resistance coefficient Eq. (3.4.85) is due completely to inner region. The leading term of the contribution from the outer region is only a constant term. At matching point r_0, we have the following relation (seeEq. (3.4.52))

$$\epsilon^{\frac{1}{2}}R_0 = r_0 = \frac{2}{\eta_0}.$$

Then the leading term of the contribution from the inner solution is (see Eq. (3.4.40)

$$f_0^i = \frac{8}{15}\ln\frac{1}{2}R_0^2 = \frac{8}{15}\left(\ln\epsilon^{-1} + \ln\frac{r_0^2}{2}\right).$$

On the other hand, from Eq. (3.4.84) we see that the leading term of the contribution from the outer solution is

$$f_0^o = \frac{8}{15}\eta_0^2 = \frac{8}{15}\left(\ln 2 + \ln\frac{2}{r_0^2}\right).$$

The terms which depend on r_0 must be cancel each other, when we combine f_0^i and f_0^o to obtain the total resistance coefficient f_0. Thus the singular term in f_0 is completely due to the inner solution, whereas the constant term $\frac{8\ln 2}{15}$ is due to the outer solution . One can easily understand this property from the lubrication theory , and here we only give the leading terms of the resistance coefficients on sphere A and sphere B. The resistance coefficient f_a on sphere A is (O'Neill and Majumdar 1970 a)[7]

$$f_a = \frac{4}{15}\frac{(2 + \lambda_1 + 2\lambda_1^2)}{(1 + \lambda_1)^3}\ln\epsilon^{-1} + O(1), \qquad (3.4.95)$$

and Eq. (3.4.95) agrees with Eq. (3.4.85) as $\lambda_1 \to 0$ (i.e. $b \to \infty$).
The resistance coefficient f_b on sphere B is (O'Neill and Majumdar (1970a))[7].

$$f_b = -\frac{4}{15}\lambda_1\frac{[2 + \lambda_1 + \lambda_1^2]}{(1 + \lambda_1)^3}\ln\epsilon^{-1} + O(1), \qquad (3.4.96)$$

and Eq. (3.4.96) agrees with Eq. (3.4.82) as $\lambda_1 \to 0$ (i.e. $b \to \infty$).

4.3. Two Rigid Spheres Moving along the Direction Perpendicular to the Line Connecting Their Centres with Arbitrary Values of the Distance between Centres

The rigid sphere A has radius a, moving with uniform velocity $(\mathcal{U}, 0, 0)$ referring to a Cartesian co-ordinate system (x, y, z), and the rigid sphere B is at rest. Let us choose the cylindrical coordinates (r, θ, z) so that the line connecting their centres lies along the z-axis. Note that (x, y, z) and (r, z) are now dimensional quantities, and sphere A lies completely in the half-space $z > 0$.

Because of the smallness of the Reynolds number, the governing equations of the flow field induced by the moving sphere A are still Eq. (3.1.1) and Eq. (2.5.2).

The boundary conditions at sphere A are

$$u = \mathcal{U}\cos\theta, \quad v = \mathcal{U}\sin\theta, \quad w = 0, \qquad (3.4.97)$$

here (u, v, w) are the cylindrical components of the velocity of the flow field, and the boundary conditions at sphere B are

$$u = v = w = 0. \qquad (3.4.98)$$

Equation Eq. (3.1.1) is satisfied when p, u, v, w are given as

$$\frac{cp}{\mu\mathcal{U}} = P\cos\theta, \quad \frac{u}{\mathcal{U}} = \frac{1}{2c}[rP + c(\chi + \psi)]\cos\theta, \qquad (3.4.99)$$

$$\frac{v}{\mathcal{U}} = \frac{1}{2}[\chi - \psi]\sin\theta, \quad \frac{w}{\mathcal{U}} = \frac{1}{2c}[zP + 2c\varphi]\cos\theta, \qquad (3.4.100)$$

where c is a constant length which will be defined later and P, φ, χ and ψ are functions of r, z satisfying

$$L_1^2 P = L_1^2 \varphi = L_2^2 \chi = L_0^2 \psi = 0, \qquad (3.4.101)$$

while the operators L_m^2 having the same definition as before.

With u, v, w given in Eq. (3.4.99) and Eq. (3.4.100), the mass-conservation equation holds provided that

$$\left[3 + r\frac{\partial}{\partial r} + z\frac{\partial}{\partial z}\right]P + c\left[\frac{\partial\psi}{\partial r} + \left(\frac{\partial}{\partial r} + \frac{2}{r}\right)\chi + 2\frac{\partial\varphi}{\partial z}\right] = 0. \qquad (3.4.102)$$

In order to facilitate the determination of the solutions, we introduce bispherical coordinates (ξ, η) defined as

$$r = \frac{c \sin\eta}{\cosh\xi - \cos\eta}, \quad z = \frac{c \sinh\xi}{\cosh\xi - \cos\eta}, \quad (0 \leq \eta \leq \pi). \qquad (3.4.103)$$

The surface defined in Eq. (3.4.103) as $\xi = \xi_0$, where ξ_0 is a non-zero constant, is a sphere of radius $c|\text{cosech}\xi_0|$ with the centre at $r = 0$, and $z = c\coth\xi_0$; the sphere lies completely in the upper half-space $z > 0$ or lower one $z < 0$ according to $\xi_0 > 0$ or $\xi_0 < 0$ respectively. Thus the spheres A and B are given by $\xi = \alpha$ and $\xi = \beta$ respectively where $\alpha > 0$ and

$$a = c \ \text{cosech}\alpha, \quad b = c|\text{cosech}\beta|, \quad d = c(\coth\alpha - \coth\beta), \qquad (3.4.104)$$

where d is the distance between the two centres of the two spheres. Equations in Eq. (3.4.104) determine α, β and the constant c of Eq. (3.4.99) and Eq. (3.4.100) uniquely. The region occupied by the fluid is determined by $\beta < \xi < \alpha, 0 \leq \eta \leq \pi$

while the points at infinity are given by $\xi = \eta = 0$. In terms of φ, P, ψ and χ, the boundary conditions Eq. (3.4.97) and Eq. (3.4.98) give

$$\frac{1}{2}rP + \chi = 0 = \frac{1}{2}zP + \varphi, \quad (\xi = \alpha), \tag{3.4.105}$$

$$\frac{1}{2}rP + \psi = 2c, \quad (\xi = \alpha), \tag{3.4.106}$$

$$\frac{1}{2}rP + \chi = \frac{1}{2}rP + \psi = \frac{1}{2}zP + \varphi = 0, \quad (\xi = \beta). \tag{3.4.107}$$

Suitable solutions of Eq. (3.4.101) in terms of bispherical coordinates were shown by O'Neill (1964) as

$$\varphi = (\cosh \xi - \sigma)^{\frac{1}{2}} \sum_{n=1}^{\infty} \left[A_n \cosh \left(n + \frac{1}{2} \right) \xi + B_n \sinh \left(n + \frac{1}{2} \right) \xi \right] P_n^1(\sigma), \tag{3.4.108}$$

$$P = (\cosh \xi - \sigma)^{\frac{1}{2}} \sum_{n=1}^{\infty} \left[C_n \cosh \left(n + \frac{1}{2} \right) \xi + D_n \sinh \left(n + \frac{1}{2} \right) \xi \right] P_n^1(\sigma), \tag{3.4.109}$$

$$\psi = (\cosh \xi - \sigma)^{\frac{1}{2}} \sum_{n=1}^{\infty} \left[E_n \cosh \left(n + \frac{1}{2} \right) \xi + F_n \sinh \left(n + \frac{1}{2} \right) \xi \right] P_n(\sigma), \tag{3.4.110}$$

$$\chi = (\cosh \xi - \sigma)^{\frac{1}{2}} \sum_{n=1}^{\infty} \left[G_n \cosh \left(n + \frac{1}{2} \right) \xi + H_n \sinh \left(n + \frac{1}{2} \right) \xi \right] P_n^2(\sigma), \tag{3.4.111}$$

where $\sigma = \cos \eta$, $P_n(\sigma)$ is the Legendre polynomial of order n and

$$P_n^m(\sigma) = (1 - \sigma^2)^{\frac{m}{2}} d^m \{ P_n(\sigma) \} / d\sigma^m$$

with $m = 1, 2$. The assumption that solutions Eq. (3.4.108)–Eq. (3.4.111) exist implies that each of the series is necessarily convergent for $\beta \le \xi \le \alpha$ and $|\sigma| \le 1$. It is therefore evident that

$$\varphi = \psi = \chi = rP = 0,$$

when $\xi = \eta = 0$ which implies that the stationary condition at infinity would then be satisfied.

Substituting Eq. (3.4.108) through Eq. (3.4.101) into the boundary conditions Eq. (3.4.105) through Eq. (3.4.107) produces the following relations expressing each of $C_n, D_n, \cdots, H_n$ in terms of A_n and B_n, one has ;

$$C_n = \overline{C}_n, \quad D_n = \overline{D}_n, \tag{3.4.112}$$

$$E_n = \overline{E}_n - 2\sqrt{2}e^{-(n+1/2)\alpha} \sinh \left(n + \frac{1}{2} \right) \beta \operatorname{cosech} \left(n + \frac{1}{2} \right) (\alpha - \beta), \tag{3.4.113}$$

$$F_n = \overline{F}_n - 2\sqrt{2}e^{-(n+1/2)\alpha} \sinh \left(n + \frac{1}{2} \right) \beta \operatorname{cosech} \left(n + \frac{1}{2} \right) (\alpha - \beta), \tag{3.4.114}$$

$$G_n = \overline{G}_n, \quad H_n = \overline{H}_n, \tag{3.4.115}$$

with $\overline{C}_n, \overline{D}_n, \cdots, \overline{H}_n$ being known functions of $A_{n-1}, A_n, A_{n+1}, B_{n-1}, B_n, B_{n+1}, \alpha$ and β. The coefficients A_n and B_n are now obtained in terms of the mass-conservation Eq. (3.4.102). We obtain the following simultaneous second-order difference equations for A_n and B_n:

$$\mathcal{D}_1(A_n, B_n) = 2\sqrt{2}\left\{e^{-(n+1/2)\alpha}\sinh\left(n+\frac{1}{2}\right)\beta\text{cosech}\left(n+\frac{1}{2}\right)(\alpha-\beta)\right.$$

$$-e^{-(n-\frac{1}{2})\alpha}\sinh\left(n-\frac{1}{2}\right)\beta\text{cosech}\left(n-\frac{1}{2}\right)(\alpha-\beta)$$

$$\left.-e^{-(n+\frac{3}{2})\alpha}\sinh\left(n+\frac{3}{2}\right)\beta\text{cosech}\left(n+\frac{3}{2}\right)(\alpha-\beta)\right\}, (n \geq 1) \qquad (3.4.116)$$

$$\mathcal{D}_2(A_n, B_n) = 2\sqrt{2}\left\{2e^{-(n+1/2)\alpha}\cosh\left(n+\frac{1}{2}\right)\beta\text{cosech}\left(n+\frac{1}{2}\right)(\alpha-\beta)\right.$$

$$-e^{-(n-\frac{1}{2})\alpha}\cosh\left(n-\frac{1}{2}\right)\beta\text{cosech}\left(n-\frac{1}{2}\right)(\alpha-\beta)$$

$$\left.-e^{-(n+\frac{3}{2})\alpha}\cosh\left(n+\frac{3}{2}\right)\beta\text{cosech}\left(n+\frac{3}{2}\right)(\alpha-\beta)\right\}, (n \geq 1) \qquad (3.4.117)$$

where

$$\mathcal{D}_1(A_n, B_n)$$

$$\equiv [(2n-1)\alpha_{n-1} - (2n-3)\alpha_n + 2\kappa]\left[\frac{(n-1)A_{n-1}}{2n-1} - \frac{nA_n}{2n+1}\right]$$

$$-[(2n+5)\alpha_n - (2n+3)\alpha_{n+1} + 2\kappa]\left[\frac{(n+1)A_n}{2n+1} - \frac{(n+2)A_{n+1}}{2n+3}\right]$$

$$+[(2n-1)(\gamma_{n-1} - \delta_{n-1}) - (2n-3)(\gamma_n - \delta_n) - 4]\left[\frac{(n-1)B_{n-1}}{2n-1} - \frac{nB_n}{2n+1}\right]$$

$$-[(2n+5)(\gamma_n - \delta_n) - (2n+3)(\gamma_{n+1} - \delta_{n+1}) + 4]\left[\frac{(n+1)B_n}{2n+1} - \frac{(n+2)B_{n+1}}{2n+3}\right], (3.4.118)$$

and $\mathcal{D}_2(A_n, B_n)$ is obtained from $\mathcal{D}_1(A_n, B_n)$ by interchanging A_n and B_n and reversing the signs in front of 2κ and 4. The constants $\alpha_n, \gamma_n, \delta_n$ and κ can be calculated by the following relations:

$$\frac{\alpha_n}{\sinh(n+\frac{1}{2})(\alpha+\beta)} = \frac{\gamma_n}{\cosh(n+\frac{1}{2})(\alpha+\beta)}$$

$$= \frac{\delta_n}{\cosh(n+\frac{1}{2})(\alpha-\beta)} = \frac{\sinh(\alpha-\beta)}{\sinh\alpha\sinh\beta\sinh(n+\frac{1}{2})(\alpha-\beta)},$$

and

$$\kappa = \frac{\sinh(\alpha + \beta)}{\sinh\alpha \sinh\beta}. \tag{3.4.119}$$

It has not been proved if it is possible to find solutions of Eq. (3.4.116) and Eq. (3.4.117) in a closed form. However A_n and B_n necessarily converge to zero for all non-zero values of α and β and hence the significant solution of the difference equations may be obtained numerically by using a successive approximations technique. Numerical solutions of Eq. (3.4.116) and Eq. (3.4.117) have been determined over a wide range of values of the ratio $\lambda = b/a$ and the minimum clearance ϵa between the spheres (O'Neill and Majumdar 1970)[8]. With these numerical solutions we can calculate the forces acting on the sphere A and sphere B, which are respectively $(-6\pi\mu a \mathcal{U} f_a, 0, 0)$ and $(-6\pi\mu b \mathcal{U} f_b, 0, 0)$ in the Cartesian co-ordinate system with (O'Neill and Majumda 1970b)[8]

$$f_a = \frac{1}{3}\sqrt{2}\sinh\alpha \sum_{n=0}^{\infty}(E_n + F_n), \tag{3.4.120}$$

and

$$f_b = \frac{1}{3}\sqrt{2}\sinh|\beta| \sum_{n=0}^{\infty}(E_n + F_n). \tag{3.4.121}$$

Finally, the method of calculating the couples acting on the two spheres, and solving the problem of a rigid sphere at rest with another sphere rotating with uniform angular velocity Ω about a diameter perpendicular to the line connecting their centres is similar to that one described above. For a detailed description of the method, the reader is referred to O'Neill and Majumda 1970b[8].

CHAPTER 4

KINETICS OF TWO SPHERES AT LOW
REYNOLDS NUMBER AND LOW STOKES NUMBER

1. Introduction

There are three kinds of forces acting on the aerosol particles viz.

1 external forces such as the gravitational force, centrifugal force, electric field force etc.;
2 forces exerted by the molecules of the medium which are divided into two parts: one of them is a very rapid fluctuating force with a molecular motion time scale, the other is a much slower fluctuating force associated with the response of the medium and can be regarded as continuous resistance to the motion of the particle;
3 interparticle interaction potential, such as the van der Waals attractive potential , Coulomb electrostatic potential which is repulsive for hydrosol system and is either repulsive or attractive for the aerosol system.

The mechanics of aerosols is to investigate the various mechanical processes of both the particle and the aerosol system under the actions of the above forces. The mechanical processes are quite various. Fuchs in his famous book " The mechanics of aerosols" [2] gave a systematic and comprehensive review. It still is a good and important reference book up to now. Although the mechanical processes of both the aerosol particles and the aerosol system are quite various, we may divide them into five divisions according a basic consideration:

1 sedimentation of aerosol particles under gravity , or under centrifugal force, or under electric field force, moreover the process of fluidization is also the content of study in which the particles are not settling, but suspend in the air flow and the velocity of the flow is equal in quantity and opposite in direction to the average settling velocity of the particles, which the particles would have, if the air were otherwise at rest;
2 coagulation of aerosol particles, including the capture processes of particles by fibre filters, by samplers, by thread of a spider web, by electric wire etc., in which the collectors are not particles themselves, but are fibres, sampling plates, cobwebs, wires, etc.;
3 mass/heat transfer from or to an aerosol particles;

4 evolution of the size distribution of aerosol particles ;

5 the mechanical properties of the aerosol system (including both the particles and the medium) such as the effective viscosity , the effective conductivity etc. Note that the Fuchs' book does not include this problem. Therefore, precisely speaking, the title of his book should be "The mechanics of aerosol particles".

It is well known that an aerosol system is a multi-particle system. Thence, all the mechanical problems pointed out above should be a multi-particle problem, in principle. This is a very complicated and difficult problem to approach, especially when the hydrodynamic interaction s of the particles are considered. Then, as the first step of the research work, people usually adopt some assumptions to simplify the problem. The simplest assumption of them is the isolated-particle approximation. Thus, the results given in sections 3.1 and 3.2 are applicable to the situation. The approximation is sufficiently good in general, if the aerosol system is extremely dilute and the volume fraction φ of the aerosol particles is much smaller than unity. In his look "The mechanics of aerosols" Fuchs gave a systematical and exhaustive summary about the isolated-particle kinetics. This is a very useful and extensive summary even at present. However, some of his view-points are not proper. For example, at the beginning of Chapter 2 of his book he said that the characteristic of the mechanics of aerosols is that among the three kinds of forces—external force, resistance from the medium, interparticle potential —, the last one is much smaller than the other two and can be neglected in most cases, thus the motions of particles become irrelevant to each other, then, the mechanics of aerosols reduces to theoretically and experimentally investigating an isolated particle moving under external forces in a damping medium. In fact, the interactions between aerosol particles are not restricted to the interparticle potential s only. The forces of the second type are type of the interactions due to the hydrodynamical interactions between particles. These forces are long-range forces. They cannot be neglected even when the aerosol system is as dilute as $\varphi \sim 10^{-2}$. For instance, the enhanced settling of the falling particle cloud in an unbound space and the hindered settling of particles in a bound space,—say, in a container—are the results of hydrodynamic interaction s between particles. We cannot understand these problems correctly without correct understanding of the hydrodynamic interactions between aerosol particles. On the other hand for the processes of coagulation of aerosol particles and the capture process by a collector, which have been discussed extensively by Fuchs in his book, the hydrodynamic interaction s between particles are especially important. There always are at least two particles or a particle with a collector involved in the coagulation or capture processes. Hence it is impossible to get a correctly understanding about these processes without correctly investigating the hydrodynamic interaction s between the particles. Lack of understanding of hydrodynamic interaction s leads to shortcomings and errors in Fuchs' book[2], and similar problems also exist in Pruppacher and Klett's book (1978)[3].

The development of low-Reynolds-number hydrodynamics during the last 20 years has pushed the mechanics of aerosols forward from isolated-particle stage to a new stage of interacting multi-particle situation, mainly to the stage of pair-interaction s. It seems to be appropriate to give a new summary about the present status of the pair-interaction s, although giving such a summary at a time when the field is rapidly growing is difficult. We will start from two-sphere kinetics in this chapter.

2. Mobility Functions for Two Spheres

The focus of attention on the mechanics of aerosols is different from that of the low-Reynolds-number hydrodynamics. The former is about how the particles move as an external force is given. In this case, a basic physical quantity is the mobility B, defined as the velocity acquired by the particle under per unit external force in a equilibrium state. Obviously, the mobility B is $(6\pi\mu a)^{-1}$ for an isolated spherical particle. Thus, it is easy to calculate the velocity U acquired by a particle on which an external force F exerts, i.e. $U = BF$. As the reversion, the problem in the low-Reynolds-number hydrodynamics is exactly the inverse problem, i.e. what the resistance is to a particle moving with a given steady velocity. The resistance of air to a particle is always equal in quantity and opposite in direction to the force exerted by the particle on the air. However, only at the state of equilibrium the resistance is equal and opposite to the external force acting on the sphere. It is indicated that for most aerosol particles the Stokes number is small, and as a first order approximation we may set it equal to zero. Hence, the inertia forces of the aerosol particles are neglected in most cases. The external forces acting on the aerosol particles always balance the resistance. Thus, the external forces exerted on the particles are always equal to the forces exerted by the particles on the air. Therefore, the mobility of a particle at zero-Stokes-number is always the velocity acquired by the particle under per unit external force. Generally speaking, the resistance coefficient is a tensor especially for the two-sphere case. Thus, the mobility is also a tensor. They are inverse matrices to each other. The detailed relation between them will be discussed in this section, and we will start our discussion from a general situation (Jeffrey and Onishi 1984)[9].

Two rigid spheres, labelled as sphere 1 and sphere 2, are immersed in infinite volume of fluid whose velocity at absence of the spheres would be the ambient velocity $U(x) = U_0 + \Omega \times x$. It is thus a superposition of a uniform stream and a rigid-body rotation. Sphere α has radius a_α and its centre is at x_α; it has angular velocity Ω_α and its mass centre has translational velocity U_α. The force F_α that sphere α exerts on the fluid is given as

$$F_\alpha = -\int_{A_\alpha} \sigma \cdot n \, dA, \qquad (4.2.1)$$

where A_α is the surface of the sphere, n is the outward normal of the particle surface and σ is the stress tensor . The force couple exerted by the sphere on the fluid, calculated relative to the centre of the sphere, is

$$C_\alpha = -\int_{A_\alpha} (x - x_\alpha) \times (\sigma \cdot n) \, dA. \qquad (4.2.2)$$

This is the antisymmetric part of the first moment of the surface stress expressed as a vector. The relations among the quantities $U_\alpha, \Omega_\alpha, F_\alpha, C_\alpha, U_0$ and Ω are the interactions we wish to study; they can be described using either a resistance matrix or a mobility matrix. If the specified quantities are the velocities of the particles and the ambient flow , we can invoke the linearity of the Stokes equation s to write

$$
\begin{vmatrix} F_1 \\ F_2 \\ C_1 \\ C_2 \end{vmatrix} = \mu \begin{vmatrix} f_{11} & f_{12} & \tilde{g}_{11} & \tilde{g}_{21} \\ f_{21} & f_{22} & \tilde{g}_{12} & \tilde{g}_{22} \\ g_{11} & g_{12} & h_{11} & h_{12} \\ g_{21} & g_{22} & h_{21} & h_{22} \end{vmatrix} \cdot \begin{vmatrix} U_1 - U(x_1) \\ U_2 - U(x_2) \\ \Omega_1 - \Omega \\ \Omega_2 - \Omega \end{vmatrix} . \tag{4.2.3}
$$

The square matrix is the resistance matrix ; it contains tensors $f_{\alpha\beta}, g_{\alpha\beta}, h_{\alpha\beta}$. The reciprocal theorem of Lorentz (1906) shows that $\tilde{g}_{\alpha\beta}$ is not an independent quantity, but has a symmetric relation with $g_{\alpha\beta}$, i.e. $\tilde{g}_{\alpha\beta} = g_{\beta\alpha}$. It might be noted that the tensors have different dimensions, and

$$
\hat{f}_{\alpha\beta} = \frac{f_{\alpha\beta}}{3\pi(a_\alpha + a_\beta)}, \tag{4.2.4a}
$$

$$
\hat{g}_{\alpha\beta} = \frac{g_{\alpha\beta}}{\pi(a_\alpha + a_\beta)^2}, \tag{4.2.4b}
$$

$$
\hat{h}_{\alpha\beta} = \frac{h_{\alpha\beta}}{\pi(a_\alpha + a_\beta)^3}, \tag{4.2.4c}
$$

where $\hat{f}_{\alpha\beta}, \hat{g}_{\alpha\beta}$ and $\hat{h}_{\alpha\beta}$ are dimensionless tensors.

Each tensor in the matrix is axisymmetric about $\mathbf{r}$, where $\mathbf{r} = x_2 - x_1$, and can be reduced to an expression containing at most two scalar functions (Batchelor 1945, Brenner 1963, 1964). Thus we can write

$$
f_{\alpha\beta} = 3\pi(a_\alpha + a_\beta) \left[f^l_{\alpha\beta} \frac{\mathbf{rr}}{r^2} + f^n_{\alpha\beta} \left(\mathbf{I} - \frac{\mathbf{rr}}{r^2} \right) \right], \tag{4.2.5a}
$$

$$
g_{\alpha\beta} = \pi(a_\alpha + a_\beta)^2 \left(g^n_{\alpha\beta} \epsilon \cdot \frac{\mathbf{r}}{r} \right), \tag{4.2.5b}
$$

$$
h_{\alpha\beta} = \pi(a_\alpha + a_\beta)^3 \left[h^l_{\alpha\beta} \frac{\mathbf{rr}}{r^2} + h^n_{\alpha\beta} \left(\mathbf{I} - \frac{\mathbf{rr}}{r^2} \right) \right], \tag{4.2.5c}
$$

where $\alpha, \beta = 1$ or 2, $r = |\mathbf{r}|$, $\mathbf{I}$ is the unit second-rank tensor. The dimensionless scalar functions $f^l_{\alpha\beta}, f^n_{\alpha\beta}, g^n_{\alpha\beta}, h^l_{\alpha\beta}, h^n_{\alpha\beta}$ depend only on two dimensionless variables s and λ, where s is the dimensionless separation of the two sphere centres scaled with the average radius, $s = 2r/(a_1 + a_2)$, and λ is the size ratio of the two spheres, $\lambda = a_2/a_1$.

We now regard the forces and couples exerted by the particles on fluid as the external forces and external couples acting on the particles when the Stokes number s of the particles are small, and write

$$
\begin{vmatrix} U_1(x_1) - U(x_1) \\ U_2(x_2 - U(x_2) \\ \Omega_1 - \Omega \\ \Omega_2 - \Omega \end{vmatrix} = \mu^{-1} \begin{vmatrix} b_{11} & b_{12} & \tilde{d}_{11} & \tilde{d}_{21} \\ b_{21} & b_{22} & \tilde{d}_{12} & \tilde{d}_{22} \\ d_{11} & d_{12} & c_{11} & c_{12} \\ d_{21} & d_{22} & c_{21} & c_{22} \end{vmatrix} \cdot \begin{vmatrix} F_1 \\ F_2 \\ C_1 \\ C_2 \end{vmatrix} . \tag{4.2.6}
$$

The square matrix is the mobility matrix ; it contains second-rank tensors $\mathbf{b}_{\alpha\beta}, \mathbf{c}_{\alpha\beta}$, $\mathbf{d}_{\alpha\beta}$. According to the reciprocal theorem of Lorentz (1906), we see that $\tilde{\mathbf{d}}_{\beta\alpha} = \mathbf{d}_{\alpha\beta}$. These tensors are also dimensional quantities, and

$$\hat{\mathbf{b}}_{\alpha\beta} = 3\pi(a_\alpha + a_\beta)\mathbf{b}_{\alpha\beta}, \qquad (4.2.7a)$$

$$\hat{\mathbf{d}}_{\alpha\beta} = \pi(a_\alpha + a_\beta)^2 \mathbf{d}_{\alpha\beta}, \qquad (4.2.7b)$$

$$\hat{\mathbf{c}}_{\alpha\beta} = \pi(a_\alpha + a_\beta)^3 \mathbf{c}_{\alpha\beta}, \qquad (4.2.7c)$$

where $\hat{\mathbf{b}}_{\alpha\beta}, \hat{\mathbf{d}}_{\alpha\beta}$ and $\hat{\mathbf{c}}_{\alpha\beta}$ are dimensionless tensors. The axisymmetry about $\mathbf{r}$ allows us that for the tensors one can use the same decompositions into at most two scalar functions:

$$\mathbf{b}_{\alpha\beta} = \frac{1}{3\pi(a_\alpha + a_\beta)} \left[A_{\alpha\beta} \frac{\mathbf{rr}}{r^2} + B_{\alpha\beta} \left(\mathbf{I} - \frac{\mathbf{rr}}{r^2} \right) \right], \qquad (4.2.8a)$$

$$\mathbf{d}_{\alpha\beta} = \frac{1}{\pi(a_\alpha + a_\beta)^2} \left(B_{\alpha\beta}^d \epsilon \cdot \frac{\mathbf{r}}{r} \right), \qquad (4.2.8b)$$

$$\mathbf{c}_{\alpha\beta} = \frac{1}{\pi(a_\alpha + a_\beta)^3} \left[A_{\alpha\beta}^c \frac{\mathbf{rr}}{r^2} + B_{\alpha\beta}^c \left(\mathbf{I} - \frac{\mathbf{rr}}{r^2} \right) \right], \qquad (4.2.8c)$$

where $\alpha, \beta = 1$ or 2, $r = |\mathbf{r}|$, $\mathbf{I}$ is the unit second-rank tensor, and the scalar functions $A_{\alpha\beta}, B_{\alpha\beta}, B_{\alpha\beta}^d, A_{\alpha\beta}^c, B_{\alpha\beta}^c$ are dimensionless functions of s and λ.

It is obvious from Eq. (4.2.3) and Eq. (4.2.6) that the dimensional resistance and mobility matrices obey the equation:

$$\begin{vmatrix} b_{11} & b_{12} & \tilde{d}_{11} & \tilde{d}_{21} \\ b_{21} & b_{22} & \tilde{d}_{12} & \tilde{d}_{22} \\ d_{11} & d_{12} & c_{11} & c_{12} \\ d_{21} & d_{22} & c_{21} & c_{22} \end{vmatrix} \cdot \begin{vmatrix} f_{11} & f_{12} & \tilde{g}_{11} & \tilde{g}_{21} \\ f_{21} & f_{22} & \tilde{g}_{12} & \tilde{g}_{22} \\ g_{11} & g_{12} & h_{11} & h_{12} \\ g_{21} & g_{22} & h_{21} & h_{22} \end{vmatrix} = \begin{vmatrix} 1 & 0 & 0 & 0 \\ 0 & 1 & 0 & 0 \\ 0 & 0 & 1 & 0 \\ 0 & 0 & 0 & 1 \end{vmatrix}. \qquad (4.2.9)$$

Using the decompositions into scalar functions, we divide this tensor matrix equation into three scalar matrix equations. This is possible because the axisymmetric translation is not coupled to axisymmetric rotation and axisymmetric motion is not coupled to asymmetric motion. Thus the dimensionless scalar functions obey

$$\begin{vmatrix} A_{11} & \frac{2}{1+\lambda} A_{12} \\ \frac{2}{1+\lambda} A_{21} & \frac{1}{\lambda} A_{22} \end{vmatrix} = \begin{vmatrix} f_{11}^l & \frac{1}{2}(1+\lambda) f_{12}^l \\ \frac{1}{2}(1+\lambda) f_{21}^l & \lambda f_{22}^l \end{vmatrix}^{-1}, \qquad (4.2.10)$$

$$\begin{vmatrix} A_{11}^c & \frac{8}{(1+\lambda)^3} A_{12}^c \\ \frac{8}{(1+\lambda)^3} A_{21}^c & \frac{1}{\lambda^3} A_{22}^c \end{vmatrix} = \begin{vmatrix} h_{11}^l & \frac{1}{8}(1+\lambda)^3 h_{12}^l \\ \frac{1}{8}(1+\lambda)^3 h_{21}^l & \lambda^3 h_{22}^l \end{vmatrix}^{-1}, \qquad (4.2.11)$$

$$
\begin{vmatrix}
B_{11} & \frac{2}{1+\lambda}B_{12} & \frac{3}{2}B_{11}^d & \frac{6}{(1+\lambda)^2}B_{21}^d \\
\frac{2}{1+\lambda}B_{21} & \frac{1}{\lambda}B_{22} & \frac{6}{(1+\lambda)^2}B_{12}^d & \frac{3}{2\lambda^2}B_{22}^d \\
\frac{3}{2}B_{11}^d & \frac{6}{(1+\lambda)^2}B_{12}^d & \frac{3}{4}B_{11}^c & \frac{6}{(1+\lambda)^3}B_{12}^c \\
\frac{6}{(1+\lambda)^2}B_{21}^d & \frac{3}{2\lambda^2}B_{22}^d & \frac{6}{(1+\lambda)^3}B_{12}^c & \frac{3}{4\lambda^3}B_{22}^c
\end{vmatrix} =
$$

$$
\begin{vmatrix}
f_{11}^n & \frac{1}{2}(1+\lambda)f_{12}^n & \frac{2}{3}g_{11}^n & \frac{1}{6}(1+\lambda)^2 g_{21}^n \\
\frac{1}{2}(1+\lambda)f_{21}^n & \lambda f_{22}^n & \frac{1}{6}(1+\lambda)^2 g_{12}^n & \frac{2}{3}\lambda^2 g_{22}^n \\
\frac{2}{3}g_{11}^n & \frac{1}{6}(1+\lambda)^2 g_{12}^n & \frac{4}{3}h_{11}^n & \frac{1}{6}(1+\lambda)^3 h_{12}^n \\
\frac{1}{6}(1+\lambda)^2 g_{21}^n & \frac{2}{3}\lambda^2 g_{22}^n & \frac{1}{6}(1+\lambda)^3 h_{21}^n & \frac{4}{3}\lambda^3 h_{22}^n
\end{vmatrix}^{-1} . \qquad (4.2.12)
$$

The development of low-Reynolds-number hydrodynamics has made it possible to calculate all the resistance scalar functions on the right-hand-side of the above three matrix equations and thus calculate all the mobility scalar functions on the left-hand-side of the matrix equations. The first exact solution to the problem of two spheres rotating with constant angular velocities about the line connecting their centres was obtained in 1915 by G.B.Jeffrey. Then the exact solution of the associated axisymmetrical problem when the spheres translate with the same velocity along the line connecting centres was obtained in 1926 by Stimson and G.B.Jeffrey. A modification of this latter analysis was carried out by Brenner (1961). A feature in common of these exact solutions is that they were constructed in terms of the bispherical coordinates which we have mentioned in Chapter 3 (see O'Neill and Majumdar 1970[8]). A detailed account of the work done on the problems of two sphere up to about 1962 was given in Happel and Brenner (1965)[10]. Several other methods for solving the equations of low-Regnolds-number flow around two spheres have been developed since then. The methods include those using reflections (Happel and Brenner 1965)[10], tangent-sphere coordinates (Cooley and O'Neill 1969), collocation methods (Ganatos, Pfeffer and Weinbaum 1978), and asymptotic methods (O'Neill and Stewartson 1967; D.J.Jeffrey 1982). In this chapter we will introduce a new method which is called as twin-multipole expansions (D.J.Jeffrey and Onishi 1984)[9] and is a development of the method of reflections. The method is chosen because it is accurate although not as inherently accurate as bispherical coordinates and it produces results in a convenient form for further usage in applications; also it can be combined easily with the results of the asymptotic methods. The method has provided a large amount of data on these mobility function s in a way in which computer subroutines can be written to evaluate these functions for any specified values of their arguments. Thus, the data it provided are comprehensive enough to meet the demands of recent developments in the dynamics of aerosol and hydrosol particles (Batchelor 1982[11] Batchelor and Wen 1982[12], Wen and Bachelor 1983, 1985[13], Wang and Wen 1990[14], Wen, Zhang and Lin 1991[15]).

Following Happel and Brenner (1965)[10], we take two sets of spherical polar coordinates $(\rho_\alpha, \theta_\alpha, \varphi)$ with a difference that θ_2 here equals their $\pi - \Theta$; let the unit vectors of the coordinates axes be $\hat{\rho}_\alpha, \hat{\theta}_\alpha, \hat{\varphi}$. The choice has an advantage that the transformation rule for spherical harmonics becomes (Hobson 1931)

$$\left(\frac{a_\alpha}{\rho_\alpha}\right)^{n+1} Y_{mn}(\theta_\alpha,\varphi) = \left(\frac{a_\alpha}{r}\right)^{n+1} \sum_{s=m}^{\infty} \binom{n+s}{s+m} \left(\frac{\rho_{3-\alpha}}{r}\right)^s Y_{ms}(\theta_{3-\alpha},\varphi), \qquad (4.2.13)$$

where $Y_{mn}(\theta,\varphi) = P_n^m(\cos\theta)\exp(im\varphi)$ and r is the distance between the centres of the spheres.

We use Lamb's general solution (Happel and Brenner 1965)[10] to write the pressure and velocity field outside the spheres as

$$p = p^{(1)} + p^{(2)}, \qquad (4.2.14a)$$

where

$$p^{(\alpha)} = \mu \sum_{m=0}^{\infty} \sum_{n=m}^{\infty} \frac{1}{a_\alpha} p_{mn}^{(\alpha)} (\frac{a_\alpha}{\rho_\alpha})^{n+1} Y_{mn}(\theta_\alpha,\varphi), \qquad (4.2.14b)$$

$$\mathbf{u} = \mathbf{u}^{(1)} + \mathbf{u}^{(2)}, \qquad (4.2.15a)$$

and

$$\mathbf{u}^{(\alpha)} = \sum_{m=0}^{\infty} \sum_{n=m}^{\infty} \left\{ \nabla \times \left[\rho_\alpha q_{mn}^{(\alpha)} (\frac{a_\alpha}{\rho_\alpha})^{n+1} Y_{mn}(\theta_\alpha,\varphi) \right] \right.$$
$$+ a_\alpha \nabla \left[v_{mn}^{(\alpha)} (\frac{a_\alpha}{\rho_\alpha})^{n+1} Y_{mn} \right] - \frac{n-2}{2n(2n-1)a_\alpha} \rho_\alpha^2 \nabla \left[p_{mn}^{(\alpha)} (\frac{a_\alpha}{\rho_\alpha})^{n+1} Y_{mn} \right]$$
$$\left. + \frac{n+1}{n(2n-1)a_\alpha} \rho_\alpha \left[p_{mn}^{(\alpha)} (\frac{a_\alpha}{\rho_\alpha})^{n+1} Y_{mn} \right] \right\}. \qquad (4.2.15b)$$

The coefficients $p_{mn}^{(\alpha)}, q_{mn}^{(\alpha)}$ and $v_{mn}^{(\alpha)}$ are functions only of r, have the dimensions of velocity and must be calculated from the boundary conditions:

$$\mathbf{u} = \mathbf{U}_\alpha(\theta_\alpha,\varphi) \qquad (\text{at} \quad \rho_\alpha = a_\alpha). \qquad (4.2.16)$$

With the solutions Eq. (4.2.14) and Eq. (4.2.15) we express the force and couple in a Cartesian system $\mathbf{i},\mathbf{j},\mathbf{e}$, where $\mathbf{e} = \mathbf{r}/r$, and the i-axis is chosen in the plane $\varphi = 0$. Then according to Happel and Brenner (1965) [10] we have

$$\mathbf{F}_\alpha = 4\pi a_\alpha \mu [p_{01}^{(\alpha)}(-l)^{3-\alpha}\mathbf{e} - p_{11}^{(\alpha)}(\mathbf{i}+i\mathbf{j})], \qquad (4.2.17)$$

$$\mathbf{C}_\alpha = 8\pi a_\alpha^2 \mu [q_{01}^{(\alpha)}(-l)^{3-\alpha}\mathbf{e} - q_{11}^{(\alpha)}(\mathbf{i}+i\mathbf{j})]. \qquad (4.2.18)$$

A sign difference from Happal and Brenner is the result of calculating the force on the fluid rather than the force on the sphere.

Based on the force Eq. (4.2.17) and Eq. (4.2.18) we may calculate all the twenty resistance functions at the right-hand-side of Eq. (4.2.10) to Eq. (4.2.12) and then all the twenty mobility function s on the left-hand-side of Eq. (4.2.10) to Eq. (4.2.12). This is a heavy task even with a computer. Fortunately, for the case of sedimentation and gravitational coagulation which we will discuss in Chapters 5 and 6 the problem can be simplified. The externally applied couple is zero for the spherical particles under gravity , i.e. $C_1 = C_2 = 0$. And the rotation of the particles is irrelevant. Thus we may put the problem of finding Ω_1 and Ω_2 aside. The fluid in which the particles are immersed is otherwise at rest. Hence $U(x) = 0$ and $\Omega = 0$. Therefore, Eq. (4.2.6) reduces to

$$U_1 = \mu^{-1}(b_{11} \cdot F_1 + b_{12} \cdot F_2),$$

$$U_2 = \mu^{-1}(b_{21} \cdot F_1 + b_{22} \cdot F_2). \qquad (4.2.19)$$

The mobility tensor s which are needed to find reduce from twelve to four , and the mobility function s reduce from twenty to eight, e.g. $A_{11}, A_{12}, A_{21}, A_{22}, B_{11}, B_{12}, B_{21}$, and B_{22}. They are functions of a dimensionless separation between the two sphere centres s and the size ratio of the two spheres λ. These two-sphere mobility function s $A_{\alpha\beta}$ and $B_{\alpha\beta}$ play a central role in the calculation of sedimentation velocity and gravitational coagulation rate of the particles.

It can be shown from exchanging the two particles that

$$A_{11}(s,\lambda) = A_{22}(s,\lambda^{-1}), \qquad A_{12}(s,\lambda) = A_{21}(s,\lambda^{-1}), \qquad (4.2.20a,b)$$

$$B_{11}(s,\lambda) = B_{22}(s,\lambda^{-1}), \qquad B_{12}(s,\lambda) = B_{21}(s,\lambda^{-1}). \qquad (4.2.20c,d)$$

And by the Lorentz reciprocal theorem (1906) we have

$$A_{12}(s,\lambda) = A_{21}(s,\lambda), \qquad B_{12}(s,\lambda) = B_{21}(s,\lambda). \qquad (4.2.21a,b)$$

Thus, among the eight mobility function s, only four functions are independent. All the eight functions can be determined, provided that the following four are known

$$A_{11} \quad \text{and} \quad B_{11} \qquad \text{within} \qquad 0 \leq \lambda \leq \infty,$$

$$A_{12} \quad \text{and} \quad B_{12} \qquad \text{within} \qquad 1 \leq \lambda \leq \infty.$$

Note that, as a consequence of the choice of the scaling factor made in Eq. (4.2.8a), $A_{\alpha\beta}$ and $B_{\alpha\beta}$ are finite for all values of λ, and

$$\mu^{-1}b_{\alpha\beta} = \begin{cases} I/6\pi\mu a_\alpha & \text{as} \quad s \to \infty, \quad \text{when} \quad \alpha = \beta, \\ 0 & \text{as} \quad s \to \infty, \quad \text{when} \quad \alpha \neq \beta, \end{cases} \qquad (4.2.22a)$$

then

$$A_{\alpha\beta}, B_{\alpha\beta} = \begin{cases} 1 & \text{as} \quad s \to \infty, \quad \text{when} \quad \alpha = \beta, \\ 0 & \text{as} \quad s \to \infty, \quad \text{when} \quad \alpha \neq \beta. \end{cases} \qquad (4.2.22b)$$

At the inner limit $s \to 2$, the standard lubrication theory (see Eq. (3.3.15) and Eq. (4.2.19)) shows that for two nearly touching spheres, the forces acting on the spheres are parallel to the line connecting their centres

$$A_{11} - \frac{4}{1+\lambda}A_{12} + \frac{1}{\lambda}A_{22} \sim \frac{(1+\lambda)^3}{2\lambda^2}\xi, \qquad (4.2.23)$$

as $\xi \to 0$, where $\xi = s - 2$. The fact that the above linear combination of $A_{\alpha\beta}$ tends to zero reflects that the resistance of a lubrication film tends to infinity. However, $A_{\alpha\beta}$ itself should not tend to zero. In fact, an applied force parallel to the line connecting the centres produces a common velocity to the two touching spheres ($s=2$) regardless of which sphere it acts on; this result gives

$$A_{11}(2, \lambda) = \frac{2}{1 + \lambda} A_{12}(2, \lambda) = \frac{1}{\lambda} A_{22}(2, \lambda). \qquad (4.2.24)$$

The far field asymptotic form for $A_{\alpha\beta}$ and $B_{\alpha\beta}$ can be obtained by the method of twin-multipole expansions (D.J.Jeffrey and Onishi 1984), viz.[9]

$$A_{11} = 1 - \frac{60\lambda^3}{(1+\lambda)^4 s^4} + \frac{32\lambda^3(15 - 4\lambda^2)}{(1+\lambda)^6 s^6}$$

$$- \frac{192\lambda^3(5 - 22\lambda^2 + 3\lambda^4)}{(1+\lambda)^8 s^8}$$

$$- \frac{256\lambda^5(70 + 375\lambda - 120\lambda^2 + 9\lambda^4)}{(1+\lambda)^{10} s^{10}} + O(s^{-12}), \qquad (4.2.25)$$

$$B_{11} = 1 - \frac{68\lambda^5}{(1+\lambda)^6 s^6} - \frac{32\lambda^3(10 - 9\lambda^2 + 9\lambda^4)}{(1+\lambda)^8 s^8}$$

$$- \frac{192\lambda^5(35 - 18\lambda^2 + 6\lambda^4)}{(1+\lambda)^{10} s^{10}} + O(s^{-12}), \qquad (4.2.26)$$

$$A_{12} = \frac{3}{2s} - \frac{2(1+\lambda^2)}{(1+\lambda)^2 s^3} + \frac{1200\lambda^3}{(1+\lambda)^6 s^7} + \frac{960\lambda^3(1+\lambda^2)}{(1+\lambda)^8 s^9}$$

$$- \frac{256\lambda^3(30 - 453\lambda^2 + 30\lambda^4)}{(1+\lambda)^{10} s^{11}} + O(s^{-13}), \qquad (4.2.27)$$

$$B_{12} = \frac{3}{4s} + \frac{1 + \lambda^2}{(1+\lambda)^2 s^3}$$

$$+ \frac{56\lambda^3(80 - 79\lambda^2 + 80\lambda^4)}{(1+\lambda)^{10} s^{11}} + O(s^{-13}). \qquad (4.2.28)$$

It is to be expected from the nature of the method of twin-multipole expansions that increasing powers in s^{-1} this series are generally associated with increasing powers in λ. It can be seen in particular from the Jeffrey and Onishis series that the terms shown in Eq. (4.2.25) to Eq. (4.2.28) are sufficient to give expressions for A_{11} and B_{11} correct to order λ^3 when $\lambda \ll 1$ given a general value of s, viz.

$$A_{11} = 1 + \lambda^3(-60s^{-4} + 480s^{-6} - 960s^{-8}) + O(\lambda^5), \qquad (4.2.29)$$

$$B_{11} = 1 + \lambda^3(-320s^{-8}) + O(\lambda^5), \qquad (4.2.30)$$

and expressions for A_{12} and B_{12} correct to order λ^2, viz.

$$A_{12} = (\frac{3}{2}s^{-1} - 2s^{-3}) + \lambda(4s^{-3}) + \lambda^2(-8s^{-3}) + O(\lambda^3), \qquad (4.2.31)$$

$$B_{12} = (\frac{3}{4}s^{-1} + s^{-3}) + \lambda(-2s^{-3}) + \lambda^2(4s^{-3}) + O(\lambda^3). \qquad (4.2.32)$$

Likewise, in view of Eq. (4.2.20) and Eq. (4.2.25), Eq. (4.2.26), $A_{22} - 1$ and $B_{22} - 1$ are of order λ when $\lambda \ll 1$, and so $A_{11} - 1$ and $B_{11} - 1$ are of order λ^{-1} when $\lambda \gg 1$.

Numerical values of $A_{11}, A_{12}, B_{11}, B_{12}$ as functions of s for arbitrary values of λ in the intermediate field can be calculated by the computer subroutine devised by Jeffrey and Onishi (1984)[9] in terms of their solution in powers of s^{-1} for $A_{\alpha\beta}$ and $B_{\alpha\beta}$. The series was summed to term $O(s^{-150})$ for $A_{\alpha\beta}$ and to terms $O(s^{-120})$ for $B_{\alpha\beta}$ and in the special case $\lambda = 1$ to term $O(s^{-220})$.

For the inner field, Jeffrey and Onishi obtained the following asymptotic form for $A_{\alpha\beta}$ and $B_{\alpha\beta}$, viz.,

$$A_{\alpha\beta} = A_{\alpha\beta}^{(1)} + A_{\alpha\beta}^{(2)} + A_{\alpha\beta}^{(3)}\xi^2\ln\xi + A_{\alpha\beta}^{(4)}\xi^2 + O(\xi^3(\ln\xi)^2), \qquad (4.2.33)$$

$$B_{\alpha\beta} = \frac{B_{\alpha\beta}^{(1)}(\ln\xi^{-1})^2 + B_{\alpha\beta}^{(2)}\ln\xi^{-1} + B_{\alpha\beta}^{(3)}}{(\ln\xi^{-1})^2 + e_1\ln\xi^{-1} + e_2} + O(\xi(\ln\xi)^3), \qquad (4.2.34)$$

where $A_{\alpha\beta}^{(1)}, A_{\alpha\beta}^{(2)}, A_{\alpha\beta}^{(3)}, A_{\alpha\beta}^{(4)}, B_{\alpha\beta}^{(1)}, B_{\alpha\beta}^{(2)}, B_{\alpha\beta}^{(3)}, e_1$ and e_2 are known functions of λ, and at the neighborhood of $\xi = 0$, Eq. (4.2.34) reduces to

$$B_{\alpha\beta} = B_{\alpha\beta}^{(1)} + \frac{B_{\alpha\beta}^{(2)} - B_{\alpha\beta}^{(1)}e_1}{\ln \xi^{-1}}. \qquad (4.2.35)$$

Numerical results have shown that the asymptotic form of the far field for $A_{\alpha\beta}$ and $B_{\alpha\beta}$ Eq. (4.2.25) through Eq. (4.2.28) can be used from $s = 3.0$, and the near field asymptotic form for $A_{\alpha\beta}$ and $B_{\alpha\beta}$ Eq. (4.2.33) and Eq. (4.2.34) can be used when $\xi \leq 0.02(s \leq 2.02)$.

3. Relative Motions of Two Spheres under the Influence of External and Interparticle Potential Forces

Having obtained the information of the two-sphere mobility function s, we can determine all the relative motions of two particles at a low Stokes number under any external force and any interparticle force, thus we can determine the relative translational motion of the two spheres with the information about $A_{\alpha\beta}$ and $B_{\alpha\beta}$.

3.1. Relative Settling Velocity of Two Spheres under Gravity

Let $\mathbf{F}_1$ and $\mathbf{F}_2$ represent gravitational forces acting on sphere 1 and sphere 2 respectively. Then $\mathbf{F}_1 = 6\pi\mu a_2 \mathbf{U}_1^{(0)}$ and $\mathbf{F}_2 = 6\pi\mu a_2 \mathbf{U}_2^{(0)}$ where $\mathbf{U}_\alpha^{(0)}(\alpha = 1$ or $2)$

is the gravitational terminal velocity in isolated state of sphere 1 or 2 respectively. From Eq. (3.1.21) we see that $U_\alpha^{(0)}$ is proportional to a_α^2 and $\rho_\alpha - \rho$ where ρ_α is the density of sphere α and ρ is the density of air. According to Eq. (4.2.19) the relative settling velocity $V_{12}(r)$ of the two sphere centres is

$$V_{12}(r) = U_2 - U_1 = 6\pi a_1 U_1^{(0)} \cdot (b_{21} - b_{11}) + 6\pi a_2 U_2^{(0)} \cdot (b_{22} - b_{21}), \qquad (4.3.1)$$

where r is the position vector of the centre of sphere 2 relative to the centre of sphere 1.

Substituting Eq. (4.2.8a) into Eq. (4.3.1) one has

$$V_{12}(r) = V_{12}^{(0)} \cdot \left\{ L(s)\frac{rr}{r^2} + M(s)\left(I - \frac{rr}{r^2}\right)\right\}, \qquad (4.3.2)$$

where $V_{12}^{(0)}$ is the relative velocity of two far separated spheres, i.e.

$$V_{12}^{(0)} = U_2^{(0)} - U_1^{(0)} = (\lambda^2\gamma - 1)U_1^{(0)}, \qquad (4.3.3)$$

and the reduced density ratio γ is defined as

$$\gamma = \frac{\rho_2 - \rho}{\rho_1 - \rho}, \qquad (4.3.4)$$

and γ may take positive or negative values. In this scenario,

$$L(s) = \frac{\lambda^2\gamma A_{22} - A_{11}}{\lambda^2\gamma - 1} + \frac{2(1 - \lambda^3\gamma)A_{12}}{(1 + \lambda)(\lambda^2\gamma - 1)}, \qquad (4.3.5)$$

$$M(s) = \frac{\lambda^2 B_{22} - B_{11}}{\lambda^2\gamma - 1} + \frac{2(1 - \lambda^3\gamma)B_{12}}{(1 + \lambda)(\lambda^2\gamma - 1)}. \qquad (4.3.6)$$

Note (i) L and M are unchanged when λ and γ are replaced by λ^{-1} and γ^{-1}, and (ii) when $\lambda = 1$ both L and M are independent of γ (because the relative velocity is then proportional to $\gamma - 1$ for all separations).

The far field asymptotic forms of L and M may be found by substituting the far field asymptotic forms of $A_{\alpha\beta}$ and $B_{\alpha\beta}$ Eq. (4.2.25) through Eq. (4.2.28) into Eq. (4.3.5) and Eq. (4.3.6). Then, we have

$$L(s) = 1 + \frac{3(1 - \lambda^3\gamma)}{(\lambda^2\gamma - 1)(1 + \lambda)s} - \frac{4(1 - \lambda^3\gamma)(1 + \lambda^2)}{(\lambda^2\gamma - 1)(1 + \lambda)^3 s^3}$$

$$- \frac{60\lambda^3(\gamma - 1)}{(\lambda^2\gamma - 1)(1 + \lambda)^4 s^4} + O(s^{-6}), \qquad (4.3.7)$$

$$M(s) = 1 + \frac{3(1 - \lambda^3\gamma)}{2(\lambda^2\gamma - 1)(1 + \lambda)s}$$

$$+ \frac{2(1 - \lambda^3 \gamma)(1 + \lambda^2)}{(\lambda^2 \gamma - 1)(1 + \lambda)^3 s^3} + O(s^{-6}). \tag{4.3.8}$$

Eq. (4.3.7) and Eq. (4.3.8) show that L and M tend to unity as $s \to \infty$, and the differences $L - 1$ and $M - 1$ decay as s^{-1}. The decay is slow. Just as we have pointed out in section 3.1, the slow decay as s^{-1} is a basic feature of the Stokes flow field.

The relative velocity field of the two spheres is not a solenoidal field, although the medium is solenoidal field. $V_{12}(r)$ tends to a solenoidal field as $s \to \infty$. In fact,

$$\nabla \cdot V_{12} = \frac{V_{12}^{(0)} \cdot r}{\frac{1}{2}(a_1 + a_2)r} W(r), \tag{4.3.9}$$

where the scalar function W is

$$W(s) = \frac{2(L - M)}{s} + \frac{dL}{ds}. \tag{4.3.10}$$

$W(s)$ cannot be zero for any finite s value. However, it tends to zero as $s \to \infty$, and the decay of $W(s)$ is faster than those of $L - 1$ and $M - 1$. The far field asymptotic form for $W(s)$ may be found by substituting Eq. (4.3.7) and Eq. (4.3.8) into Eq. (4.3.10). Then, we have

$$W(s) = \frac{120\lambda^3(\gamma - 1)}{(\lambda^2 \gamma - 1)(1 + \lambda)^4 s^5} + \frac{24\lambda^3}{(1 + \lambda)^6 s^7} \left\{ \frac{27(\gamma - \lambda^2)}{\gamma\lambda^2 - 1} - 80 \right\}$$

$$+ \frac{12000\lambda^3(\gamma\lambda^3 - 1)}{(\lambda^2 \gamma - 1)(1 + \lambda)^7 s^8} + \frac{64\lambda^3}{(1 + \lambda)^8 s^9} \left\{ \frac{100(\gamma\lambda^4 - 1)}{\lambda^2 \gamma - 1} \right.$$

$$\left. + \frac{63(\gamma - \lambda^4) - 405(\gamma - 1)^2}{\lambda^2 \gamma - 1} \right\} + O(s^{-10}). \tag{4.3.11}$$

$W(s)$ decays as s^{-5} owing to the cancellation in the terms s^{-2} and s^{-4}, and as s^{-7} if $\gamma = 1$.

Substituting the near field asymptotic forms of $A_{\alpha\beta}$ and $B_{\alpha\beta}$—Eq. (4.2.33) and Eq. (4.2.34)—into Eq. (4.3.5) and Eq. (4.3.6) one has

$$L = L_1\xi + L_2\xi^2 \ln\xi + L_3\xi^2 + O(\xi^2(\ln\xi)^2) \qquad \text{as} \quad \xi \to 0, \tag{4.3.12}$$

$$M = \frac{M_0(\ln\xi^{-1})^2 + M_1\ln\xi^{-1} + M_2}{(\ln\xi^{-1})^2 + e_1(\ln\xi^{-1}) + e_2} + O(\xi(\ln\xi)^3) \qquad \text{as} \quad \xi \to 0. \tag{4.3.13}$$

In the neighbourhood of $\xi = 0$, Eq. (4.3.13) reduces to

$$M = M_0 + \frac{M_1 - M_0 e_1}{\ln\xi^{-1}}. \tag{4.3.14}$$

The coefficients in Eq. (4.3.12) and Eq. (4.3.3) are given by

$$\left. \begin{array}{l} L_1 = c_2 A_{22}^{(2)} - c_1 A_{11}^{(2)} + c_3 A_{12}^{(2)} \\ L_2 = c_2 A_{22}^{(3)} - c_1 A_{11}^{(3)} + c_3 A_{12}^{(3)} \\ L_3 = c_2 A_{22}^{(4)} - c_1 A_{11}^{(4)} + c_3 A_{12}^{(4)} \end{array} \right\}, \tag{4.3.15}$$

$$
\left.
\begin{aligned}
M_0 &= c_2 B_{22}^{(1)} - c_1 B_{11}^{(1)} + c_3 B_{12}^{(1)} \\
M_1 &= c_2 B_{22}^{(2)} - c_1 B_{11}^{(2)} + c_3 B_{12}^{(2)} \\
M_3 &= c_2 B_{22}^{(3)} - c_1 B_{11}^{(3)} + c_3 B_{12}^{(3)}
\end{aligned}
\right\}. \tag{4.3.16}
$$

where

$$
c_1 = (\lambda^2\gamma - 1)^{-1}, \quad c_2 = \lambda^2\gamma(\lambda^2\gamma - 1)^{-1}, \quad c_3 = 2(1 - \lambda^3\gamma)(1+\lambda)^{-1}(\lambda^2\gamma - 1)^{-1}. \tag{4.3.17}
$$

The constant term in the longitudinal scalar function L should be zero, i.e. $L^{(0)} = c_2 A_{22}^{(1)} - c_1 A_{11}^{(1)} + c_3 A_{12}^{(1)} = 0$. If we multiply Eq. (4.3.5) by $(\lambda^2\gamma - 1)$, Eq. (4.3.5) becomes

$$
\begin{aligned}
(\lambda^2\gamma - 1)L &= \lambda^2\gamma A_{22} - A_{11} + \frac{2(1 - \lambda^3\gamma)}{1+\lambda} A_{12} \\
&= \left(-A_{11} + \frac{4}{1+\lambda} A_{12} - \frac{1}{\lambda} A_{22}\right) \\
&\quad + \left(\lambda^3\gamma - 1\right)\left(\frac{1}{\lambda} A_{22} - \frac{2}{1+\lambda} A_{12}\right). \tag{4.3.18}
\end{aligned}
$$

The first term on the right-hand-side of Eq. (4.3.18) tends to zero as $\xi \to 0$ due to that the lubrication resistance becomes infinity (see Eq. (4.2.23)). The second term on the right-hand-side of Eq. (4.3.18) also tends to zero as $\xi \to 0$ due to the impenetrability of two rigid spheres (see Eq. (4.2.24)). Thus the constant term in L must be zero and L also tends to zero as $\xi \to 0$. However, the transverse scalar function M is not zero as $\xi \to 0$, since M_0 is not equal to zero. In fact, as sphere 2 contacts sphere 1, the centre of sphere 2 can still rotate about the centre of sphere 1. Thus the transverse component of the relative velocity of the centre of sphere 2 with respect to the centre of sphere 1 must not be zero. This conclusion has been verified by the theory and experiment made by Nir and Acrivos (1973)[16].

3.2. Relative Velocity of Two Spheres Due to Interparticle Potential

The interparticle potential force can also cause a relative motion between two spheres. For the case of central interparticle potential , this force can cause only a translational relative motion between the spheres. The van der Waals attractive potential is a central interparticle potential . Thus, we can calculate the relative velocity of the two spheres caused by the van der Waals potential force with the information of the mobility function s $A_{\alpha\beta}$ and $B_{\alpha\beta}$.

Suppose that the central interparticle potential is $\Phi_{12}(r)$, then, the force acting on sphere 2 is $-\nabla\Phi_{12}(r)$, and that acting on sphere 1 is $\nabla\Phi_{12}(r)$. According to Eq. (4.2.19), we may write the relative velocity $\mathbf{V}_{12}^{\Phi}$ due to the interparticle potential, as

$$
\begin{aligned}
\mathbf{V}_{12}^{\Phi} = \mathbf{U}_2^{\Phi} - \mathbf{U}_1^{\Phi} &= -\mu^{-1}(\mathbf{b}_{22} + \mathbf{b}_{11} - \mathbf{b}_{21} - \mathbf{b}_{12}) \cdot \nabla\Phi_{12}(r) \\
&= -\mathbf{B}^{\Phi} \cdot \nabla\Phi^{12}(r), \tag{4.3.19}
\end{aligned}
$$

where the composite mobility tensor $\mathbf{B}^{\bullet}$ is given as $\mu^{-1}(\mathbf{b}_{22} + \mathbf{b}_{11} - \mathbf{b}_{21} - \mathbf{b}_{12})$.

In aerosol dispersions the interparticle potential force is the attractive van der Waals potential force, and if the particle carry electrical charges, it may be either repulsive or attractive. In hydrosols the interparticle potential force is the sum of an attractive van der Waals force and a repulsive electrical double layer force.

The van der Waals force between two isolated particles was first calculated by Hamaker (1937) who assumed pairwise additivity of the intermolecular attractions[17].

The intermolecular potential force is a very short-range force and it decays rapidly at the molecular scale. Suppose that $d\tau_1$ denotes a volume element in sphere 1 with q_1 atoms per unity volume, and $d\tau_2$ denotes a volume element in sphere 2 with q_2 atoms per unit volume. Then, the attractive molecular potential between these two volume elements is

$$d\Phi_{12} = \frac{-q_1 q_2 \lambda_w}{r^6} d\tau_1 d\tau_2, \tag{4.3.20}$$

where λ_w is the London–van der Waals constant. Eq. (4.3.20) indicates that the force is extremely short-ranged, and it decays as r^{-7}. However, Hamaker showed, the cumulative effect was a long-range interparticle potential with scales of the particle size . Integrating Eq. (4.3.20),[17]

$$\Phi_{12} = -\frac{A}{6}\left[\frac{8\lambda}{(s^2 - 4)(1 + \lambda)^2}\right.$$
$$\left. + \frac{8\lambda}{s^2(1 + \lambda)^2 - 4(1 - \lambda)^2} + \ln\frac{(s^2 - 4)(1 + \lambda)^2}{s^2(1 + \lambda)^2 - 4(1 - \lambda)^2}\right], \tag{4.3.21}$$

where A is the composite Hamaker constant for the materials which compose the two spheres and the surrounding fluid medium. Its value should be determined experimentally or by more rigorous theory (Lifshitz (1955)). When the spheres are very close, $\xi \ll 1$ and $\xi < \lambda$, this expression reduces to

$$\Phi_{12} = \frac{-A\lambda}{3(1 + \lambda)^2 \xi}. \tag{4.3.22}$$

The Hamaker calculation neglected electromagnetic retardation and hence is valid only for separations less than the London wavelengh λ_L, which is typically $0.1\mu m$. Retardation was included by Shenkel and Kitchener (1960) who reported the following best fit to their numerical integrations for $\xi \ll 1$:

$$\Phi_{12} = -\frac{A\lambda}{3(1 + \lambda)^2 \xi(1 + 11.2d/\lambda_L)} \qquad \text{when} \qquad \frac{d}{\lambda_L} \leq \pi, \quad (4.3.23a)$$

$$\Phi_{12} = -\frac{4 \times 10^{-3} A\lambda}{(1 + \lambda)^2 \xi}\left\{\frac{6.5}{d/\lambda_L} - \frac{0.305}{(d/\lambda_L)^2} + \frac{0.0057}{(d/\lambda_L)^3}\right\} \qquad \text{when} \qquad \frac{d}{\lambda_L} > \pi, \quad (4.3.23b)$$

where d is the gap between the spheres.

4. Relative Brownian Diffusion of Two Spheres and the Equation for the Pair-Distribution Function

Brownian motion is one of the basic features of aerosol particles. Nearly all the problems of the mechanics of aerosols are influenced by it.

The displacement induced by Brownian motion is important for most aerosol particles. For the case of Aitken nuclei , the root mean square Brownian displacement in 1 second is of order 10μm, and is larger than the gravitational displacement by two orders. Tables 2 and 3 in Chapter 1 show that the Brownian transport process is the leading one for the case of Aitken nuclei . For the case of large nuclei , the root mean square Brownian displacement in 1 second is of order $10^0\mu$m, and is comparable to the gravitational displacement. Only for the case of giant nuclei , the Brownian motion becomes to be of secondary importance to other transport processes (see Tables 2 and 3 of Chapter 1). It follows that the locations and orientations and relative positions of most aerosol particles are random and must be described by a statistical method. Thus, the thermal motion of aerosol particles has consequences for almost all problems of the mechanics of aerosols, and the phenomenon of Brownian motion itself, which was first discovered by Robert Brown, an English botanist, in 1827, is still an inexhaustible source of surprises and new questions for investigation.

The understanding of Brownian motion began in 1905 when as a result of Einstein's work it was realized that the reason why the mean-square velocity of a suspended particle observed under microscope is much less than the theoretical value obtained from equipartition of energy, viz. $3kT/m$ for a particle of mass m in a fluid medium at temperature T, is that the rapid fluctuations in particle velocity cannot be resolved by eye. Einstein pointed out that a quantity which could be more readily observed and compared to the theoretical value is the rate of increase of mean-square displacement of a particle, or, equivalently in at least some circumstances, the diffusivity defined as the flux of particle number per unit concentration gradient.

The classical theory of Brownian diffusion developed by Einstein (1905, 1906) and many later investigators is concerned with random migration of isolated colloidal particles or large solute molecules due to their interaction with molecules of the suspending fluid. The results of this theory are applicable to very dilute solutions or suspensions in which the particles on average are far apart from each other. When the suspension is not extremely dilute, the hydrodynamic interaction of particles will affect their migration, and modifications should be made to the classical Einstein's theory. This is the new development in the theory of Brownian motion achieved in the last twenty years, and we shall consider this effects later in this section. As a starting point we now turn back to the classic Einstein theory.

Suppose that the position of the volume center of a particle subjecting to the Brownian motion is given at some initial instant, and that the displacement and velocity of the volume centre after a time t are expressed as $\mathbf{X}(t)$ and $\mathbf{v}(t)$, where

$$\mathbf{X}(t) = \int_0^t \mathbf{v}(t')\mathrm{d}t'. \tag{4.4.1}$$

In circumstances such that the velocity $\mathbf{v}$ is a stationary random function of time, it follows from the central limit theorem that the probability distribution of $\mathbf{X}$ tends to the normal or Gaussian form as $t \to \infty$. We may see this formally by dividing the range of t into a number of equal intervals Δt, whence $\mathbf{X}(t)$ becomes the sum of a series of terms with equal means and equal variances. If now we choose Δt to be sufficiently large compared to the correlative time τ_0 of the Brownian motion , and then let $t/\Delta t \to \infty$, the conditions for application of the central limit theorem are satisfied. Obviously, the Brownian correlative time τ_0 should be very small for the central limit theorem to be applicable. That is just the case for the aerosol particles. For instance, τ_0 is of order 10^{-8} to 10^{-9} seconds for the case of Aitken nuclei , and is of order 10^{-5} to 10^{-7} seconds for the case of large nuclei , and is of order 10^{-3} to 10^{-5} seconds for the case of giant nuclei . Therefore, the probability distribution of $\mathbf{X}$ of an aerosol particle must be a Gaussian distribution.

A Gaussian form for the probability density function $P(\mathbf{x})$ satisfies the diffusion equation.

$$\frac{\partial P}{\partial t} = \nabla \cdot \left(\frac{1}{2} \frac{d < \mathbf{XX} >}{dt} \cdot \nabla P \right),$$

and the tensor coefficient

$$\frac{1}{2} \frac{d < \mathbf{XX} >}{dt} = \int_0^t \mathbf{R}(\xi_1) d\xi_1 \qquad (4.4.2)$$

is termed the diffusivity, where $\mathbf{R}(\xi) = < \mathbf{v}(t)\mathbf{v}(t + \xi_1) >$ and the angle brackets denote the ensemble average . Provided that the integral in Eq. (4.4.2) converges as $t \to \infty$, the diffusivity attains the asymptotic value

$$\mathbf{D} = \int_0^\infty \mathbf{R}(\xi_1) d\xi_1 \qquad (4.4.3)$$

after a correlative time τ_0, where

$$\tau_0 = \int_0^\infty |\mathbf{R}(\xi_1)| d\xi_1 / |\mathbf{R}(0)| \qquad (4.4.4)$$

and $|\mathbf{R}|$ denotes the trace of $\mathbf{R}$. The asymptotic value Eq. (4.4.3) is usually the only relevant value of the diffusivity in practical diffusion problems since τ_0 is a very small time. We note that since the principle of equipartition of energy gives

$$|\mathbf{R}(0)| = 3kT/m$$

for a particle of mass m in an equilibrium situation, the Brownian correlative time τ_0 is equal to $m|\mathbf{D}|/3kT$.

The argument used by Einstein (1905) to determine the Brownian diffusivity of an isolated particle suspended in fluid is ingenious and simple, and involves a hypothesized equilibrium system. Einstein considered a situation in which a steady external force derived from a potential $\Phi(\mathbf{X})$ acts on the particle and drives it towards an impermeable boundary against the random movement away from the boundary due to the thermal agitation. In this state of thermodynamic equilibrium involving the suspended particle and the fluid molecules at temperature T, the probability density function for the position of the particle is given by the Boltzmann distribution as

$$P(\mathbf{X}) = P_0 \exp(-\Phi/kT), \qquad (4.4.5)$$

where P_0 is a constant determined by the normalization condition $\int P(\mathbf{X})d\mathbf{X} = 1$. Alternatively, we can follow Einstein (1905) more closely and use the fact that, in a system with various compositions, uniformity of the chemical potential of each component (the components in the case being the particle and the fluid which suspends the particles), as well as of the temperature, is a necessary condition for thermodynamic equilibrium . In a dilute suspension the local chemical potential of the particle at the presence of the applied force is approximately

$$kT\ln P(\mathbf{X}) + \Phi(\mathbf{X}) = const.,$$

where Eq. (4.4.5) is recovered.

Now in this equilibrium system the mean particle flux due to the movement under the action of the applied force balances that due to Brownian diffusion down the probability gradient. Provided the particle is so small that the fluid flow about it is governed by the linear low-Reynolds-number equation of motion, the velocity imparted to the particle by a steady force $\mathbf{F}$ is $\mu^{-1}\mathbf{b} \cdot \mathbf{F}$, where the second-rank tensor $\mathbf{b}$ is the particle mobility. Hence the local flux balance in the equilibrium state is represented by

$$-\mu^{-1}P\mathbf{b} \cdot \nabla\Phi - \mathbf{D} \cdot \nabla P = 0. \qquad (4.4.6)$$

Substituting for P from Eq. (4.4.5), the diffusivity due to Brownian motion has a uniform value

$$\mathbf{D} = \mu^{-1}kT\mathbf{b}. \qquad (4.4.7)$$

For a rigid spherical particle of radius a the mobility tensor is isotropic and may be obtained from the Stokes resistance law, giving the classical formula

$$\mathbf{D} = \frac{kT}{6\pi\mu a}\mathbf{I}, \qquad (4.4.8)$$

where $\mathbf{I}$ is the unit isotropic tensor and μ is the viscosity of the fluid which suspends the particles.

This Einstein's formula for $\mathbf{D}$ has been derived for the postulated equilibrium situation; but since the Brownian agitation and the steady applied force lead to independent and superposable movements of the particle in a linear low-Reynolds-number system, the expression for the diffusive flux is equally valid at the absence of the applied force.

It should be noted for the future reference that the particle flux due to Brownian migration is the same here as if a certain steady force acted on the particle (this force being equal in magnitude and opposite in direction to the external force $-\nabla\Phi$ that, in the equilibrium situation, produces a convective flux which balances the diffusive flux), the same as if a steady force

$$\mathbf{F} = -kT\nabla\ln P(\mathbf{X}) \qquad (4.4.9)$$

acting on the particle. When the probability density of the particle position, is nonuniform, the mean Brownian velocity of a particle, is non-zero simply as a consequence of the fact the particle is more likely to have come from a direction in which the probability density increases, other than from one in which it decreases; and it is equivalent, so far as its effect on the diffusive flux is concerned, to the action of the steady force Eq. (4.4.9) on the particle. The equivalent steady force Eq. (4.4.9) is sometimes termed the thermodynamic force.

This classical Einstein's argument has been reproduced here because it is the basis of our investigation of the effects of multi-particle interactions. The Einstein's formula holds when the particle velocity $\mathbf{v}$ is a stationary random function of t. That is certainly true if the particle is free from interactions with other particles and if any external force acting on it is constant. But when there are interactions between particles which depend on the relative positions of the particles, the statistical properties of the velocity of one particle might change slowly with time simply as a consequence of change of the spatial configuration of particles. For instance, as two particles are very close their relative Brownian vanishes, but as a joint pair, it still responds to thermal forces and possesses a Brownian motion with a mass and hydrodynamic resistance for the pair containing two particles. Whereas after some time when the particles have wandered far apart each of them would have the Brownian motion as isolated particles. It is conceivable that a theory which takes full account of the effect of change of the particle configuration on the Brownian movement could be devised, but here we shall evade the difficulties by supposing that the change in the particle configuration during the correlative time τ_0 is negligibly small. In these circumstances the velocities of a number of interacting particles are approximately stationary random functions of time over a time interval of order τ_0; and a simple generalization of the argument above to multi-particle statistics shows that the joint probability distribution of the displacement vectors of N interacting particles tends to the Gaussian form and that the corresponding diffusion coefficient tensor of the form

$$\frac{1}{2}\frac{\mathrm{d}<\mathbf{X}_i\mathbf{X}_j>}{\mathrm{d}t} = \int_0^t \mathbf{R}_{ij}(\xi_1)\mathrm{d}\xi_1, \tag{4.4.10}$$

where i and j refer to two different particles, tend to constant values like Eq. (4.4.3), viz.

$$\mathbf{D}_{ij} = \int_0^\infty \mathbf{R}_{ij}(\xi_1)\mathrm{d}\xi_1, \tag{4.4.11}$$

which are functions of the particle configuration, and where $\mathbf{R}_{ij}(\xi_1) = <\mathbf{v}_i(t)\mathbf{v}_j(t+\xi_1)>$.

The essence of our assumption is that the particle configuration effectively remains constant during the time τ_0 that characterizes the diffusion process. If for purposes of estimation we suppose that each particle migrates with the scalar diffusivity $\tau_0 kT/m$ (and that any other motion-producing agency changes the particle configuration more slowly), the change in the distance between two particles in time interval τ_0 is of order $\tau_0(kt/m)^{1/2}$, or $(mkT)^{1/2}/6\pi\mu a$ in the case of rigid spheres of radius a where μ is the fluid viscosity . Being expressed as a fraction of a particle radius, this relative displacement of two spheres in time interval τ_0 is

$$(mkT)^{1/2}/6\pi\mu a^2,$$

which is less than 10^{-2} in the case of spheres of radius 1μm in air at temperature 20^0C and is normally negligible. If two particles are very close to each other, and ϵ is the minimum distance between the two surfaces, it would be more logical to compare the relative displacement of the two particles during time interval τ_0 with the ϵ; but then the relative diffusivity differs from that assumed by just a factor of order ϵ/a (as will be seen later, and the relation $(mkT)^{1/2}/6\pi\mu a^2$ is again an estimate of the change in the distance between the two centres in time interval τ_0 expressed as a fraction of the relevant configuration length which is now ϵ. It appears that our assumption of a constant configuration during the time interval τ_0 that characterizes the diffusion process will be accurate in common circumstances.

Consider now a group of m particles in a homogeneous suspension (where $m > 1$ but is not very large) whose separations are within a few particle radii and so are close enough to interact with each other hydrodynamically. The chance of another particle being close to the group is small, in a dilute suspension, and we may therefore regard the group as being isolated from other particles. We now hypothesize an equilibrium situation in which a steady external interactive force with mutual potential energy $\Phi(\mathbf{X}_1, \mathbf{X}_2, \cdots, \mathbf{X}_m)$ is applied to the particles of the group, where $\mathbf{X}_1, \mathbf{X}_2, \cdots, \mathbf{X}_m$ are the position vectors of the centres of the m particles, and Φ depends only on the relative position vectors $\mathbf{X}_2 - \mathbf{X}_1, \mathbf{X}_3 - \mathbf{X}_1, \cdots, \mathbf{X}_m - \mathbf{X}_1$ and not on the location of the group in space. There is in addition the interactive force with a potential energy $U(\mathbf{X}_1, \mathbf{X}_2, \cdots, \mathbf{X}_m)$ which in our case of hard spheres is infinite as $\epsilon \leq 0$ and zero as $\epsilon > 0$. Since the group of particles is independent of other particles in the suspension we again have the Boltzmann form

$$P(\mathbf{X}_1, \mathbf{X}_2, \cdots, \mathbf{X}_m) = P_0 \exp\left(\frac{-\Phi - U}{kT}\right) \qquad (4.4.12)$$

for the joint probability distribution of the position vectors of the m particles.

The diffusive probability flux of one particle relative to others in this equilibrium situation is equal in magnitude and opposite in direction to the convective flux produced by the steady applied force. Provided the two fluxes are independent, as may be assumed in view of linearity of the hydrodynamic equations for the motion due to particles under the external force, it follows that when the external force is suddenly removed there is a flux due to diffusion which is the same as if each particle were under a steady force equal in magnitude and opposite in direction to that derived from Φ, where Φ is given in Eq. (4.4.12), that is, the same as if the k-th particle were under the steady force

$$\mathbf{F}_k = -kT \frac{\partial \ln P(\mathbf{X}_1, \mathbf{X}_2, \cdots, \mathbf{X}_m)}{\partial \mathbf{X}_k} \qquad (k = 1, 2, \cdots, m). \qquad (4.4.13)$$

Such values of $\mathbf{X}_k$ as the k-th particle contacts another particle must be excluded since $P(\mathbf{X}_1, \mathbf{X}_2, \cdots, \mathbf{X}_m)$ may be discontinuous there). Thus when the probability density function for the particle configuration is known, the relative diffusive flux can be calculated. Such a calculation involves a consideration of the hydrodynamic resistance to relative motion of neighbouring particles under the action of the force Eq. (4.4.13), and we now take it up for the case of a group containing two particles.

In the case of a statistically homogeneous suspension of particles which is being deformed, a non-uniform probability distribution of the relative positions of particles may be produced by the bulk flow . The effect of Brownian motion of the particles is then to tend to restore the uniformity.

When the suspension is dilute, the chance of a particle having m neighbours within a few diameters at any instant is of order φ^m, where $\varphi(\ll 1)$ is the volume fraction of the particles . We shall consider here the relative diffusion of an effectively isolated pair of spherical particles with a given vector separation. For a homogeneous suspension the joint probability density function of the positions of sphere 1 of radius a_1 and of sphere 2 of radius a_2 is of the form

$$P(\mathbf{X}_1, \mathbf{X}_2) = n_1 n_2 p_{12}(\mathbf{r}), \qquad (4.4.14)$$

where $\mathbf{X}_1$ is the position vector of the centre of sphere 1, $\mathbf{X}_2$ is that of sphere 2, $\mathbf{r} = \mathbf{X}_2 - \mathbf{X}_1$ and n_1, n_2 are the (uniform) number densities of the two types of spheres respectively. The pair-distribution function $p_{12}(\mathbf{r})$ will satisfy the condition

$$p_{12}(\mathbf{r}) \to 1 \quad \text{as} \quad r \to \infty$$

at absence of long-range order in the suspension.

According to the above results the relative diffusive flux due to the Brownian motion is the same as if spheres 1 and 2 moved under the action of steady applied forces $\mathbf{F}_1$ and $\mathbf{F}_2$ respectively, in view of Eq. (4.4.13),

$$\left. \begin{array}{l} \mathbf{F}_1 = -kT\frac{\partial \ln P(\mathbf{X}_1, \mathbf{X}_2)}{\partial \mathbf{X}_1} = kT\nabla \ln p_{12}(\mathbf{r}) \\ \mathbf{F}_2 = -kT\frac{\partial \ln P(\mathbf{X}_1, \mathbf{X}_2)}{\partial \mathbf{X}_2} = -kT\nabla \ln p_{12}(\mathbf{r}) \end{array} \right\}. \qquad (4.4.15)$$

Now when two couple-free spheres are under external forces $\mathbf{F}_1$ and $\mathbf{F}_2$ in fluid which is at rest at infinity, they acquire velocities $\mathbf{U}_1$ and $\mathbf{U}_2$ given by Eq. (4.2.19). The diffusive flux of sphere 2 relative to sphere 1 is then

$$(\mathbf{U}_2 - \mathbf{U}_1)P(\mathbf{X}_1, \mathbf{X}_2) = -\mu^{-1}kT(\mathbf{b}_{11} + \mathbf{b}_{22} - \mathbf{b}_{12} - \mathbf{b}_{21}) \cdot \nabla(n_1 n_2 p_{12}), \qquad (4.4.16)$$

and we may define the relative Browinan diffusivity of two spheres with separation vector $\mathbf{r}$ as

$$\mathbf{D}_{12}(\mathbf{r}) = \mu^{-1}kT(\mathbf{b}_{11} + \mathbf{b}_{22} - \mathbf{b}_{12} - \mathbf{b}_{21}). \qquad (4.4.17)$$

This is the diffusivity that is needed in an investigation of the pair-distribution function $p_{12}(\mathbf{r})$ in a dilute homogeneous suspension of spheres subjected to a bulk deforming motion.

Substituting the expressions for the mobility tensor s $\mathbf{b}_{\alpha\beta}$ given in Eq. (4.2.8a) into Eq. (4.4.17) we find (Batchelor 1976)[18]

$$\mathbf{D}_{12}(\mathbf{r}) = D_{12}^{(0)}\left\{ G(r)\frac{\mathbf{rr}}{r^2} + H(r)\left(\mathbf{I} - \frac{\mathbf{rr}}{r^2} \right) \right\}, \qquad (4.4.18)$$

where

$$D_{12}^{(0)} = \frac{kT}{6\pi\mu}\left(\frac{1}{a_1} + \frac{1}{a_2}\right),\tag{4.4.19}$$

and the longitudinal scalar function G and the transverse scalar function H are

$$G(s,\lambda) = \frac{\lambda A_{11} + A_{22}}{1+\lambda} - \frac{4\lambda A_{12}}{(1+\lambda)^2},\tag{4.4.20a}$$

$$H(s,\lambda) = \frac{\lambda B_{11} + B_{22}}{1+\lambda} - \frac{4\lambda B_{12}}{(1+\lambda)^2},\tag{4.4.20b}$$

respectively, and $s = 2r/(a_1 + a_2), \lambda = a_2/a_1$.

We note from the inversion relations of $A_{\alpha\beta}, B_{\alpha\beta}$ Eq. (4.4.20) and Eq. (4.2.21) that

$$G(s,\lambda) = G(s,\lambda^{-1}), H(s,\lambda) = H(s,\lambda^{-1}).\tag{4.4.21}$$

When the two spheres are far apart they move independently, corresponding to

$$G \to 1, H \to 1 \quad\text{and}\quad \mathbf{D}_{12} \to D_{12}^{(0)}\mathbf{I} \quad\text{as } s \to \infty.\tag{4.4.22}$$

The far field asymptotic form for G and H as $s \to \infty$ can be found from Eq. (4.2.25) to Eq. (4.2.28) correct to order of s^{-12}. From Eq. (4.2.23) and Eq. (4.4.20a) we may see that the near field asymptotic form for G is

$$G = G_1\xi + O(\xi^2\ln\xi),\tag{4.4.23}$$

where $G_1 = \frac{(1+\lambda)^2}{2\lambda}$. The fact that the longitudinal scalar function G tends to zero as ξ tends to zero reflects the fact that the lubrication resistance tends to infinity as $\xi \to 0$. The near field asymptotic form for H can be obtained by substituting Eq. (4.2.34) into Eq. (4.4.20b), then

$$H = \frac{H_0(\ln\xi^{-1})^2 + H_1\ln\xi^{-1} + H_2}{(\ln\xi^{-1})^2 + e_1\ln\xi^{-1} + e_2} + O(\xi(\ln\xi)^3),\tag{4.4.24}$$

where

$$\left.\begin{array}{rcl} H_0 &=& c_1 B_{11}^{(1)} + c_2 B_{22}^{(1)} - c_3 B_{12}^{(1)} \\ H_1 &=& c_1 B_{11}^{(2)} + c_2 B_{22}^{(2)} - c_3 B_{12}^{(2)} \\ H_2 &=& c_1 B_{11}^{(3)} + c_2 B_{22}^{(3)} - c_3 B_{12}^{(3)} \end{array}\right\},\tag{4.4.25}$$

and $c_1 = \lambda/1 + \lambda, c_2 = 1/1 + \lambda, c_3 = 4\lambda/(1+\lambda)^2$. In the neighbourhood of $\xi = 0$, the simple near field asymptotic form for H is

$$H = H_0 + \frac{H_1 - H_0 e_1}{\ln\xi^{-1}}.\tag{4.4.26}$$

The fact that the transverse scalar function H is not equal to zero as $\xi \to 0$ reflects that the centre of the sphere 2 may rotate about the centre of the sphere 1 when the sphere 2 contacts the sphere 1.

The values of G and H in the intermediate field can be obtained from Jeffrey and Onishi's mobility function s $A_{\alpha\beta}$ and $B_{\alpha\beta}$ (1984)[9], and have been calculated by Batchelor and Wen (1982) for rigid spheres[12].

Failing in understanding the hydrodynamic interaction s of the particles led to some mistakes in Fuchs' book (1964)[2] and Pruppacher and Klett's book (1978)[3]. For example, they still followed Smoluchowski 's isolated particle theory in their books, and thus thought that the relative Brownian diffusivity remained constant even when the gap between the two spheres tended to zero. Obviously, the assumption of the constant Brownian diffusivity of sphere 2 relative to sphere 1 cannot be correct. In fact, it is impossible for the finite Brownian energy to break through the lubrication resistance as $\xi \to 0$. The scalar functions G and H cannot be identically equal to unity, and must be functions of the vector separation $\mathbf{r}$ of the two spheres when we consider the hydrodynamic interaction between the two spheres. This is one of the essential difference between isolated-particle theory and multi-particle theory. The latter is a new development in the Brownian motion theory made in the last twenty years by Batchelor (1976,[18], 1983) and by other workers like Deutch and Oppenheim (1971), Murphy and Aguirre (1972), Aguirre and Murphy (1973), Felderhof (1978), Hanna, Hess and Klein (1982), Jones and Burfield (1982), Jones (1979), McQuarrie and Deutch (1980) and Pusey and Tough (1982).

We come now to the other aspect of the problem, viz. concern with the statistical structure of the aerosol dispersion. The position of an aerosol particle should be described by a statistical method, since, it undergoes a Brownian motion . All the quantity to be solved in the mechanics of aerosols should be a statistical average quantity. Thus to find a microstructure of aerosol dispersions is one of the key points in the mechanics of aerosols. In a dilute aerosol system, this is the pair-distribution function $p_{ij}(\mathbf{r})$ (see Eq. (4.4.14)), defined as the probability that the centre of sphere j lies within a unity volume at position $\mathbf{r}$ relative to the centre of sphere i. The pair-distribution function is governed by a conservative equation of the Fokker–Planck type. Colloidal scientists have tended either to overlook this need or to dismiss it on the grounds that usually they are considering systems in which effects of convection are weak relative to those of Brownian motion (see, for instance, Peterson and Fixman 1963, Reed and Anderson 1976 and Dickinson 1980). In a sedimenting system (or a deforming system), effects of motion due to gravity (or deforming motion) are not always small in practice. Thus, solving the conservative equation of Fokker–Planck type is normally an unavoidable part of the investigation of properties of a dynamical system which is not in thermodynamic equilibrium .

For this differential equation we need to know the velocity of the particle centre of species j relative to the particle centre of species i. For the relative velocity $\mathbf{V}_{ij}(\mathbf{r})$ due to gravity alone we have the expression Eq. (4.3.2), and that due to central inter-particle forces we have the expression Eq. (4.3.19) (in which all the subscripts 1 and 2 should be replaced by i and j). The total relative velocity due to the deterministic forces is thus

$$\mathbf{V}_{ij} - \mathbf{D}_{ij} \cdot \nabla(\Phi_{ij}/kT),$$

in which the composite mobility tensor $\mathbf{B}^{\circledast}$ has been replaced by $\mathbf{D}_{ij}/kT$.

Taking account also of relative diffusion we find that the differential equation expressing the conservation of pairs made up of a sphere of species i and another of species j is

$$\frac{\partial p_{ij}}{\partial t} = -\nabla \cdot (\mathbf{V}_{ij}p_{ij}) + \nabla \cdot \{p_{ij}\mathbf{D}_{ij} \cdot \nabla(\Phi_{ij}/kT)\} + \nabla \cdot (\mathbf{D}_{ij} \cdot \nabla p_{ij}). \qquad (4.4.27)$$

The expressions for the dimensionless separation s, the size ratio λ and the reduced density ratio γ are defined by

$$s = \frac{2r}{a_i + a_j}, \quad \lambda = \frac{a_j}{a_i}, \quad \gamma = \frac{\rho_j - \rho}{\rho_i - \rho}, \qquad (4.4.28)$$

where $i, j = 1, 2, \cdots, m$, if there are m different species of rigid spherical particle in the aerosol dispersions.

For a statistically homogeneous aerosol system, the outer boundary condition is

$$p_{ij}(\mathbf{r}) \to 1 \quad \text{as} \quad r \to \infty. \qquad (4.4.29)$$

For a stable system the scalar product of the radial vector $\mathbf{r}$ and the total flux of sphere j

$$\mathbf{r} \cdot [\{\mathbf{V}_{ij}p_{ij} - \mathbf{D}_{ij} \cdot \nabla\{\Phi_{ij}/kT\}\}p_{ij} - \mathbf{D}_{ij} \cdot \nabla p_{ij}]$$

should be zero at the inner boundary $r = a_i + a_j$. Now $\mathbf{r} \cdot \mathbf{V}_{ij} \to 0$ as $r \downarrow (a_i + a_j)$, as a consequence of the rigidity of the spheres. The total flux at $r = a_i + a_j$ reduces to

$$\mathbf{r} \cdot [\{-\mathbf{D}_{ij} \cdot \nabla\{\Phi_{ij}/kT\}\}p_{ij} - \mathbf{D}_{ij} \cdot \nabla p_{ij}].$$

It will be noted from Eq. (4.4.23) that, as another consequence of the sphere rigidity,

$$\mathbf{r} \cdot \mathbf{D}_{ij} \sim r - (a_i + a_j) \quad \text{as} \quad r \downarrow a_i + a_j;$$

and so, in cases in which the limit value of $|\nabla \Phi_{ij}|$ as $r \downarrow a_i + a_j$ is finite, the condition to be satisfied at $r = a_i + a_j$ reduces to

$$\mathbf{r} \cdot \mathbf{D}_{ij} \cdot \nabla p_{ij} = 0, \qquad (4.4.30)$$

and imposes a restriction on the rapidity with which p_{ij} varies near $r = a_i + a_j$.

On the other hand, for an unstable system, the sphere j will be removed from the system at the inner boundary $r = a_i + a_j$, and the particles undergo coagulation . Thus the inner boundary condition is

$$p_{ij}(\mathbf{r}) = 0 \quad \text{at} \quad r = a_i + a_j. \qquad (4.4.31)$$

In a steady state, the sum of the three divergence terms on the right-hand-side of Eq. (4.4.27) are zero. The form of the solution for $p_{ij}(r)$ depends on the relative magnitudes of the three terms, which are measured by the Péclet number $\mathcal{P}_{ij}$, defined as

$$\mathcal{P}_{ij} = \frac{\frac{1}{2}(a_i + a_j)V_{ij}^{(0)}}{D_{ij}^{(0)}}, \qquad (4.4.32)$$

which measures the relative magnitudes of the effects of convection and diffusion on p_{ij}, and in which $V_{ij}^{(0)}$, $D_{ij}^{(0)}$ are representative magnitudes of $\mathbf{V}_{ij}$ and $\mathbf{D}_{ij}$ given by Eq. (4.3.3) (for the sedimenting dispersions) and Eq. (4.4.19) respectively. The Q_{ij} number is defined as

$$Q_{ij} = \frac{\frac{1}{2}(a_i + a_j)V_{ij}^{(0)}kT}{D_{ij}^{(0)}\Phi_{ij}^{(0)}}, \qquad (4.4.33)$$

which measures the relative magnitudes of the effects of convection and interparticle potential on p_{ij}, and $\Phi_{ij}^{(0)}$ is the representative magnitude of the interparticle potential and is the Hamaker constant A for the case of an unstable system.

From the Markov process theory we may derive a differential equation similar to Eq. (4.4.27) which is called as the second Kolmogorov differential equation. The diffusion term is $\nabla \cdot \nabla \cdot (\mathbf{D}_{ij}p_{ij})$ which is different from the gradient diffusion transport term $\nabla \cdot (\mathbf{D}_{ij} \cdot \nabla p_{ij})$ in Eq. (4.4.27). These two differential equations coincide with each other only provided that the relative Brownian diffusivity is constant, i.e. $\mathbf{D}_{ij} \equiv D_{ij}^{(0)}\mathbf{I}$. Unfortunately the diffusivity depends on the vector separation $\mathbf{r}$ under the hydrodynamic interaction s of particles. Chandrasekhar (1943) showed that the gradient diffusion approximation could be used only when the displacement $|\Delta \mathbf{r}|$ was larger than the diffusion displacement $(D_{ij}^{(0)}\tau_0)^{1/2}$[19]. In that case the second Kolmogorov differential equation reduces to the Fokker–Planck equation. Suppose the average radius of aerosol particles to be 1μm, then we have.

$$\frac{(D_{ij}^{(0)}\tau_0)^{1/2}}{|\Delta \mathbf{r}|} \sim 10^{-2} \ll 1.$$

Thus, generally speaking, Eq. (4.4.27) is valid for the case of aerosol particles.

Another modification to the classical Einstein's theory of the Brownian motions that are needed when rigid spherical particles are close enough to interact hydrodynamically is to the gradient diffusivity in an inhomogeneuos aerosol dispersion.

Batchelor (1976) showed in a monodisperse system that there was a dependence of the gradient Brownian diffusivity on the concentration of particles. The dependence on concentration is a consequence of interactions of the particles, both hydrodynamic interaction s transmitted from one moving particle to another through the fluid and direct interactions which in the simple case of rigid uncharged spheres reduce to an exclusion of overlapping particles. Eq. (4.4.9) is still valid, but the effective chemical potential ζ (i.e. $kT \ln P(\mathbf{X})$ now depends on the interaction potential, in a way which is familiar from the theory of ideal solutions (analogous to the theory of ideal gases), and the relation between the steady thermodynamic driving force $\mathbf{F}$ and the resulting average velocity of the particles is also different because the particles interfere with each other hydrodynamically, exactly as in the problem of sedimentation of particles under gravity which will be discussed in detail next chapter.

In the case of a dispersion of uncharged rigid identical spheres which occupy a small volume fraction φ, it turns out that the effect of forces between particles is to make the thermodynamic driving force greater by a factor $1 + 8\varphi + O(\varphi^2)$ and to make the mobility less by a factor $1 - 6.55\varphi + O(\varphi^2)$ (because a movement of particles in one direction is accompanied by a counter-current of fluid which hinders the movement), giving a diffusivity equal to (Batchelor 1976)[18]

$$\mathbf{D} = \frac{kT}{6\pi\mu a}(1 + 8\varphi + O(\varphi^2))(1 - 6.55\varphi + O(\varphi^2))\mathbf{I}$$

$$= \frac{kT}{6\pi\mu a}(1 + 1.45\varphi)\mathbf{I}, \qquad (4.4.34)$$

correct to the order of φ. It appears that the enhancement of the diffusivity due to the greater availability of particle sites in regions of lower concentration is a little greater than the reduction due to hydrodynamic hindrance to the movement of particles.

In 1974, some extensive measurements on the concentration dependence of various properties of a suspension of compact DNA molecules (molecular weight 1.9×10^6, effective diameter about $0.03\mu m$) by the new and much more accurate technique of spectral analysis of fluctuations in the light scattered by the particles from an incident laser beam give the value of the fractional rate of change of gradient diffusivity with respect to φ as 1.2 ± 0.4 (Newman, Swinney, Berkowitz and Day), which is in fair agreement with the above theoretical value. Measurements on the gradient diffusivity of stabilized silica spheres (with $a = 0.021\mu m$) determine the coefficient of ϕ in Eq. (4.4.34) as 1.3 ± 0.3, which is also in fair agreement with the theoretical value 1.45.

On the basis of the extensive calculations of sedimentation in a dilute polydisperse system of interacting spheres (Batchelor and Wen 1982)[12], Batchelor (1983) generalized the above gradient diffusion theory to the gradient diffusion theory in a polydisperse system of interacting spheres.

When m different species of small particles are dispersed in a fluid where a (small) spatial gradient of concentration of particles of type j (with radius a_j, volume fraction φ_j) exists, as a consequence of Brownian motion of the particles, by a flux of particles of type i (with radius a_i, volume fraction φ_i). The flux and the gradient are linearly related, and the tensor gradient diffusivity $_g\mathbf{D}_{ij}$ is the proportionality constant. When the volume fraction of the particles $_g\mathbf{D}_{ij}$ is approximately a linear function of the volume fractions $\varphi_1, \varphi_2, \cdots, \varphi_m$, with coefficients which depend on the interactions between particle pairs, and given by (Batchelor 1983)[20]

$$_g\mathbf{D}_{ij} = _g\mathbf{D}_i^{(0)}\left\{\delta_{ij} + \beta_{ij}\varphi_i\left(\frac{1 + \lambda_{ij}}{2}\right)^3 + \delta_{ij}\sum_{k=1}^m K'_{ik}\varphi_k + K''_{ij}\varphi_i\right\}, \qquad (4.4.35)$$

correct to the order of φ, where $\delta_{ij} = 1$ when $i = j$ and zero otherwise, $\lambda_{ij} = a_j/a_i$, $_g\mathbf{D}_i^{(0)}$ is the gradient diffusivity tensor of independent particles of species i,

$$_g\mathbf{D}_i^{(0)} = \frac{kT}{6\pi\mu a_i}\mathbf{I}. \qquad (4.4.36)$$

The coefficient β_{ij} in Eq. (4.4.35) represents the effect of interactions of particle pairs on the thermodynamic driving force. β_{ij} is (a numerical multiple of) the 'second virial coefficient' giving a correction to the perfect-gas law due to the finite size and range of interaction of the gas molecules, or to the ideal-solution expression for the osmotic pressure in the solutions of non-electrolytes. Explicitly, and is given by

$$\beta_{ij} = 8 - \alpha_{ij}$$

where

$$\alpha_{ij} = 3 \int_2^\infty \left\{ \exp\left(-\frac{\Phi_{ij}}{kT}\right) - 1 \right\} s^2 ds, \qquad (4.4.37)$$

and $\alpha_{ij}\varphi_j$ can be interpreted as the mean number of j-th particles to be found within a sphere of radius $R(>> a_i + a_j)$ centred at a given particle i, minus the number of j-th particle in dispersion which is of the same concentration and satisfies $\Phi_{ij} = 0$. The coefficient β_{ij} thus can be calculated when the interparticle potential of the force exerted between two spheres of the same species ($i = j$) and between two spheres of different species ($i \neq j$) are known.

The coefficients K'_{ij} and K''_{ij} in Eq. (4.4.35) are the bulk mobility coefficients. It can be shown that K'_{ij} and K''_{ij} are the same as the sedimentation coefficients S'_{ij} and S''_{ij} of the polydisperse system of interacting spheres at small Péclet number , which we shall discuss in detail in Chapter 5. Then we have

$$K'_{ij} = S'_{ij}, \qquad K''_{ij} = S''_{ij}, \qquad (4.4.38)$$

where the coefficient S'_{ij} and S''_{ij}, and then the coefficient K'_{ij} and K''_{ij} depend on λ_{ij} only and

$$K'_{ii} + K''_{ii} = S, \qquad (4.4.39)$$

where S is the sedimentation coefficient in a monodisperse system of interacting spheres, and is given with reasonable accuracy by the approximate relation

$$S = -6.55 + 0.44\alpha, \qquad (4.4.40)$$

where

$$\alpha = 3 \int_2^\infty \left\{ \exp\left(-\frac{\Phi}{kT}\right) - 1 \right\} s^2 ds, \qquad (4.4.41)$$

and $S = -6.55$ in the case of $\Phi = 0$.

Then, the diagonal element of the gradient diffusivity matrix is[20]

$$_g D_{ii} =_g D_i^{(0)} \{1 + (\beta + S)\varphi_i + \sum_{k(\neq i)} K'_{ik}\varphi_k\}, \qquad (4.4.42)$$

and the off-diagonal element is[20]

$$_g D_{ij} =_g D_i^{(0)} \varphi_i \left\{ \beta_{ij} \left(\frac{1 + \lambda_{ij}}{2}\right)^3 + K''_{ij} \right\} \quad (i \neq j). \qquad (4.4.43)$$

For the case of a monodisperse system $\varphi_k = 0(k \neq i)$ and Eq. (4.4.42) reduces to[20]

$$D = D^{(0)}\{1 + (\beta + S)\varphi + O(\varphi^2)\}, \tag{4.4.44}$$

where $\beta = 8 - \alpha$ and $\beta + S \sim 1.45 - 0.56\alpha$, and $\beta + S = 1.45$ in the case of $\Phi = 0$, then $\alpha = 0$, thus Eq. (4.4.43) becomes

$$D = D^{(0)}\{1 + 1.45\varphi\}, \tag{4.4.45}$$

correct to the order of φ. The Eq. (4.4.34) is thus recovered.

CHAPTER 5

SEDIMENTATION OF AEROSOL PARTICLES

1. Sedimentation in a Dilute Monodisperse Stable System

The term 'Sedimentation' conventionally refers to a large number of aerosol particles falling under gravity through air which is otherwise at rest. It is actually a process of particles which external forces exert on and are separated from the medium. Besides gravitational settling, it also includes the separating process in an ultra-centrifuge due to a centrifugal force, sedimentation of charged particles in an electric field, and the related process of fluidization in chemical and energy source industry. In an extremely dilute system, the problem reduces to an isolated-particle problem, and has been solved by Stokes (1851), (see Eq. (3.1.21) for the terminal velocity of an isolated spherical particle caused by gravity). When an aerosol system is not so dilute that the hydrodynamic interaction between particles must be taken into account, we have a problem of multi-particle system. This is a classical problem studied in extensive literature since Smoluchowski (1912). The total research effort over the years has been massive, but it has proved to be difficult to obtain quantitative theoretical results, mainly because of lack of an appropriate means to analyze the motion of many interacting particles. Only in recent years, have we obtained an answer which is regarded as reliable.

The whole problem can be divided into two main divisions. The first one involves investigation on a cloud of particles settling in unbound space. The other involves investigation on particles settling in bound space, for example, falling in a container. In the first case, the settling speed will generally be larger than for an isolated particle due to the direct hydrodynamic interaction between particles. In the second case, the settling speed is found to be less than for an isolated particle due to the global influences (the existence of the container), and the phenomenon is often referred to as 'hindered settling'. Fuchs (1964) reviewed the work before his time in his book "The mechanics of aerosols"[2], and pointed out that it was exceedingly difficult to obtain an exact solution for the second problem. Now the progresses have been made just in this respect since the seventies. In this chapter we will make a review of it.

The difficulty in determination of the hydrodynamic interaction of the particles originates first from the slowness of the velocity disturbance in the fluid due to an isolated falling particle, which decreases to zero at increasing distance, whereas secondly from the random arrangement of the particles in a real dispersion. The magnitude of the fluid velocity at distance r from a single sphere of radius a falling with speed U_0 varies asymptotically as $U_0 a/r$, and so a naive attempt to sum over the contributions to the velocity at one point from an indefinitely large number of falling spheres in a homogeneous dispersion leads to a series or an integral which diverges strongly. This is the first main obstacle for obtaining the exact solution, while the second obstacle is to determine the statistical distribution of the particles, i.e. to find the solution for the pair-distribution function .

Investigation on sedimentation in a monodisperse system can make the second difficulty disappear. Because in this case, the relative gravitational convective term is zero, and the interparticle potential term is assumed to be zero, the problem then reduces to a pure Brownian diffusion problem. Thus, the solution for the pair-distribution equation Eq. (4.4.27) in a steady state with boundary conditions Eq. (4.4.29) (the uniform outer boundary condition), and Eq. (4.4.30) (the zero-flux inner boundary condition), must be a uniform distribution, since in this case, the Brownian motion can smooth out any non-uniform distribution. Thus for the case of monodisperse dispersion we can concentrate our effort to overcome the first obstacle.

Previous investigations fall into three groups, corresponding to the assumptions made about the arrangement of the spherical particles in the dispersion and the nature of their interactions. In the first group we suppose that the centres of the spheres lie according to some regular geomatrical patterns with the length scale of the array being of order $a\phi^{-1/3}$, where ϕ is the fraction of the total volume occupied by the spheres. These calculations yield a result that for a dilute dispersion ($\phi \ll 1$) the fractional reduction of the fall speed due to particle interactions is proportional to $\phi^{1/3}$. The constant of proportionality is of order unity and varies with the type of assumed arrangement (simple cubic, body-centred cubic, rhombohedral, etc.)

In the second group we use the 'cell' model of the interaction effects. The assumption here is that the average hydrodynamic effect on one sphere at presence of all the other spheres in the dispersion is equivalent to that of a boundary, which is usually taken as a spherical surface enclosing the sphere under consideration. The radius of this outer spherical boundary is usually chosen as $a\phi^{-1/3}$. The motion of the fluid in the cell satisfies the no-slip condition at the surface of the rigid sphere, and some suitably chosen conditions at the stationary artificial outer boundary of the cell. One simple choice is that the fluid velocity is zero there. All these calculations with a cell radius proportional to $\phi^{-1/3}$ give a fractional reduction of the fall speed which is proportional to $\phi^{1/3}$ for $\phi \ll 1$. The constant of proportionality is again of order unity, but is not the same as that found for a regular array of spheres.

Investigations in the third group have employed statistically analytical methods in attempt to determine the hindered settling of a random distribution of spheres in a dilute dispersion (Burgers 1942; Pyun and Fixman 1964). Burgers tried a variety of ways to overcome the difficulty caused by lack of absolute convergence of the sum over the effects of an indefinitely large number of falling spheres to the given sphere, both for a random distribution and for a regular arrangement. His sequence of papers is remarkable. Burgers recognized the arbitrariness of some of his summation procedures, and was uncertain whether he had found correct expressions. Pyun and Fixman employed generally similar statistical approaches, and improved Burger's calculations in one detail respect but failed in following Burgers in another respect.

For a random distribution of spheres Burgers and Pyun and Fixman found that the average change in the speed was proportional to ϕ, it is a result which is so different in form from that found either for a regular array of sphere or from the cell model . The result obtained from statistical methods was not generally accepted at that time in the hydrosol literatures, however, Fuchs (1964) thought in his book "The mechanics of aerosols" that the statistical theory did contain essentially the right approach and achieve the right result, although Fuchs pointed out that to obtain an exact solution from statistical theory was an extremely difficult problem. Not until Batchelor (1972), did it have got a satisfying answer, and in this section we will mainly introduce Batchelor's renormalization approach and the corresponding result (1972)[21].

We consider a statistically homogeneous dispersion of N identical rigid spherical particles ($N \gg 1$), which is allowed to stand in a container of volume V. The volume flux of a material (which may be either fluid or solid) per unit area of any stationary plane surface in the dispersion defines a local velocity vector whose statistical ensemble mean is uniform over the whole volume , and the axes of reference are so chosen that this mean velocity is zero. Inertia forces on either the solid particles and the fluid will be neglected due to the smallness of the particle in the dispersion.

We shall denote the set of position vectors of the centres of N spheres by $\mathcal{C}_N$, which is a $3N$ dimensional stochastic field, and term it a configuration of N spheres . The probability density of the configuration is $P(\mathcal{C}_N)$, meaning that the probability of sphere centres being located simultaneously in the volume element $\delta \mathbf{r}_1, \delta \mathbf{r}_2, \cdots \delta \mathbf{r}_N$ about the point in the $3N$ space

$$\mathbf{x} + \mathbf{r}_1, \mathbf{x} + \mathbf{r}_2, \cdots \mathbf{x} + \mathbf{r}_N$$

e.g.

$$P(\mathcal{C}_N)\delta \mathcal{C}_N = P(\mathbf{x} + \mathbf{r}_1, \mathbf{x} + \mathbf{r}_2, \cdots \mathbf{x} + \mathbf{r}_N)\delta \mathbf{r}_1 \delta \mathbf{r}_2 \cdots \delta \mathbf{r}_N.$$

The N spheres in the volume V are identical and so we have the normalization relation

$$\int P(\mathcal{C}_N)\delta \mathcal{C}_N = N!. \tag{5.1.1}$$

The probability density of the position vector of a single sphere centre will be written as

$$P(\mathcal{C}_1) = P(\mathbf{x} + \mathbf{r}) = N/V = n, \tag{5.1.2}$$

and

$$P(\mathcal{C}_2) = P(\mathbf{x} + \mathbf{r}_1, \mathbf{x} + \mathbf{r}_2) = N(N-1)/V^2 \cong n^2. \tag{5.1.3}$$

We also introduce the conditional probability density $P(\mathcal{C}_N|\mathbf{x})$ which refers to a dispersion in which there are $N + 1$ spheres in V; $P(\mathcal{C}_N|\mathbf{x})\delta \mathcal{C}_N$ is the probability of a configuration of N spheres whose centres are found in the region $\delta \mathcal{C}_N$ about $\mathcal{C}_N$, given that there is a sphere centred at the point $\mathbf{x}$. A connection between the conditional and unconditional probabilities is provided by the identity

$$P(\mathcal{C}_N) = P(\mathbf{x} + \mathbf{r}_k)P(\mathcal{C}_{N-1}|\mathbf{x} + \mathbf{r}_k) = nP(\mathcal{C}_{N-1}|\mathbf{x} + \mathbf{r}_k), \tag{5.1.4}$$

and the connection between the one-sphere conditional probability and the two-sphere conditional one is given by

$$P(\mathcal{C}_N|\mathbf{x}) = P(\mathbf{x} + \mathbf{r}_k|\mathbf{x})P(\mathcal{C}_{N-1}|\mathbf{x}, \mathbf{x} + \mathbf{r}_k). \tag{5.1.5}$$

When the dispersion is dilute, it will be assumed that there is no long-range order in the dispersion, and that the probabilities of sphere centres being at points which are separated from each other far enough compared to the sphere radius a are independent of each other. In particular we have

$$P(\mathcal{C}_N|\mathbf{x}) \approx P(\mathcal{C}_N), \tag{5.1.6}$$

when each of components of $\mathcal{C}_N$ is at a distance from $\mathbf{x}$ which is large compared to a.

We may now express the average of a quantity $G(\mathbf{x}, \mathcal{C}_N)$ which is associated with a point $\mathbf{x}$ in the dispersion (G necessarily being defined for points both in the fluid and within the rigid spheres) and which is determined by the configuration for N spheres, as

$$\overline{G} = \frac{1}{N!} \int G(\mathbf{x}, \mathcal{C}_N)P(\mathcal{C}_N)\mathrm{d}\mathcal{C}_N. \tag{5.1.7}$$

We shall also concern the average of a quantity $H(\mathbf{x}, \mathcal{C}_N)$, which is associated with a sphere whose centre is at the point $\mathbf{x}$; this requires the conditional probability density and is

$$\overline{H} = \frac{1}{N!} \int H(\mathbf{x}, \mathcal{C}_N)P(\mathcal{C}_N|\mathbf{x})\mathrm{d}\mathcal{C}_N. \tag{5.1.8}$$

If H expresses the settling velocity of a given sphere whose centre is at $\mathbf{x}$ when the configuration of the centres of the surrounding spheres has the form $\mathcal{C}_N$, Eq. (5.1.8) can then be used to calculate the mean settling speed of the test sphere.

The above Eq. (5.1.7) and Eq. (5.1.8) can be simplified if the dispersion is dilute. The probability that just one of the sphere centres of the configuration $\mathcal{C}_N$ is located at a distance from the reference point $\mathbf{x}$ which is a multiple of a is of order ϕ, and so is a small quantity ($\phi \ll 1$). Likewisely the probability that two sphere centres are simultaneously at such a distance from $\mathbf{x}$ is of order ϕ^2. It follows that, in any case where the quantity G(or H) decreases to zero sufficiently fast with increase of the distance r of a single rigid sphere from the reference point $\mathbf{x}$, the average Eq. (5.1.7) (or Eq. (5.1.8)) may be calculated, with an error of order ϕ^2 as if the surrounding configuration would contain just one sphere; 'sufficiently fast' means as rapidly as $(a/r)^{3+\epsilon}$, where $\epsilon > 0$, in order to ensure absolute convergence of the integration. This approximation corresponds to using the identity Eq. (5.1.4) and the approximate relation Eq. (5.1.6) to replace $P(\mathcal{C}_N)$ in the integrand of Eq. (5.1.7.) by

$$P(\mathbf{x} + \mathbf{r}_k)P(\mathcal{C}_{N-1})$$

for that part integration of $\mathbf{r}_k$ for which r_k/a (where $r_k = |\mathbf{r}_k|$) is of order unity and to ignoring the influence of the spheres on the value of G. Thus we have a simplified expression

$$\overline{G} = \sum_{k=1}^{N} \frac{1}{N!} \int P(\mathcal{C}_{N-1})\mathrm{d}\mathcal{C}_{N-1} \int G(\mathbf{x}, \mathbf{x} + \mathbf{r}_k)P(\mathbf{x} + \mathbf{r}_k)\mathrm{d}\mathbf{r}_k$$

$$+O(\phi^2) = \int G(\mathbf{x}, \mathbf{x} + \mathbf{r})P(\mathbf{x} + \mathbf{r})\mathrm{d}\mathbf{r} + O(\phi^2), \tag{5.1.9}$$

and similarly we have

$$\overline{H} = \int H(\mathbf{x}, \mathbf{x} + \mathbf{r})P(\mathbf{x} + \mathbf{r}|\mathbf{x})\mathrm{d}\mathbf{r} + O(\phi^2). \tag{5.1.10}$$

However, it must be noted that not all the quantities represented by G or H occurring in the problem under discussion satisfy the required condition, that they decrease to zero sufficiently fast with increase of the distance of a single rigid sphere from the point $\mathbf{x}$. In particular, the velocity at the point $\mathbf{x}$ induced by one falling sphere with centre at $\mathbf{x} + \mathbf{r}$ behaves as r^{-1} when $r/a \gg 1$. It is definitely not permissible in this case to disregard contributions to an average defined in Eq. (5.1.7) from spheres at distances large compared to a. The approximation Eq. (5.1.9) (or Eq. (5.1.10)) for the average would be a divergent integral if $G(\mathbf{x}, \mathbf{x} + \mathbf{r})$ (or $H(\mathbf{x}, \mathbf{x} + \mathbf{r})$) were to be interpreted as the velocity at $\mathbf{x}$ caused by a sphere with centre at $\mathbf{x} + \mathbf{r}$. On the other hand, without simplifying the average to an expression involving only a small number of spheres it does not seem to be possible to make any progress for obtaining analytical solutions. This is the central difficulty of the problem to which we have already referred.

The purpose of this section is to determine the mean velocity of a test sphere, which is represented formally by

$$\overline{\mathbf{U}} = \frac{1}{N!} \int \mathbf{U}(\mathbf{x}_0, \mathcal{C}_N)P(\mathcal{C}_N|\mathbf{x}_0)\mathrm{d}\mathcal{C}_N. \tag{5.1.11}$$

The centre of the test sphere whose velocity is $\mathbf{U}(\mathbf{x}_0, \mathcal{C}_N)$ but $\mathbf{U}_0$ in isolated state falling in infinite fluid, is situated at $\mathbf{x}_0$. According to the Faxen theorem , $\mathbf{U}(\mathbf{x}_0, \mathcal{C}_N)$ could be decomposed into $\mathbf{U}_0 + \mathbf{V}$, where

$$\mathbf{V}(\mathbf{x}_0, \mathcal{C}_N) = \mathbf{u}(\mathbf{x}_0, \mathcal{C}_N) + \frac{1}{6}a^2\{\nabla^2\mathbf{u}(\mathbf{x}, \mathcal{C}_N)\}_{\mathbf{x}=\mathbf{x}_0}. \tag{5.1.12}$$

Here $\mathbf{u}(\mathbf{x}, \mathcal{C}_N)$ is the velocity distribution of the fluid that would exist in the dispersion if the test sphere were replaced by fluid of the same viscoscity μ while the configuration $\mathcal{C}_N$ is unchanged.

Note that the Faxen theorem is valid only for the case where there are no boundaries to the fluid. The expression Eq. (5.1.12) takes account of the velocity distribution in the fluid near $\mathbf{x}_0$ due to the motion of all spheres other than the test spheres, but it is incomplete because the force acting on the surface of the test sphere needs to be accompanied by image systems on the surfaces of all other spheres in order to ensure the no-slip condition to be satisfied at those boundaries. The effect of the presence of the test sphere is to induce a velocity at distance r in the surrounding fluid and its order of magnitude is U_0a/r and the velocity gradient is of order U_0a/r^2, etc. A rigid sphere of radius a whose centre is at distance r from that of the test sphere as a consequence will acquire an additional translational velocity, and in addition there will be changes in stress distribution at its surface due to the ambient velocity gradient which have a net force dipole with a magnitude of order $\mu U_0 a^4/r^2$. This new force dipole will in turn induce a change in the velocity distribution near the test sphere, and in particular the test sphere will be given an additional translational velocity of order $U_0 a^4/r^4$. All other spheres in the dispersion will have a similar effect on the test sphere. The translational velocity of the test sphere with centre being at x_0 should thus be written as

$$\mathbf{U}(x_0, \mathcal{C}_N) = \mathbf{U}_0 + \mathbf{V}(x_0, \mathcal{C}_N) + \mathbf{W}(x_0, \mathcal{C}_N), \qquad (5.1.13)$$

where $\mathbf{W}$ represents the effect of the image systems of the Stokeslets at the surface of the test sphere. Note that (1) these Stokeslets include both the ones related to the gravitational force on the test sphere and those required to cancel the velocity field $\mathbf{u}$, (2) the contribution to $\mathbf{W}$ from another sphere at distance r varies asymptotically as r^{-4}, showing that the sum of the separate contributions from an indefinitely large number of spheres located with uniform probability density would converge, and (3) $\mathbf{W}$ is defined precisely as the difference between $(\mathbf{U} - \mathbf{U}_0)$ and the velocity of $\mathbf{V}$ given by Eq. (5.1.12).

With the relation Eq. (5.1.12) and Eq. (5.1.13) in mind it is possible to see how to overcome the central difficulty referred at the beginning of this section. Replacing $\overline{\mathbf{U}}$ we now evaluate the two mean quantities $\overline{\mathbf{V}}$ and $\overline{\mathbf{W}}$. For the latter we may use the dilute approximation represented by Eq. (5.1.10), because the value of $\mathbf{W}$ due to one sphere of the configuration at distance r from x_0 decreases rapidly as $r/a \to \infty$, so that only those spheres of the configuration which are within a distance of order a from x_0 have significant effects. Thus we have

$$\begin{aligned}\overline{\mathbf{W}} &= \tfrac{1}{N!} \int \mathbf{W}(x_0, \mathcal{C}_N) P(\mathcal{C}_N|x_0) d\mathcal{C}_N \\ &= \int \mathbf{W}(x_0, x_0 + \mathbf{r}) P(x_0 + \mathbf{r}|x_0) d\mathbf{r} + O(\phi^2),\end{aligned} \qquad (5.1.14)$$

where

$$\begin{aligned}\mathbf{W}(x_0, x_0 + \mathbf{r}) &= \mathbf{U}(x_0, x_0 + \mathbf{r}) - \mathbf{U}_0 - \mathbf{u}(x_0, x_0 + \mathbf{r}) \\ &\quad - \tfrac{1}{6} a^2 \{\nabla^2 \mathbf{u}(x, x + \mathbf{r})\}_{x=x_0},\end{aligned} \qquad (5.1.15)$$

All the difficulties concerning the non-convergence now are incorporated in the contribution

$$\overline{\mathbf{V}} = \overline{\mathbf{V}'} + \overline{\mathbf{V}''}$$

where

$$\overline{\mathbf{V}'} = \frac{1}{N!} \int \mathbf{u}(x_0, \mathcal{C}_N) P(\mathcal{C}_N|x_0) d\mathcal{C}_N, \qquad (5.1.16a)$$

$$\overline{\mathbf{V}''} = \frac{1}{N!} \int \frac{1}{6} a^2 \{\nabla^2 \mathbf{u}(x, \mathcal{C}_N)\}_{x=x_0} P(\mathcal{C}_N|x_0) d\mathcal{C}_N. \qquad (5.1.16b)$$

It is not possible to reduce either of these two integrals to an integral with respect to the position of only one sphere, because $|\mathbf{u}|$ behaves as a/r at distance r from one falling sphere and $a^2|\nabla^2 \mathbf{u}|$ behaves as a^3/r^3, the latter decreases to zero as $r/a \to \infty$, so is only just too slow for an absolute convergence of such an integral. We therefore try to evaluate the integrals in Eq. (5.1.16) with the aid of the mean which is known exactly from some overall condition or constraint in the specification of the problem and its value at x_0 has the same long-range dependence on the presence of a sphere at $x_0 + r$ as the velocity of the test sphere; and once we find the difference between U and the mean an integral like Eq. (5.1.11) can then legitimately be reduced to an integral like Eq. (5.1.10).

One obvious result is that the mean value of the velocity at a point in the dispersion is zero, and it is expressed formally as

$$\overline{u} = \frac{1}{N!} \int u(x, \mathcal{C}_N) P(\mathcal{C}_N) d\mathcal{C}_N = 0, \qquad (5.1.17)$$

where the local velocity u has its ordinary physical meaning in each of the fluid and solid parts of the dispersion. The contribution V' may thus be written as

$$\overline{V'} = \frac{1}{N!} \int u(x_0, \mathcal{C}_N) \{ P(\mathcal{C}_N | x_0) - P(\mathcal{C}_N) \} d\mathcal{C}_N. \qquad (5.1.18)$$

Here $u(x_0, \mathcal{C}_N)$ denotes the velocity at the point x_0 in the presence of the configuration of N spheres represented by $\mathcal{C}_N$ and the two terms within the curly brackets are the two probability distributions for $\mathcal{C}_N$; the first, $P(\mathcal{C}_N | x_0)$, represents the distribution of $\mathcal{C}_N$ if the centre of another sphere was known to be at x_0 (but note that such a sphere is not present so far as the value of $u(x_0, \mathcal{C}_N)$ is concerned), and the second, $P(\mathcal{C}_N)$, is the unconditional distribution of $\mathcal{C}_N$, involving some configurations for which the point x_0 lies within a rigid sphere.

We may assume that the difference $P(x_0 + r_k | x_0) - P(x_0 + r_k)$ tends to zero sufficiently fast as $r_k/a \to \infty$, then the integral Eq. (5.1.18) is now in a form which reduces the integral to that for the dilute dispersion Eq. (5.1.10) by using identities Eq. (5.1.4), Eq. (5.1.5) and the approximate relation Eq. (5.1.6). Then we have

$$\overline{V'} = \sum_{k=1}^{N} \frac{1}{N!} \int \int u(x_0, \mathcal{C}_N) \{ P(x_0 + r_k | x_0)$$

$$- P(x_0 + r_k) \} P(\mathcal{C}_{N-1}) dr_k d\mathcal{C}_{N-1} + O(\phi^2)$$

$$= \int u(x_0, x_0 + r) \{ P(x_0 + r_k | x_0) - P(x_0 + r) \} dr + O(\phi^2), \qquad (5.1.19)$$

where the leading approximation to $u(x_0, \mathcal{C}_N)$ is $u(x_0, x_0 + r_k)$ when only the sphere centre at $x_0 + r_k$ is close to x_0.

In order to evaluate the contribution $\overline{V''}$ we introduce the deviatoric stress tensor d_{ij} which is defined in both the fluid and solid parts of the dispersion and has the Newtonian form $2\mu e_{ij}$ in the fluid, where e_{ij} is the rate-of-strain tensor . Now $d_{ij}(x)$ is a stationary random function of positions in a statistically homogeneous dispersion, and so has a constant mean. It follows that the mean of $\partial d_{ij}/\partial x_{ij}$ is zero, as

$$\frac{1}{N!} \int \frac{\partial d_{ij}(x, \mathcal{C}_N)}{\partial x_j} P(\mathcal{C}_N) d\mathcal{C}_N = 0, \qquad (5.1.20)$$

in which the differentiation of d_{ij} with respect to $\mathbf{x}$ is carried out while C_N keeps fixed and the position vectors of N sphere centres can be chosen as $\mathbf{x}_0 + \mathbf{r}_1, \mathbf{x}_0 + \mathbf{r}_2, \cdots, \mathbf{x}_0 + \mathbf{r}_k, \cdots, \mathbf{x}_0 + \mathbf{r}_N$. We shall divide the range of integration with respect to $\mathbf{r}_k$ into two parts, one is specified by $|\mathbf{x}_0 + \mathbf{r}_k - \mathbf{x}| \leq a$ (so that $\mathbf{x}$ lies inside or on the surface of the k-th sphere) and makes a contribution to the integral in Eq. (5.1.20) and by using identity Eq. (5.1.4), it may be written as

$$\int\int_{|\mathbf{x}_0 + \mathbf{r}_k - \mathbf{x}| \leq a} \frac{\partial d_{ij}(\mathbf{x}, C_N)}{\partial x_{ij}} P(\mathbf{x}_0 + \mathbf{r}_k) P(C_{N-1}|\mathbf{x}_0 + \mathbf{r}_k) d\mathbf{r}_k dC_{N-1}.$$

The integral of $\partial d_{ij}/\partial x_{ij}$ with respect to $\mathbf{r}_k$ over the volume of the k-th sphere can be transformed into an integral of d_{ij} over the surface of this volume by the Gauss theorem. Note that when $|\mathbf{x}_0 + \mathbf{r}_k - \mathbf{x}| > a$ for all k, we may put

$$\frac{\partial d_{ij}}{\partial x_j} = \mu \nabla^2 u_i,$$

thus we have in place of Eq. (5.1.20)

$$\sum_{k=1}^{N} \frac{n}{N!} \int \mathbf{f}(\mathbf{x}_0 + \mathbf{r}_k, C_{N-1}) P(C_{N-1}|\mathbf{x}_0 + \mathbf{r}_k) dC_{N-1}$$

$$+ \frac{1}{N!} \int_{|\mathbf{x}_0 + \mathbf{r}_k - \mathbf{x}| > a, all \ k} \mu \nabla^2 \mathbf{u}(\mathbf{x}, C_N) P(C_N) dC_N = 0, \qquad (5.1.21)$$

where $\mathbf{f}(\mathbf{x}_0 + \mathbf{r}_k, C_N)$ is the total force exerted by the deviatoric stress on the surface of a rigid sphere centred at $\mathbf{x}_0 + \mathbf{r}_k$ in the presence of $N - 1$ other spheres. The first of these integrals is independent of $\mathbf{r}_k$, and so we can put $\mathbf{r}_k = 0$; and the second is independent of $\mathbf{x}$, so we can put $\mathbf{x} = \mathbf{x}_0$. The contribution $\overline{V''}$ may now be written as

$$\overline{V''} = \frac{1}{N!} \int_{r > a, all \ k} \frac{1}{6} a^2 \{\nabla^2 \mathbf{u}(\mathbf{x}, C_N)\}_{\mathbf{x} = \mathbf{x}_0} \{P(C_N|\mathbf{x}_0) - P(C_N)\} dC_N$$

$$- \frac{1}{(N-1)!} \int \frac{na^2}{6\mu} \mathbf{f}(\mathbf{x}_0, C_{N-1}) P(C_{N-1}|\mathbf{x}_0) dC_{N-1}. \qquad (5.1.22)$$

The argument which leads to the reduction of Eq. (5.1.18) into the form of the dilute approximation Eq. (5.1.19) may also be applied to the first integral in Eq. (5.1.22). And for the second integral in Eq. (5.1.22) we note that the leading approximation to the mean value of $na^2 f/6\mu$ for $\phi \ll 1$ is $(-1/2)\phi \mathbf{U}_0$ (the drag due to the viscous stress at the surface of a single falling sphere emersed in an infinite fluid is two-thirds of the total drag force, see Eq. (3,1,19)). The approximate expression for $\overline{V''}$ is therefore

$$\overline{V''} = \int_{r > a} \frac{1}{6} a^2 \{\nabla^2 \mathbf{u}(\mathbf{x}, \mathbf{x}_0 + \mathbf{r})\}_{\mathbf{x} = \mathbf{x}_0} \{P(\mathbf{x}_0 + \mathbf{r}|\mathbf{x}_0)$$

$$- P(\mathbf{x}_0 + \mathbf{r})\} d\mathbf{r} + \frac{1}{2} \phi \mathbf{U}_0 + O(\phi^2). \qquad (5.1.23)$$

The final expression for the mean velocity of a sphere in the mono-dispersion is

$$\overline{\mathbf{U}} = \mathbf{U}_0 + \overline{\mathbf{V}'} + \overline{\mathbf{V}''} + \overline{\mathbf{W}}, \qquad (5.1.24)$$

where $\overline{\mathbf{V}'}$ is given approximately by Eq. (5.1.19), $\overline{\mathbf{V}''}$ by Eq. (5.1.23), and $\overline{\mathbf{W}}$ by Eq. (5.1.14), in all the three cases an error is of order ϕ^2.

For the case of monodispersion, we have indicated that the pair-distribution function will be uniform for $r \geq 2a$ and will be zero for $r < 2a$, which arises from the fact that rigid spheres do not overlap. Thus

$$P(\mathbf{x}_0 + \mathbf{r}|\mathbf{x}_0) - P(\mathbf{x}_0 + \mathbf{r}) = \left\{ \begin{array}{ll} 0 & if \ r \geq a, \\ -n & if \ r < 2a. \end{array} \right\} \qquad (5.1.25)$$

We shall need to make a use of the expression Eq. (3.1.15) for the fluid velocity $\mathbf{u}$ at $\mathbf{x} + \mathbf{r}$ in the presence of only a single rigid sphere of radius a with centre instantaneously being at $\mathbf{x}$ which falls through the infinite fluid with velocity $\mathbf{U}_0$. And obviously $\mathbf{u} = \mathbf{U}_0$ for $r \leq a$. The corresponding expression for $\nabla_r^2 \mathbf{u}$ at points in the fluid is derived from Eq. (3.1.15)

$$\nabla_r^2 \mathbf{u} = \mathbf{U}_0 \frac{3a}{2r^3} - \mathbf{r} \frac{\mathbf{r} \cdot \mathbf{U}_0}{r^2} \frac{9a}{2r^3}.$$

The first term on the right-hand-side of expression $\mathbf{W}$ Eq. (5.1.15) can be calculated in terms of the result of two-sphere kinetics, i.e. Eq. (4.2.19) using the values of mobility function with $\lambda = 1$, obtained in section 4.2.

Substituting all the above relations into Eq. (5.1.24), it yields

$$\overline{\mathbf{V}'} = -\frac{22}{3}\pi a^3 n \mathbf{U}_0 = -\frac{11}{2}\phi \mathbf{U}_0, \qquad (5.1.26)$$

$$\overline{\mathbf{V}''} = \frac{1}{2}\phi \mathbf{U}_0, \qquad (5.1.27)$$

and

$$\overline{\mathbf{W}} = -1.55\phi \mathbf{U}_0. \qquad (5.1.28)$$

Thus we arrive at the Batchelor (1972) sedimentation theory for a monodisperse system, viz.[21]

$$\mathbf{U} = \mathbf{U}_0(1 - 6.55\phi). \qquad (5.1.29)$$

Now it is interesting to investigate the physical meanings and the magnitudes of the various contributions. The downward flux of a volume of solid material in the dispersion is accompanied by a corresponding net upward flux of a fluid volume; this change in fluid environment for one sphere causes the mean settling speed to differ from the value it would have in infinite clear fluid by an amount $-\phi \mathbf{U}_0$. The falling spheres also drag down with them the adjoining fluid volume, and this downward flux of fluid volume is also accompanied by an equal upward flux of volume. Thus, the corresponding contribution to the change on mean settling velocity is $-4.5\phi \mathbf{U}_0$.

(These two contributions associated with conservation of the volume flux comprise the term denoted by $\overline{\mathbf{V}'}$.) The motion of the spheres generates collectively a velocity distribution in the fluid such that the second derivative of the velocity (or $\nabla^2 \mathbf{u}$, precisely) has a non-zero mean, and this property of the environment for a particular sphere changes its mean velocity by an amount $(+1/2)\phi \mathbf{U}_0$. Finally, when the test sphere is near one of the other spheres in the dispersion, the interaction between these two spheres gives the test sphere a translational velocity which is significantly different from that estimated from the velocity distribution of the fluid in the absence of the test sphere. So the previous three contributions need to be supplemented; this causes a further change of the mean settling speed to be equal to $-1.55\phi \mathbf{U}_0$.

The largest contribution to the change of the mean settling speed thus comes from the upward current which compensates for the downward flux of fluid volume in the inaccessible shells surrounding the rigid spheres; and this contribution is a reduction of the settling speed.

Measurement on the average velocity of a particular sphere in a dispersion over a long time is difficult, and many observers have sought other kinds of average of the particle velocity, such as the speed of fall of the relatively sharp 'top' of the particle cloud in a vessel. The available data of various kinds have been examined by Maude and Whitemore (1958) who concluded that the observed mean settling velocity of identical spheres moving with small Reynolds number is best represented by relation of the form

$$\overline{\mathbf{U}} = \mathbf{U}_0(1 - \phi)^s$$

in which the value of the constant S 'is uncertain but is approximately 5'. In 1977 Garside and Al-Dibouni indicated that a value of $S = 5.1$ most accurately reflects the observed data for small Reynolds numbers. The agreement between this empirical relation in the case $\phi \ll 1$ (exactly speaking, $\phi < 0.08$) and the theoretical result Eq. (5.1.29) is not worse than one would expect from the general scatter of the observations. In section 5.3 we will discuss the reason for the discrepancy between the theoretical value $S = -6.55$ (Batchelor 1972[21]) and the observed value ($S \sim -5$).

Sometimes and for some person there is a persistent belief that the decrease in the settling speed in a dilute suspension should be proportional to $\phi^{1/3}$, despite the fact that such a relation predicts appreciably large decrease in settling speed than those obtained from most experiments. It is worthwhile to consider in general the reasons why these methods yield a result which is different in form from that obtained by Batchelor's statistical theory (1972).

We may note first that any method of calculations by the regular array model or 'cell' model which represents each falling sphere wholly as a point force $\mathbf{G}$ applied to the fluid is certainly on dimensional grounds to find that the additional velocity near one sphere due to the presence of all the others is proportional to $\mathbf{G}/\mu R$, and the length R associated with the particle arrangement is the only length entering the problem. The force $\mathbf{G}$ can be written as $6\pi \mu a \mathbf{U}_0$, and so the change in the falling velocity of the chosen sphere up to the first order in ϕ is predicted to be of the form

$$\mathbf{U} - \mathbf{U}_0 \propto a\mathbf{U}_0/R.$$

In the case of a regular array of the spheres, R can be taken as the distance between neighbouring spheres, whence a/R is proportional to $\phi^{1/3}$. Likewise in the case of a cell model , R can only be a linear dimension of the cell and then, with the same volume fraction of solid material in the cell as in the real dispersion, a/R is proportional to $\phi^{1/3}$. In all the published investigations which adopt either a regular arrangement of the spheres or a cell model , it appears that each sphere was represented by a point force and the radius a entered the analysis only through either the identity connecting $\mathbf{G}$ and $\mathbf{U}_0$ or R as a function of ϕ. A result of the form

$$\mathbf{U} - \mathbf{U}_0 \propto \phi^{\frac{1}{3}} \mathbf{U}_0$$

is consequently natural.

We can see also more directly how this result comes about and why it differs from the result for a random distribution of the spheres. The key point of the explanation is that, since the fluid velocity due to a single moving sphere varies with distance r as $G/\mu r$ (or $U_0 a/r$), any method which selects and takes account only of one particular value of the distance between spheres, the average spacing R being the unique choice, as do both the regular array model and the cell model , will find a resultant induced velocity at the position of one sphere which is of order $U_0 a/R$. On the other hand, for a random distribution of spheres such that the position vector of the centre of a sphere takes all possible values with equal probability, there is no direct significance or preference associated with the average particle separation. The result for random distribution of spheres is different, because a test sphere samples all accessible positions in the fluid and the change in fall speed due to the presence of the other spheres is determined primarily by the fact that, since the average velocity over a large volume of the dispersion is zero, the integral of the fluid velocity over all possible positions for the test sphere is equal in magnitude and negative to the integral over the inaccessible region, which occupies a fraction of the total volume of order ϕ and in which the velocity is of order U_0.

The Batchelor's (1972) theory has already achieved great success. Many experiments for monodisperse dispersion have proved that the observed data do have a ϕ-decrease in mean fall speed, as Cheng and Schachman (1955)[22], Maude and Whitemore (1958), Garside and Al-Dibouni (1977), Kops-Werkhoven and Fijnaut (1981), Buscall et al. (1982)[23], Tackie et al. (1983), etc. discussed. However, the sedimentation theory is still faced with a challenge even for the dilute monodisperse system . Having summarized the results of many experiments in monodisperse system s, Barnea and Mizrahi (1973) gave a different empirical formula, viz.

$$\overline{\mathbf{U}} = \mathbf{U}_0 \frac{(1 - \phi)^2}{(1 + \phi^{\frac{1}{3}}) \exp(5\phi/3(1 - \phi))}.$$

In the dilute limit this formula gives a $\phi^{1/3}$-decrease in mean fall speed. Then Davis and Acrivos (1985) insisted on that the regular array model seemed to be still correct in that case. They thought there existed in some suspensions a microscale 'structure' whose origin is unknown[24] at present.

Although Davis and Acrivos' argument of regular-array model is still not a convincing one, the existence of two different kinds of monodisperse system experiments requires that people carry out experiments more carefully, as well as theoretical work on the problem of sedimentation in a monodisperse dispersion. Perhaps there is still a long way before the problem is eventually solved.

2. Solution of the Equation for the Pair-Distribution Function in a Polydisperse Stable System

Starting from this section we will consider the problem of polydisperse system. The calculation of sedimentation in a polydisperse system is much more complicated than in a monodisperse system because the pair-distribution function is no longer uniform. The relative motion of two unequal spheres under gravity in a polydisperse system is not zero, the interparticle force between two particles should be included in most cases, the pair-distribution function is thus made non-uniformity . Then, the second obstacle—to find the statistical structure in the dispersion—cannot be avoided again like in the monodisperse-system case, and must be determined before calculating the sedimentation in an unequal-sized system. This work was first done systematically by Batchelor (1982)[11] and Batchelor and Wen(1982)[12]. They calculated the pair-distribution function s for a wide range of special conditions, including large values of the Péclet number (negligible effect of the Brownian motion); small values of the Péclet number; and extreme values of the ratio of the radii and the reduced density of the two spheres. In some cases they also calculated the effect of the interparticle force on the statistical structure. We will introduce the results of Batchelor (1982)[11] and Batchelor and Wen (1982)[12] in this section.

2.1. Large Péclet Nnumber (with $\Phi_{ij} = 0$)

According to the definition Eq. (4.4.32) of the Péclet number $\mathcal{P}_{ij}$, the expression Eq. (3.1.21) for the Stokes terminal velocity , and the expression Eq. (4.4.19) for $D_{ij}^{(0)}$, the value of the Péclet number of the aerosol particles under the conditions $T = 293^0 K, \lambda = 1/2, \rho_p = 1\text{gm/cm}^3$ can be given by the following formula

$$\mathcal{P}_{ij} \sim 1.9 \times a_i^4 (\mu m).$$

From this formula we may see that the values of $\mathcal{P}_{ij}$ for the aerosol particles with a radius larger than one micron are larger than unity. The values of $\mathcal{P}_{ij}$ for aerosol particles with a radius larger than 3 microns are larger than 100. From Eq. (4.4.32) and Eq. (4.4.33), there is a connection between $\mathcal{P}_{ij}$ and Q_{ij}, viz.

$$Q_{ij} = \mathcal{P}_{ij} \frac{kT}{A}. \tag{5.2.1}$$

The magnitude of kT/A is of order 10^{-1} for aerosol particles (and is of order unity for hydrosol particles). Therefore, for most of the giant nuclei, we have $\mathcal{P}_{ij} \gg 1, Q_{ij} \gg 1$.

When $\mathcal{P}_{ij} \gg 1$ and $Q_{ij} \gg 1$ for $r > a_i + a_j$ the equation Eq. (4.4.27) for the pair-distribution function is approximately

$$\frac{\partial p_{ij}}{\partial t} + \nabla.(\mathbf{V}_{ij}p_{ij}) = 0. \tag{5.2.2}$$

(Retention of the time derivative creates no additional difficulty in this particular case.) This equation can be solved by the same kind of procedure as was used in the case of dispersion of identical force-free spheres in a bulk linear deforming motion at high Péclet number (Batchelor and Green 1972). We note from Eq. (4.3.2) and Eq. (4.3.9) that $\nabla.\mathbf{V}_{ij}$ can be written as

$$\nabla.\mathbf{V}_{ij} = -\frac{\mathbf{r} \cdot \mathbf{V}_{ij}}{rq}\frac{dq(r)}{dr}, \tag{5.2.3}$$

where

$$\frac{d\log q}{dr} = \frac{W(r)}{\frac{1}{2}(a_i + a_j)L(r)}, \tag{5.2.4}$$

and so Eq. (5.2.2) becomes

$$\left(\frac{\partial}{\partial t} + \mathbf{V}_{ij} \cdot \nabla\right)\left\{\frac{p_{ij}(\mathbf{r},t)}{q(r)}\right\} = 0. \tag{5.2.5}$$

Thus $p_{ij}(\mathbf{r})/q(r)$ is a constant for a 'material' point on a trajectory in r-space. In the case of a trajectory coming from infinity, where $p_{ij} = 1$, the value of that constant is $1/q(\infty)$ and as we have

$$p_{ij}(\mathbf{r},t) = q(r)/q(\infty),$$

at all points on that trajectory and thence at all points on all such trajectories. The scalar function $q(r)$ according to its definition Eq. (5.2.4), includes an arbitrary constant. However, it can not affect the values of p_{ij}, hence, we can put $q(\infty) = 1$, then we have

$$p_{ij}(\mathbf{r},t) = q(r). \tag{5.2.6}$$

In the case of a trajectory which does not extend to infinity, Eq. (5.2.5) still gives the relation between the values of p_{ij} at any two points on the trajectory, but the relation between the values on two different trajectories is not determined by Eq. (5.2.2). The occurrence of trajectories of finite length has been investigated by Wacholder and Sather (1974). They found for each value of λ there is a range of γ for which some of the trajectories do not extend to infinity (except for $\lambda = 1$, when the range contracts to zero). One bound (higher bound) of the range of values of γ for which some of the trajectories are finite is given by $\gamma = \lambda^{-2}$ (this being the value for which two widely separated spheres have zero relative velocity). The other bound (lower bound) varies with λ in a way which is known only numerically. In the important practical case of spheres of uniform density ($\gamma = 1$), and for any value of λ, all the trajectories extend to infinity.

When some of the trajectories in a flow field end finite, it is necessary to include the effect of the relative Brownian diffusion of the two spheres in the governing equation in order to make the pair-distribution function fully determinative. The effect of a small diffusivity here has a singular perturbing effect, and the required analysis has not yet been worked out. Since sedimenting systems in which some of the trajectories end finite are not typical, we shall set this problem aside and consider the high-Péclet- number form of the pair-distribution function only for cases in which all the trajectories extend to infinity and Eq. (5.2.6) consequently holds for all relative positions of the two spheres.

Returning to Eq. (5.2.6) and Eq. (5.2.4), we find

$$\log p_{ij}(s) = \int_s^\infty \frac{W(s)}{L(s)} ds = \int_s^\infty \left\{ \frac{2(L-M)}{sL} + \frac{1}{L}\frac{dL}{ds} \right\} ds, \qquad (5.2.7)$$

which has the note-worthy property of the dependence on the separation magnitude s alone. A spherical symmetric pair-distribution function is generated, despite the directional character of the sedimentation of two spheres, essentially because $\mathbf{r} \cdot \mathbf{V}_{ij}$ and $\nabla \cdot \mathbf{V}_{ij}$ have the same dependence on the direction of $\mathbf{r}$.

The asymptotic form for p_{ij} (for $s \gg 1$) may be found while $W(s)$ is given in Eq. (4.3.11) and $L(s)$ and $M(s)$ are obtained from Eq. (4.2.25–28), to order of s^{-6} as

$$p_{ij} = 1 + \frac{30(\gamma-1)\lambda^3}{(\gamma\lambda^2-1)(1+\lambda)^4}s^{-4}$$

$$+ \frac{72(\gamma-1)(\gamma\lambda^3-1)^3}{(\gamma\lambda^2-1)^2(1+\lambda)^5}s^{-5}$$

$$+ \frac{4\lambda^3}{(1+\lambda)^6}\left\{ \frac{45(\gamma-1)(\gamma\lambda^3-1)^2}{(\gamma\lambda^2-1)^3} \right.$$

$$\left. + \frac{27(\gamma-\lambda^2)}{\gamma\lambda^2-1} - 80 \right\}s^{-6} + O(s^{-7}). \qquad (5.2.8)$$

In an important case ($\gamma = 1$), the first two non-uniform terms in Eq. (5.2.8) vanish and $p_{ij} - 1$ falls off rapidly as s^{-6}, when $s \gg 1$.

At the other end of the range of s values ($s \to 2$), we find from the asymptotic expansion for L and M (for $\xi \to 0$) Eq. (4.3.12) Eq. (4.3.14) that

$$p_{ij} \sim \frac{const.}{\xi^\alpha(\log \xi^{-1})^\beta} \qquad \text{as } \xi \to 0 \qquad (5.2.9)$$

where

$$\alpha = \frac{L_1 - M_0}{L_1}, \quad \beta = \frac{M_1 - M_0 e_1}{L_1}. \qquad (5.2.10)$$

The parameter α is usually positive, and there is usually a singularity in p_{ij} at $\xi = 0$.

Figure 1 of Chapter 5 shows an example of p_{ij} as function of s for $\lambda = 1/2$ and various values of γ.

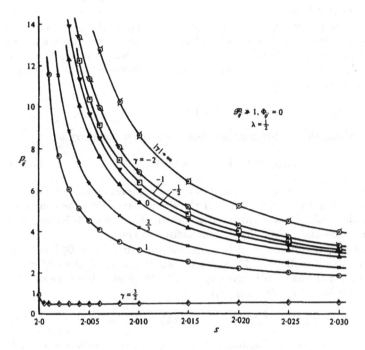

Fig. 1. The pair-distribution function at large Péclet number for $\lambda = 1/2$ and various values of γ (Batchelor and Wen 1982)[12]

The pair-distribution function is unchanged when λ and γ are replaced by λ^{-1} and γ^{-1}.

The nontypical negative values of $p_{ij} - 1$ for $\lambda = 1/2, \gamma = 3/2$ are associated with the proximity of the point representing this combination of values to the lower boundary of the region with trajectories of finite lengths. In terms of the asymptotic form Eq. (5.2.9) and Eq. (5.2.10), the parameter α decreases when the lower boundary of the region with trajectories of finite length is approached from below and passes though zero, and then α becomes negative before the boundary is reached.

2.2. The Limits of $\lambda \to 1, \gamma \to 1, D_{ij}^{(0)} \to 0$ (with $\Phi_{ij} = 0$)

Taking these three limit s we arrive at the case of a dispersion of spheres of the same radius and density which sediment with zero diffusivity due to the Brownian motion. It appears however that the pair-distribution function corresponding to this multiple limit is not unique, and depends on which of the three limit s is taken the last. It is worthwhile to examine the different limiting forms of $p_{ij}(\mathbf{r})$, because they are relevant to practical situations in which one of the three limit s is realized less perfectly than the other two.

First take the case in which the limit $D_{ij}^{(0)} \to 0$ is taken the last. This may be regarded as a mathematical description of an accurately monodisperse system sedimenting with a small but non-zero diffusivity of the spheres. When $\lambda = 1$ and $\gamma = 1$ we have $\mathbf{V}_{ij} = 0$, and so the Eq. (4.4.27) in a steady-state (neglecting again the interparticle force) reduces to

$$\nabla \cdot (\mathbf{D}_{ij} \cdot \nabla p_{ij}) = 0.$$

The solution satisfying the condition of zero flux at the inner boundary is

$$p_{ij}(r) = 1 \quad \text{for} \quad r > a_i + a_j, \tag{5.2.11}$$

regardless of how small $D_{ij}^{(0)}$ may be. This is of course the solution for the dominant Brownian motion—dominant here because of the circumstance that the convection term is zero for identical spheres. This solution was adopted in section 5.1 for calculating the mean sedimentation velocity in a monodisperse system .

Suppose now that the limit $\gamma \to 1$ is taken the last. If we put $D_{ij}^{(0)} = 0$, the equation for p_{ij} reduces to Eq. (5.2.2), and since when $\lambda = 1$, all the trajectories extend to infinity and the solution is given by Eq. (5.2.7). And if $\lambda = 1$ we see from Eq. (4.3.5), Eq. (4.3.6) and Eq. (4.2.20) that

$$L(r) = (A_{11} - A_{12})_{\lambda=1}, \quad M(r) = (B_{11} - B_{12})_{\lambda=1}. \tag{5.2.12}$$

The pair-distribution function is thus independent of γ, and it is unnecessary to take the limit $\gamma \to 1$. The function $p_{ij}(\mathbf{r})$ has been calculated for this case by Feuillebois (1980). The effect of varying γ here is to vary the speed with which one of the two spheres moves relative to the other along its trajectory but not to change the trajectory itself.

Finally, suppose that the limit $\lambda \to 1$ is taken the last with $D_{ij}^{(0)} = 0$ the equation for p_{ij} is again Eq. (5.2.2) and since when $\gamma = 1$ all trajectories extend to infinity the solution is again Eq. (5.2.7). Now in the neighbourhood of $\lambda = 1$ we may put

$$A_{11}(\lambda) = A_{11}(1) + (\lambda - 1)(\partial A_{11}/\partial \lambda)_{\lambda=1} + O((\lambda - 1)^2),$$

and, in view of the relation Eq. (4.2.20),

$$A_{22}(\lambda) = A_{11}(1) - (\lambda - 1)(\partial A_{11}/\partial \lambda)_{\lambda=1} + O((\lambda - 1)^2),$$

and similarly for B_{11} and B_{22}. When $\gamma = 1$ we then have

$$\lim_{\lambda \to 1} L(r) = \left(A_{11} - \frac{\partial A_{11}}{\partial \lambda} - \frac{3}{2} A_{12} \right)_{\lambda=1},$$

$$\lim_{\lambda \to 1} M(r) = \left(B_{11} - \frac{\partial B_{11}}{\partial \lambda} - \frac{3}{2} B_{12} \right)_{\lambda=1}, \qquad (5.2.13)$$

and these functions of r must be substituted into the expression Eq. (5.2.7) for p_{ij}. The trajectories have different shapes in the two cases $\lambda = 1, \gamma \to 1$, and $\gamma = 1, \lambda \to 1$, so this is the essential reason for the different forms of the pair-distribution function in the two cases. The existence of the difference between these two limit s is obvious when the asymptotic form Eq. (5.2.8) is considered.

Figure 2 of Chapter 5 shows $p_{ij}(r)$ for the three different limit orders corresponding to different ways of approaching the state with equal sphere radii, sphere densities, and zero Brownian diffusivity.

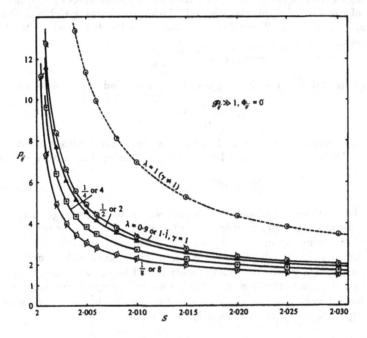

Fig. 2. Pair-distribution functions corresponding to different ways of taking the limit s $\lambda \to 1, \gamma \to 1, D_{ij}^{(0)} \to 0$ (Batchelor and Wen 1982)[12]

The axis $p_{ij} = 1$ in Figure 2 of Chapter 5 shows p_{ij} when $D_{ij}^{(0)} \to 0$ is taken the last. The values of p_{ij} in this case are equal to unity and are the smallest ones among the three different limit s. The limit of p_{ij} when $\lambda \to 1$ is taken the last is slightly complicated. There are four curves for p_{ij} in Figure 2 of Chapter 5 as functions of s as found in numerical integration for the cases $\gamma = 1, \lambda = 1/8, 1/4, 1/2$ and 0.9. These four curves apply equally to the cases $\gamma = 1, \lambda = 8, 4, 2$ and 1.1 respectively, and it is evident from the trends of the curves that the curve corresponding to $\gamma = 1, \lambda \to 1$ must be close to that for $\gamma = 1, \lambda = 0.9$. The values of p_{ij} in this limit case are thus larger than the first one, but are smaller than the corresponding points on the curve with $\lambda = 1$ and γ arbitrary (but $\gamma \neq 1$). The latter is corresponding to the limit for p_{ij} when $\gamma \to 1$ is taken the last, and is shown in Figure 2 of Chapter 5 as a broken curve.

The case in which the limit $\lambda \to 1$ is taken the last provides a mathematical description of a dispersion of spheres which sediment at high Péclet number with accurately uniform density and a small variation of radius. The variation of radius must not be too small, because it would be difficult then to satisfy in practice the condition of high Péclet number. We can see the practical limitations by considering a dispersion of equal-density spheres, some of which have radius a and some have radius $a + \delta a$, where $\delta a \ll a$. The difference between the free-fall speeds of the two types of sphere is

$$|\mathbf{U}_2^{(0)} - \mathbf{U}_1^{(0)}| \approx \frac{2\delta a}{a} U_0,$$

where U_0 is the free-fall speed of either type of the spheres, and so the Péclet number is

$$\mathcal{P} = \frac{2aU_0}{D_0} \frac{\delta a}{a},$$

where D_0 is the relative diffusivity at large separations. For particles of relative density 2gm/cm³ in water or of density 1gm/cm³ in air at room temperature we have

$$\mathcal{P} = 16 \frac{\delta a}{a} (a \; \mu\mathrm{m})^4, \qquad (5.2.14)$$

showing that for particles not smaller than 1.6μm with a tolerable range of 10% in radius, the Péclet number would be large (i.e. 10 or more). We also see incidentally that for particles of radius 1μm, a size difference of a few percent is sufficient to give $\mathcal{P}$ a value of about unity, which suggests that in an experiment with a supposedly monodisperse system (in which the Péclet number should be equal to zero) the pair-distribution cannot possess the form Eq. (5.2.11) which is only appropriate to the exactly identical spheres. It is important to make an accurate observation of the size variation in dispersions of spheres of radius 1μm or larger.

2.3. Structural Equilibrium

When the spheres are truly identical, both in density and in radius, the pair-distribution function may still be non-uniform if the interparticle potential, Φ_{ij} is considered. In these circumstances retention of the effect of interparticle forces causes no difficulty. The steady state solution of Eq. (4.4.27) that satisfies the outer boundary condition Eq. (4.4.29) and the inner boundary condition Eq. (4.4.30) is then the equilibrium obeying the Boltzmann distribution

$$p_{ij}(\mathbf{r}) = \exp\{-\Phi_{ij}(\mathbf{r})/kT\}. \qquad (5.2.15)$$

Thus at positions near $r = a_i + a_j$ where $\Phi_{ij} < 0$ (the value of Φ_{ij} at large r being zero) there is an excess density of sphere pairs and at positions where $\Phi_{ij} > 0$ there is a deficiency. The consequences for the mean speed of fall of particles are that two identical spheres fall more quickly when they are closer to each other than when they are far apart.

We present here a simplified model for the interparticle potential Φ_{ij}. The force between two neighbouring colloidal particles in a stable dispersion is usually resulted by the van der Waals attractive force and an electrostatic repulsive force between the charges held at the surfaces of the two particles and screened by the presence of counter-ions in the double layer surrounding each particle. The dependence of the potential of the van der Waals force on the distance between two spheres is fairly well established, and a commonly used expression for the potential in the case where two spheres are of the same radius a in a nearly-touching position is

$$\Phi_{vdw} = -\frac{A}{12\xi(1 + 11.2\xi a\lambda_L^{-1})}, \tag{5.2.16}$$

where A is the composite Hamaker constant which depends on the atomic composition of both the particles and the fluid medium, and λ_L is the London wavelength representing the retardation effect usually taken as $0.1\mu m$. This expression is believed to be applicable when $\xi a < \lambda_L/\pi (= 0.032\mu m)$. At large separations, where retardation effects are stronger, but with the gap between the two spheres still small compared to the radius ($\xi \ll 1$), the potential is given approximately by

$$-\frac{10^{-3}A}{\xi}\left\{\frac{6.50}{\xi a\lambda_L^{-1}} - \frac{0.305}{(\xi a\lambda_L)^2} + \frac{0.0057}{(\xi a\lambda_L)^3}\right\}. \tag{5.2.17}$$

These two expressions were put forward first by Schenkel and Kitchener (1960), and have been used by many later workers. The electrostatic potential is less well established, and it also varies widely with the nature of the dispersions. The two main controlling parameters in hydrosol dispersions are the charge density at the surface of a particle or equivalently the electric potential at the surface, and the thickness of the double layer surrounding each particle. The potential of the Coulomb repulsion between the two spheres varies relatively slowly for sphere separations smaller than a double layer thickness and falls off exponentially when exceeds this thickness. Roughly speaking, the surface electric potential determines the magnitude of the potential barrier and the double layer thickness determines the location of the barrier. The absolute sphere size appears in expressions Eq. (5.2.16) and Eq. (5.2.17), so at least three independent parameters are needed to represent the resultant interparticle force in a stable dispersion.

We here present only a few exploratory calculations for spheres of equal size which show the general nature and magnitude of the effect of interparticle force on sedimentation velocities, with this modest objective in mind we thus adopted the following simplified form for the Coulomb repulsion potential

$$\Phi_{Coul} = \begin{cases} \Phi_0(\gg kT) & \text{for } 0 < \xi < \xi_0, \\ 0 & \text{for } \xi \geq \xi_0, \end{cases} \tag{5.2.18}$$

where $\xi_0 a$ can be identified roughly as the double layer thickness. The Coulomb barrier is assumed here to be high enough to exclude any particle pair with a spacing smaller than $\xi_0 a$, and at large spacings the Coulomb force is zero. For the potential of the van der Waals force the expression Eq. (5.2.16) will be assumed to hold over the whole range of (small) values of ξ, despite the fact that for some part of that range $\xi a > 0.032 \mu m$ Eq. (5.2.17) is more appropriate. And since the magnitude of the van der Waals potentials falls off much more rapidly than Eq. (5.2.16) or Eq. (5.2.17) as soon as the gap between the spheres is no longer small compared to the smaller of the two sphere radii, we assume that the potential jumps to zero at $\xi = 0.2$.

The resultant interparticle force potential is thus specified now by two parameters, viz. ξ_0 and a/λ_L. This is not put forward as a realistic form of the potential, but it has the merit of allowing an examination of the effect of varying the important parameter ξ_0 which represents the double layer thickness without too many numerical complications. For sphere separations in the range $0 < \xi < \xi_0$ the Coulomb repulsion is dominant and the pair-distribution function is zero, whereas for $\xi_0 < \xi < 0.2$. The interparticle force is attractive, the pair-distribution will be given by Eq. (5.2.15) and there are more sphere pairs than at absence of interparticle forces and finally $p_{ij} = 1$ for $\xi \geq 0.2$. Varying ξ_0 from zero to 0.2, thus changes the balance between the effect on sedimentation velocities of an excess number of close pairs resulted from attractive forces and that of a deficiency of close pairs resulted from the Coulomb repulsion over a wide range. The depth of the secondary minimum of Φ at $\xi = \xi_0$ has a strong effect on the maximum of the pair-distribution function and, with our simple model is determined directly by ξ_0. The pair-distribution function has a very high maximum for small values of ξ_0 (ξ_0 less than about 0.02 for $a = 0.1 \mu m$, and less than about 0.01 for $a = 1 \mu m$), reminding us that such small values of ξ_0 are incompatible with our basic premise that the dispersion under consideration is stable.

Note that the above repulsive double layer potential between two charged spheres only exists in hydrosol dispersions. However, it is also important for the aerosol scientists to understand this problem, since in aerosol mechanics, experiments with hydrosol dispersion are also used to simulate the sedimentation in aerosol system (see Fuchs' book "The mechanics of aerosols"(1964))[2].

2.4. Small Péclet Number

According to the formula for calculating the Péclet number ($\mathcal{P}_{ij} \sim 1.9 \times a^4(\mu m)$, see the beginning of this section), the values of the Péclet number of aerosol particles with radius equal to $0.5 \mu m$ are only 0.1, and the values of $\mathcal{P}_{ij}$ of aerosol particles with radius equal to $0.1 \mu m$ are merely 10^{-4}. Therefore, for all Aitken nuclei and most large nuclei , their Péclet number is much less than unity.

When $\mathcal{P}_{ij} \ll 1$, the convection term in Eq. (4.4.27) is in general small compared to the diffusion term. Thus we recover again the Boltzmann distribution Eq. (5.2.15). We shall investigate here an improved approximation to $p_{ij}(\mathbf{r})$ which takes some account of the effect of convection in a steady state. It will be seen later that a knowledge on this improved approximation is needed for a correct prediction of the mean speed if sedimentation of each particle species is in the limit $\mathcal{P}_{ij} \to 0$.

The appropriate form of perturbation of the Boltzmann distribution at small values of $\mathcal{P}_{ij}$ is

$$p_{ij}(\mathbf{r}) = \exp\{-\Phi_{ij}(\mathbf{r})/kT\}\{1 + \mathcal{P}_{ij}p_{ij}^{(1)}(\mathbf{r}) + O(\mathcal{P}_{ij}^2)\}. \tag{5.2.19}$$

If we substitute this expression for p_{ij} into the (steady-state form of) Eq. (4.4.27) and neglect the terms of order $\mathcal{P}_{ij}^2$ we are left with

$$\mathcal{P}_{ij}\nabla_r \cdot (e^{-\Phi_{ij}/kT}\mathbf{D}_{ij} \cdot \nabla_r p_{ij}^{(1)}) = \nabla_r \cdot (\mathbf{V}_{ij}e^{-\Phi_{ij}/kT}). \tag{5.2.20}$$

The scalar function $p_{ij}^{(1)}$ is evidently axially symmetric about the vertical direction, and we try a form for it

$$p_{ij}^{(1)}(\mathbf{r}) = \frac{\mathbf{r} \cdot \mathbf{V}_{ij}^{(0)}}{rV_{ij}^{(0)}}Q(s). \tag{5.2.21}$$

Substituting into Eq. (5.2.20), after some straightforward work we find Eq. (5.2.20) is satisfied by this expression for $p_{ij}^{(1)}$ provided $Q(s)$ is a solution of

$$\frac{\mathrm{d}}{\mathrm{d}s}(s^2 G \frac{\mathrm{d}Q}{\mathrm{d}s}) - \frac{\mathrm{d}(\Phi_{ij}/kT)}{\mathrm{d}s}s^2 G \frac{\mathrm{d}G}{\mathrm{d}s} - 2HQ$$

$$= s^2 W - \frac{\mathrm{d}(\Phi_{ij}/kT)}{\mathrm{d}s}s^2 L. \tag{5.2.22}$$

The outer boundary condition to be imposed on Q is

$$Q \to 0 \quad \text{as} \quad s \to \infty, \tag{5.2.23a}$$

and the inner boundary conditions Eq. (4.4.30) reduces to

$$G \mathrm{d}Q/\mathrm{d}r = 0 \quad \text{at} \quad s = 2. \tag{5.2.23b}$$

At the small Péclet number the applied forces $\mathbf{F}_i^{(0)}$ and $\mathbf{F}_j^{(0)}$ represents perturbations of an equilibrium system dominated by the Brownian and interparticle forces and we may expect that p_{ij} is a linear function of both $\mathbf{F}_i^{(0)}$ and $\mathbf{F}_j^{(0)}$. This linearity is obscured here by the inclusion of the factor $\lambda^2\gamma - 1$ in the definitions of scalar functions L and M, but it may be seen that $(\lambda^2\gamma - 1)L$ and $(\lambda^2\gamma - 1)M$ are linear functions of r and hence that $(\lambda^2\gamma - 1)Q$ is also linear in γ. Thus the perturbation term in Eq. (5.2.19) is a linear function of γ, or, equivalently a linear function of $\mathbf{F}_i^{(0)}$ and $\mathbf{F}_j^{(0)}$ as expected.

Leaving aside the factors in Eq. (5.2.20) containing Φ_{ij}, which are significantly different from unity only when the gap between the two spheres is very small, Eq. (5.2.20) or Eq. (5.2.22) is essentially a pure diffusion equation, but more complicated than the usual type because the diffusivities are different for the radial and transverse directions and are functions of r. The right-hand-side represents a source, which owes its existence to that the relative trajectories are not volume preserving (i.e. $\nabla.\mathbf{V}_{ij} \neq 0$). The scalar function $W(s)$ representing the magnitude of $\nabla.\mathbf{V}_{ij}$ varies as s^{-5} at large values of s in general, and as s^{-7} in the particular case $\gamma = 1$, and the particular integral of Eq. (5.2.22) is correspondingly of order s^{-3} in general and of order s^{-5} for spheres of equal density. Both Q and H approach unity as $s \to \infty$, and the complementary function for Q is of order s^{-2} when $s \gg 1$, as would be expected for a solution with the dipole structure Eq. (5.2.21). Consequently Q may be written as

$$Q = \frac{K}{s^2} + \sum_{n=3}^{\infty} \frac{Q_n}{s^n}, \qquad (5.2.24)$$

when $s \gg 1$, where K is unknown and Q_n may be found in terms of K by substituting Eq. (5.2.24) into Eq. (5.2.22) with $\Phi_{ij} = 0$. For further information about $Q(s)$ it will be necessary to solve Eq. (5.2.22) numerically.

Figure 3 of Chapter 5 shows the numerically calculated values of Q as a function of s for $\gamma = 1, \Phi_{ij} = 0$ and $\lambda = 1/8, 1/4, 1/2, 0.9$, and those for $\gamma = 0, \Phi_{ij} = 0$ and $\lambda = 1/8, 1/4, 1/2, 1, 2, 4, 8$ are shown in Figure 4 of Chapter 5

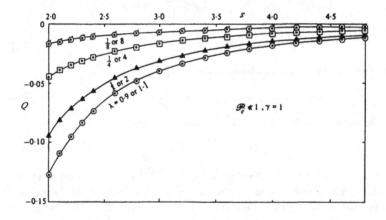

Fig. 3. The perturbation of the pair distribution function at small Péclet number, for $\gamma = 1$ and various values of λ with $\Phi_{ij} = 0$ (Batchelor and Wen 1982)[12]

The function $Q(s)$ for any other values of γ can be obtained from these two sets of curves, because $(\lambda^2 - 1)Q$ is linear in γ, viz.

$$(\gamma\lambda^2 - 1)Q(s,\lambda,\gamma) = Q'(s,\lambda) + \gamma Q''(s,\lambda), \qquad (5.2.25)$$

where

$$Q'(s,\lambda) = -Q(s,\lambda,\gamma)_{\gamma=0}, \\
Q''(s,\lambda) = Q(s,\lambda,\gamma)_{\gamma=0}, \\
+(\lambda^2-1)Q(s,\lambda,\gamma)_{\gamma=1}.$$

(5.2.26)

Again there is a difference between the pair-distribution function for $\lambda = 1$ with γ

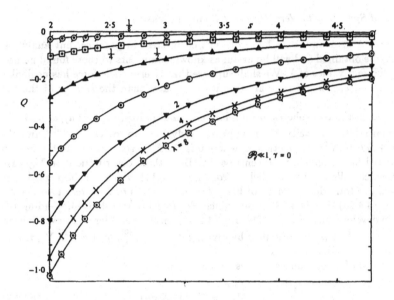

Fig. 4. The perturbation of the pair-distribution function at small Péclet number, for $\gamma = 0$ and various values of λ. The same set of curves applies to $|\gamma| \to \infty$ when λ is replaced by λ^{-1} with $\Phi_{ij} = 0$ (Batchelor and Wen 1982)[12]

having any value except unity and the p_{ij} for $\gamma = 1$ with λ tending to unity. It will be noticed that, whereas the values of Q for a given s and $\gamma = 1$ seem to tend to zero as γ becomes large, those for $\gamma = 0$ are actually increasing as λ changes from 1 to 8, although according to the theoretical result $Q \to 0$ as $\lambda \to \infty$ for a given s. In general, W and hence also Q are of order λ^{-3} when $\lambda \gg 1$, but $\gamma = 0$ is a special case owing to the occurrence of the products $\lambda^2\gamma$ and $\lambda^3\gamma$ in the expressions for L and M, hence W and Q are of order λ^{-1} when $\lambda \gg 1$ in the case $\gamma = 0$. The convergence of Q to zero as $\lambda \to \infty$ is thus rather slow when $\gamma = 0$. The trend of the curves in Figure 4 of Chapter 5 suggests that at values of λ larger than 8 the curves would begin to approach the axis $Q=0$.

It has been seen that the pair-distribution function is spherically symmetric in both limit s $\mathcal{P}_{ij} \to \infty$ and $\mathcal{P}_{ij} \to 0$, although for quite different physical reasons (and in the case $\mathcal{P}_{ij} \to \infty$ the spherical symmetry was established only when $\Phi_{ij} = 0$). The declination from the spherical symmetry is small in the case $\mathcal{P}_{ij} \ll 1$ because diffusive smoothing effect is then strong, and it is small for the case $\mathcal{P}_{ij} \gg 1$ (with $\Phi_{ij} = 0$) owing to that $\nabla.\mathbf{V}_{ij}$ has the same dependence on the direction of $\mathbf{r}$ as $\mathbf{r}.\mathbf{V}_{ij}$. There is no reason to expect that for a general value of $\mathcal{P}_{ij}$ the declination from spherical symmetry will be small (although $p_{ij}(\mathbf{r})$ is necessarily symmetrical about the direction of $\mathbf{g}$).

2.5. Pairs of Spheres with Very Different Radii or Densities

The cases $\lambda \ll 1$ and $\lambda \gg 1$ are of limited interest in practice, but analytical results may be obtained and will be useful as known end points to those found numerically for different values of λ. We shall suppose that interparticle force has negligible effect when $r - (a_i + a_j)$ is comparable with or greater than the smaller of the two sphere radii.

When one of the two spheres has a relatively small radius, say $\lambda = a_j/a_i \ll 1$, the gravitational relative velocity $\mathbf{V}_{ij}$ is approximately the sum of $\mathbf{U}_j^{(0)} - \mathbf{U}_i^{(0)}$ and the value of $\mathbf{u} + (1/6)a_j^2\nabla^2\mathbf{u}$ at position $\mathbf{x}_j$ where $\mathbf{u}$ is the fluid velocity due to the motion of the isolated bigger sphere with centre at $\mathbf{x}_i$. Both these contributions to $\mathbf{V}_{ij}$ have zero divergence. We see this explicitly from Eq. (4.3.11) which can be regarded as giving an asymptotic development of $W(s)$ for small values of λ correct to order λ^2, just as Eq. (4.4.25)–Eq. (4.4.28), from which Eq. (4.3.11) was derived, yielding the asymptotic developments Eq. (4.2.29)–Eq. (4.2.32) correct to order λ^2 or even better. Thus, as making $\nabla.\mathbf{V}_{ij}$ dimensionless by divising it over $2\mathbf{V}_{ij}^{(0)}/(a_i+a_j)$, $\nabla.\mathbf{V}_{ij}$ is zero up to order λ^2, for all values of $\mathbf{r}$.

The equation for p_{ij} reduces in these circumstances to

$$\frac{\partial p_{ij}}{\partial t} + \mathbf{V}_{ij}.\nabla p_{ij} = \nabla.(\mathbf{D}_{ij}.\nabla p_{ij}), \qquad (5.2.27)$$

that is, to the equation for convection and diffusion in an incompressible fluid, except for the values of $s - 2$ which are small compared to λ. There is now no source terms, nor any non-uniformity of p_{ij} specified by the boundary conditions, and so $p_{ij}(\mathbf{r}) \approx 1$. Furthermore, since the term omitted at Eq. (5.2.27) is $p_{ij}\nabla.\mathbf{V}_{ij}$, $\nabla.\mathbf{V}_{ij}$ is known to be of order λ^3 when $\lambda \ll 1$, the solution of the full equation is

$$p_{ij}(\mathbf{r}) = 1 + O(\lambda^3), \qquad (5.2.28)$$

when $\lambda \ll 1$, for a given value of $\mathcal{P}_{ij}$. This holds at all values of the Péclet number. The asymptotic development of p_{ij} in Eq. (5.2.8) for the case of large Péclet number is consistent with Eq. (5.2.28). By definition we have $p_{ij}(\mathbf{r}) = p_{ji}(-\mathbf{r})$, that is $p_{ij}(\mathbf{r},\gamma,\lambda) = p_{ji}(-\mathbf{r},\gamma^{-1},\lambda^{-1})$, where it follows from Eq. (5.2.28) that

$$p_{ij}(\mathbf{r}) = 1 + O(\lambda^{-3}), \qquad (5.2.29)$$

when $\lambda \gg 1$, for any given value of the Péclet number .

The limiting case $\gamma \to 0$ and $|\gamma| \to \infty$ are likewise worth consideration. It is necessary here to only put $\gamma = 0$ or $|\gamma| \to \infty$ in the expression for $L(s)$ and $M(s)$ Eq. (4.3.5) and Eq. (4.3.6), and to use the resulting expression for $W(s)$ in the calculations of $p_{ij}(\mathbf{r})$. The order of the two limit operations $\gamma \to 0, \lambda \to \infty$, is significant, since the products $\lambda^2\gamma$ and $\lambda^3\gamma$ occur in Eq. (4.3.5) and Eq. (4.3.6).

3. Sedimentation in a Dilute Polydisperse Stable System

The majority of theoretical and experimental investigations of the sedimentation process has focused on the monodisperse dispersion, whereas in most cases of practical importance, the dispersions contain particles with different sizes, different densities, even different shapes. The less on polydisperse system is due to its complexity of observations of sedimentation , as well as the calculations for sedimentation in a polydisperse suspension. Especially there was not any theoretical work on sedimentation in a polydisperse dispersion until the sixties. Since then, there is some work in which people use theoretical models, like the Smith's cell model (1965, 1966, 1967), the Lockett and Al-Habbooby's model (1973, 1974), Mirza and Richardson's improved L and A-H model, and the Selim et al's model (1983) etc. In each of the theoretical models mentioned above, the results for monodispersion are used to predict the effects of particle -particle interactions in polydisperse systems , or else an ad hoc modification of the monodisperse results was introduced. However, for general systems which contain large variations in particle sizes or densities, we would not expect that polydisperse settling phenomena could be accurately described by this type of approach (Davis and Acrivos 1985)[24].

On the other hand , based on the calculated pair-distribution function (see the above section), Batchelor (1982)[11] and Batchelor and Wen (1982)[12] first developed an exact theoretical solution for sedimentation in a polydisperse suspension by the statistical method. In this section we will describe Batchelor (1982) and Batchelor and Wen 's (1982) results[11,12].

Let us consider a statistically homogeneous dilute polydisperse system remaining in a container, and there are m different species of rigid spherical particles in this dispersion. The radius, density, number density, volume fraction, and velocity in isolation of each particle of species i are denoted by

$$a_i, \rho_i, n_i, \phi_i, \mathbf{U}_i^{(0)} \quad (i = 1, 2, \cdots, m)$$

respectively. A uniform body force per unit mass g (which will be described as gravity although it might also represent centrifugal force) acts on the dispersion, so

$$\mathbf{U}_i^{(0)} = \frac{2a_i^2(\rho_i - \rho_o)}{9\mu}\mathbf{g} \quad (i = 1, 2, \cdots, m), \tag{5.3.1}$$

and

$$\mathbf{U}_j^{(0)} = \gamma\lambda^2\mathbf{U}_i^{(0)}. \tag{5.3.2}$$

The mean velocity of the particles of species i correct to order $\phi (= \phi_1 + \phi_2 + \cdots + \phi_m$, and $\phi \ll 1$) depends on pair interactions only, and, since it must be vertical like $\mathbf{U}_i^{(0)}$, may be written as

$$\langle \mathbf{U}_i \rangle = \mathbf{U}_i^{(0)} (1 + \sum_{j=1}^{m} S_{ij} \phi_j) \quad (i = 1, 2, \cdots, m). \tag{5.3.3}$$

The dimensionless sedimentation coefficient S_{ij} is a complicated function of the size ratio λ, the reduced density ratio γ, the Péclet number of the relative motion of an i-particle and a j-particle, the Q_{ij} number measuring the relative magnitude of the effect of gravity - induced motion and the effect of interparticle potential Φ_{ij}. When $i = j$, S_{ij} is equal to the sedimentation coefficient S for a monodisperse system , and $S = -6.55$ according to the Batchelor 1972 theory[21], $S \sim -5$——6 according to the existing experimental data. The purpose of this section is to show how S_{ij} may be determined from the interaction of two spheres of different sizes and densities, and try explain the reason of the difference between Batchelor's 1972 theory and the observed results in a monodisperse system .

In a dilute dispersion under consideration the probability of more than one particle being found in the neighbourhood of a given particle is negligible. The interactions to be considered are thus those arising when a particle of species i on which a force $\mathbf{F}_i$ acts, finds in its neighbourhood, a particle of species j on which a force $\mathbf{F}_j$ acts. The forces $\mathbf{F}_i$ and $\mathbf{F}_j$ include not only the applied gravitational force but also a mutual interparticle force which depends on $\mathbf{r}$ and tends to zero as $\mathbf{r} \to \infty$. The additional velocity of the particle of species i due to the presence of the particle of species j is then (see Chapter 4),

$$\mu^{-1}(\mathbf{b}_{11}.\mathbf{F}_i + \mathbf{b}_{12}.\mathbf{F}_j) - \frac{\mathbf{F}_i^{(0)}}{6\pi \mu a_i}, \tag{5.3.4}$$

where the superscript (0) indicates the corresponding value for large separations, that is , for a particle effectively isolated from others. Only gravity contributes to $\mathbf{F}_i^{(0)}$.

It is evident that the mean additional velocity of an i-species particle due to the presence of other particles involves an integral of an expression like Eq. (5.3.4) over all positions of the particles of species j with the pair-distribution function $n_j p_{ij}(\mathbf{r})$ as a weight factor. The integral is not absolutely convergent (see section 5.1). And the ways proposed by Batchelor (1972) for overcoming the divergence difficulty is now generalized to the case of polydisperse system, we then have[11,12]

$$\langle \Delta \mathbf{U}_i \rangle = \sum_{j=1}^{m} \phi_j \left\{ \left(\frac{1+\lambda}{2\lambda} \right)^3 (\mathbf{U}_i^{(0)}.\mathbf{J}' + \mathbf{U}_j^{(0)}.\mathbf{J}'' + \mathbf{K}_{ij}) \right.$$

$$\left. - \left(1 + \frac{3}{\lambda} + \frac{1}{\lambda^2} \right) \mathbf{U}_j^{(0)} \right\}, \tag{5.3.5}$$

where

$$\mathbf{J}' = \frac{3}{4\pi} \int_{s \geq 2} \left\{ A_{11} \frac{\mathbf{ss}}{s^2} + B_{11} \left(\mathbf{I} - \frac{\mathbf{ss}}{s^2} \right) - \mathbf{I} \right\} p_{ij}(\mathbf{s}) d\mathbf{s}, \tag{5.3.6}$$

$$\mathbf{J}'' = \frac{3}{4\pi} \frac{2\lambda}{1+\lambda} \int_{s\geq 2} \left[\left\{ A_{12} \frac{\mathbf{ss}}{s^2} + B_{12} \left(\mathbf{I} - \frac{\mathbf{ss}}{s^2} \right) \right\} p_{ij}(\mathbf{s}) \right.$$

$$\left. - \left\{ \frac{3}{4s} \left(\mathbf{I} + \frac{\mathbf{ss}}{s^2} \right) + \frac{1+\lambda^2}{(1+\lambda)^2 s^3} \left(\mathbf{I} - \frac{3\mathbf{ss}}{s^2} \right) \right\} \right] d\mathbf{s}, \qquad (5.3.7)$$

$$\mathbf{K}_{ij} = \frac{3}{4\pi} \int_{s\geq 2} \left[\left\{ A_{11} \frac{\mathbf{ss}}{s^2} + B_{11} \left(\mathbf{I} - \frac{\mathbf{ss}}{s^2} \right) \right\} \cdot \frac{(\mathbf{F}_i - \mathbf{F}_i^{(0)})}{6\pi\mu a_i} \right.$$

$$\left. + \frac{2\lambda}{1+\lambda} \left\{ A_{12} \frac{\mathbf{ss}}{s^2} + B_{12} \left(\mathbf{I} - \frac{\mathbf{ss}}{s^2} \right) \right\} \frac{(\mathbf{F}_j - \mathbf{F}_j^{(0)})}{6\pi\mu a_j} \right] p_{ij}(\mathbf{s}) d\mathbf{s}. \qquad (5.3.8)$$

The right-hand-side of Eq. (5.3.5) is linear in the forces $\mathbf{F}_i, \mathbf{F}_j$, for a given pair-distribution function , and it is convenient to recognize the separate contributions made by gravity $\langle \Delta \mathbf{U}_i \rangle^{(G)}$ and by interparticle forces on the two spheres $\langle \Delta \mathbf{U}_i \rangle^{(I)}$. The expression of $\langle \Delta \mathbf{U}_i \rangle$ in terms of the three integrals in Eq. (5.3.5) demands this decomposition.

When $\mathbf{F}_i$ and $\mathbf{F}_j$ represent gravitational forces, and so are independent of $\mathbf{s}$, we have

$$\mathbf{F}_i = \mathbf{F}_i^{(0)} = 6\pi\mu a_i \mathbf{U}_i^{(0)}, \mathbf{F}_j = \mathbf{F}_j^{(0)} = 6\pi\mu a_j \mathbf{U}_j^{(0)}, \mathbf{U}_j^{(0)} = \gamma\lambda^2 \mathbf{U}_i^{(0)}, \text{ and } \mathbf{K}_{ij} = 0.$$

Hence

$$\langle \Delta \mathbf{U}_i \rangle^{(G)} = \mathbf{U}_i^{(0)} \cdot \sum_{j=1}^{m} \phi_j \{ (\frac{1+\lambda}{2\lambda})^3 (\mathbf{J}' + \gamma\lambda^2 \mathbf{J}'') - \gamma(\lambda^2 + 3\lambda + 1)\mathbf{I} \}. \qquad (5.3.9)$$

On the other hand, when $\mathbf{F}_i$ and $\mathbf{F}_j$ represent mutual interparticle forces which tend to zero as $r \to \infty$, we have

$$\mathbf{F}_i = \mathbf{F}_j, = \mathbf{F}_{ij}(r) \text{ say, and } \mathbf{F}_i^{(0)} = \mathbf{F}_j^{(0)} = 0,$$

and

$$\mathbf{K}_{ij} = \frac{1}{8\pi^2 \mu a_i} \int_{s\geq 2} \left\{ \left(A_{11} - \frac{2}{1+\lambda} A_{12} \right) \frac{\mathbf{ss}}{s^2} \right.$$

$$\left. + \left(B_{11} - \frac{2}{1+\lambda} B_{12} \right) \left(\mathbf{I} - \frac{\mathbf{ss}}{s^2} \right) \right\} \cdot \mathbf{F}_{ij} p_{ij}(\mathbf{s}) d\mathbf{s}. \qquad (5.3.10)$$

The notation of interparticle forces causing a mean drift of the particles of one species may seem strange at first sight. A mean drift would of course be impossible, if the pair-distribution is spherically symmetric or has a uniform form, since $\mathbf{F}_{ij}(\mathbf{r}) = -\mathbf{F}_{ij}(-\mathbf{r})$ and the integral in Eq. (5.3.10) is then zero; but as we have seen in section 5.2 the pair-distribution function is spherically symmetric or uniform only in the two extreme ends, i.e. for the case of $\mathcal{P}_{ij} \to \infty$ and $\mathcal{P}_{ij} \to 0$. In general, when $\mathcal{P}_{ij}$ is neither infinity nor zero, the pair-distribution function is asymmetrical about the horizontal plane through $s = 0$ in a polydisperse system, and a mean drift of particles of one species may then occur.

In a particular case where a central interparticle force is derived from a potential Φ_{ij} which is a function of r alone (e.g. a van der Waals force), we have

$$\mathbf{F}_{ij} = -\nabla_{x_i}\Phi_{ij}(r) = \frac{2}{a_i + a_j}\frac{\mathbf{s}}{s}\frac{d\Phi_{ij}}{ds}, \qquad (5.3.11)$$

and the corresponding contribution to $\langle \Delta \mathbf{U}_i \rangle$ is seen from Eq. (5.3.5), Eq. (5.3.10) to be

$$\langle \Delta \mathbf{U}_i \rangle^{(I)} = \frac{1}{8\pi^2 \mu a_i^2}\sum_{j=1}^{m}\phi_j \frac{(1+\lambda)^2}{4\lambda^3}\int_{s\geq 2}\left(A_{11}\right.$$

$$\left.-\frac{2}{1+\lambda}A_{12}\right)\frac{\mathbf{s}}{s}\frac{d\Phi_{ij}}{ds}p_{ij}(\mathbf{s})d\mathbf{s}. \qquad (5.3.12)$$

When the pair-distribution function is asymmetric about the horizontal plane through $s = 0$, there will be another direct contribution to the mean particle velocities due to the Brownian thermodynamics, just like the interparticle force. This asymmetry of $p_{ij}(\mathbf{r})$ at small Péclet number can readily be seen as a consequence of the combined effects of convection and diffusion represented in Eq. (4.4.27). Let us consider for example a case in which $\gamma\lambda^2 > 1$ and the particle of species j falls downward relative to the particle of species i. The exclusion of the trajectories from the region $r < a_i + a_j$ crowds the trajectories together in the upper half-space, corresponding to $\nabla.\mathbf{V}_{ij} < 0$, whereas in the mirror image in the lower half-space the trajectories move apart from each other and $\nabla.\mathbf{V}_{ij} > 0$. The negative $\nabla.\mathbf{V}_{ij}$ in the upper half-space and positive ones in the lower half-space act as a diffusive source and sink respectively, giving the whole distribution of $p_{ij}(\mathbf{r})$ the character of diffusion from an upwardly directed dipole source at the centre of the field. The diffusive effects are dominant at small Péclet number , and negligible at large Péclet number, and for finite Péclet number we must expect to find larger values of p_{ij} in the upper half-space than in the lower half-space.

The probability flux of a particle of species j relative to a particle of species i that results from diffusive levelling thus in general has a vertical component which is mostly downward. The steady–state gradients of p_{ij} are of course feeble in the case of small Péclet number, but the product of small gradient by the relatively large Brownian diffusivity is not a small quantity. The final step in the discussion is to recognize that a relative flux of the two kinds of particles in a certain direction implies the existence of an absolute flux (i.e. a flux relative to the container walls) of each of the two species which must be taken into account in the calculation of mean particle-velocities.

We now make a precise calculation of the direct contribution to the velocity of the particles of species i due to the Brownian diffusion (the indirect contribution being that resulting from the influence of the Brownian motion on the pair-distribution function) for arbitrary values of λ, γ and the Péclet number.

In Chapter 4, it was shown that if two particles with labels i and j are alone in an infinite fluid and the joint probability density of the positions of their centres is

$$P(\mathbf{x}_i, \mathbf{x}_j) = n_i n_j p_{ij}(\mathbf{r}),$$

where $\mathbf{r} = \mathbf{x}_j - \mathbf{x}_i$ then the diffusive flux of the spheres due to the Brownian motion is the same as if i and j spheres move under the action of a steady Brownian thermodynamics $\mathbf{F}_{ij}^{(B)}(\mathbf{r})$ and $-\mathbf{F}_{ij}^{(B)}(\mathbf{r})$ respectively, where $\mathbf{F}_{ij}^{(B)}$ is given by Eq. (4.4.15), viz.

$$\mathbf{F}_{ij}^{(B)} = -kT\frac{\partial \log P(\mathbf{x}_i, \mathbf{x}_j)}{\partial \mathbf{x}_i} = kT\nabla_r \log p_{ij}(\mathbf{r}).$$

Then ane i sphere acquires a velocity (see Eq. (4.2.19))

$$\mu^{-1}(\mathbf{b}_{11}.\mathbf{F}_{ij}^{(B)} - \mathbf{b}_{12}.\mathbf{F}_{ij}^{(B)}) = kT\mu^{-1}(\mathbf{b}_{11} - \mathbf{b}_{12}).\nabla_r \log p_{ij}(\mathbf{r}). \qquad (5.3.13)$$

The mean velocity of species i due to the relative Brownian diffusion of i- and j-particles is now obtained by integrating Eq. (5.3.13.) over all values of $\mathbf{r}$ with $n_j p_{ij}(\mathbf{r})$ as a weight function. The specific Brownian contribution to the change in mean velocity of a particle of species i due to all pair interactions is thus

$$\langle\Delta\mathbf{U}_i\rangle^{(B)} = kT\sum_{j=1}^{m}\frac{1}{4}(a_i + a_j)^2 n_j \int_{s\geq 2} \mu^{-1}(\mathbf{b}_{11} - \mathbf{b}_{12}).\nabla_s p_{ij} \mathrm{d}s. \qquad (5.3.14)$$

The divergence problem of the integral in Eq. (5.3.14.) also may be seen not to exist by writing

$$
\begin{aligned}
\int_{s\geq 2} \mu^{-1}(\mathbf{b}_{11} - \mathbf{b}_{12}).\nabla p_{ij}\mathrm{d}s &= \int_{s\geq 2} \nabla.\{\mu^{-1}(\mathbf{b}_{11} - \mathbf{b}_{12})(p_{ij} - 1)\}\mathrm{d}s \\
&- \int_{s\geq 2} \nabla.\mu^{-1}(\mathbf{b}_{11} - \mathbf{b}_{12})(p_{ij} - 1)\mathrm{d}s.
\end{aligned}
\qquad (5.3.15)
$$

The asymptotic forms of $A_{11}, A_{12}, \cdots$ given in Chapter 4 show that $\nabla.(\mathbf{b}_{11} - \mathbf{b}_{12})$ is of order s^{-5} for large s, and so the second integral on the right-hand-side is convergent when the low-Péclet-number approximation is used for $p_{ij} - 1$. The first integral may be transformed to a sum of the two surface integrals, the one over a surface 'at infinity ' which is zero because $p_{ij} - 1$ is of order s^{-4} when s is sufficiently large, no matter how small the Péclet number may be, like other such low-Péclet-number approximations to convection-diffusion equation neglecting the convective term are not valid for indefinitely large values of s. As s increases the second derivative diffusion term in Eq. (4.4.27) ultimately becomes smaller than the first-derivative convective term, no matter how small the Péclet number may be, the high-Péclet-number Eq. (5.2.2) in which only convection term is retained, becomes the appropriate approximation to Eq. (4.4.27) at sufficiently large values of s, the solution for $p_{ij} - 1$ then is of order s^{-4} (see Eq. (5.2.8.))), and the other over the surface $s = 2$ is zero because

$$\mu^{-1}\mathbf{s}.(\mathbf{b}_{11} - \mathbf{b}_{12}) = \frac{s}{6\pi\mu a_i}\left(A_{11} - \frac{2}{1+\lambda}A_{12}\right) \to 0 \quad \text{as} \quad s \to 2$$

in view of Eq. (4.2.24.).

We may thus rewrite Eq. (5.3.14) as

$$\langle\Delta\mathbf{U}_i\rangle^{(B)} = \frac{3}{4\pi a_i}\sum_{j=1}^{m}\frac{\phi_j D_{ij}^{(0)}}{2\lambda^2}\int_{s\geq 2}\left\{\frac{A_{11} - B_{11}}{s} + \frac{1}{2}\frac{\mathrm{d}A_{11}}{\mathrm{d}s}\right.$$

$$- \frac{2(A_{12} - B_{12})}{(1+\lambda)s} - \frac{1}{1+\lambda}\frac{\mathrm{d}A_{12}}{\mathrm{d}s} \Bigg\} \frac{\mathbf{s}}{s}(1 - p_{ij})\mathrm{d}s. \tag{5.3.16}$$

The final results will be expressed in terms of the sedimentation coefficient S_{ij} defined by Eq. (5.3.3). Except where the contrary is stated, the formulae allow for the effects of three different forces acting on each particle, (a) the gravitational force (b) the interparticle force, (c) the Brownian thermodynamical force. All the three make additive direct contributions to the change in the mean velocity $\langle \Delta U_i \rangle$, and so the sedimentation coefficient as indicated by the notation,

$$S_{ij} = S_{ij}^{(G)} + S_{ij}^{(I)} + S_{ij}^{(B)}. \tag{5.3.17}$$

And all the three forces have a further indirect effect on the mean velocity through their influence on the pair-distribution function —the statistical structure of the dispersion.

For the direct contribution to $\langle \Delta U_i \rangle^{(G)}$ due to gravity we have the expression Eq. (5.3.9). To convert this to a contrition to S_{ij} we note that, since p_{ij} is symmetrical about the vertical axis, each of the tensors $\mathbf{J}'$ and $\mathbf{J}''$ given by Eq. (5.3.6) and Eq. (5.3.7) has one of its principal axis along the vertical direction, whence

$$S_{ij}^{(G)} = \left(\frac{1+\lambda}{2\lambda}\right)^3 \left(\frac{\mathbf{g}\cdot\mathbf{J}'\cdot\mathbf{g}}{g^2} + \gamma\lambda^2\frac{\mathbf{g}\cdot\mathbf{J}''\cdot\mathbf{g}}{g^2}\right) - \gamma(\lambda^2 + 3\lambda + 1). \tag{5.3.18}$$

For the interparticle force contribution $\langle \Delta U_i \rangle^{(I)}$, we have Eq. (5.3.12) and again the symmetry of p_{ij} about the vertical direction may be invoked to justify the applicability of retaining only the vertical component of Eq. (5.3.12) whence

$$S_{ij}^{(I)} = \frac{3}{8\pi}\frac{\gamma\lambda^2 - 1}{\mathcal{P}_{ij}}\frac{(1+\lambda)^2}{4\lambda^2}\int_{s\geq 2}\Big(A_{11}$$

$$- \frac{2}{1+\lambda}A_{12}\Big)\frac{\mathbf{s}\cdot\mathbf{V}_{ij}^{(0)}}{sV_{ij}^{(0)}}\frac{\mathrm{d}(\Phi_{ij}/kT)}{\mathrm{d}s}p_{ij}(\mathbf{s})\mathrm{d}\mathbf{s}. \tag{5.3.19}$$

For the Brownian thermodynamical contribution $\langle \Delta U_i \rangle^{(B)}$ we have the expression Eq. (5.3.16). The symmetry of p_{ij} about the vertical direction shows that only the vertical component of Eq. (5.3.16) is non-zero hence that

$$S_{ij}^{(B)} = \frac{-3}{4\pi}\frac{\gamma\lambda^2 - 1}{\mathcal{P}_{ij}}\frac{(1+\lambda)^2}{4\lambda^2}\int_{s\geq 2}\Bigg\{\frac{A_{11} - B_{11}}{s}$$

$$+ \frac{1}{2}\frac{\mathrm{d}A_{11}}{\mathrm{d}s} - \frac{2(A_{12} - B_{12})}{(1+\lambda)s} - \frac{1}{1+\lambda}\frac{\mathrm{d}A_{12}}{\mathrm{d}s}\Bigg\}\frac{\mathbf{s}\cdot\mathbf{V}_{ij}^{(0)}}{sV_{ij}^{(0)}}(1 - p_{ij})\mathrm{d}\mathbf{s}. \tag{5.3.20}$$

3.1. Large Péclet Number and $\Phi_{ij} = 0$

The pair-distribution function found in this case is spherically symmetric (see Eq. (5.2.7)). In these circumstances $\langle \Delta U_i \rangle^{(I)}$ and $\langle \Delta U_i \rangle^{(B)}$ are zero, thus the only contribution to the sedimentation coefficient is $S_{ij}^{(G)}$. The integration over the surface of a sphere of radius s in the expressions $\mathbf{J}'$ and $\mathbf{J}''$ (Eq. (5.3.6) and Eq. (5.3.7)) can be carried out. The sedimentation coefficient is then[11,12]

$$S_{ij}(\lambda,\gamma) = \int_2^\infty \left[\left(\frac{1+\lambda}{2\lambda} \right)^3 (A_{11} + 2B_{11} - 3)p_{ij} \right.$$

$$\left. + \frac{1}{4}(1+\lambda)^2 \left\{ (A_{12} + 2B_{12})p_{ij} - \frac{3}{s} \right\} \right] s^2 ds - \gamma(\lambda^2 + 3\lambda + 1). \qquad (5.3.21)$$

Figure 5 of Chapter 5 shows the numerical results of the variation of S_{ij} with respect to γ for various values of λ, and the variations of S_{ij} with respect to λ for given γ is shown in Figure 6 of chapter 5.

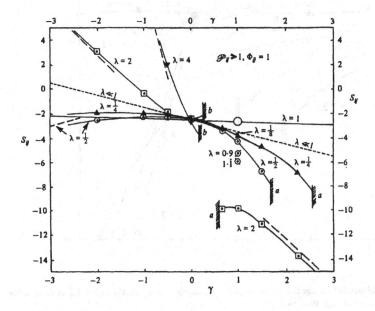

Fig. 5. The sedimentation coefficient S_{ij} at large Péclet number as a function of γ for various values of λ. (From Batchelor and Wen 1982)[12]

Nearly all the behaviours of $S_{ij}(\gamma)$ are non-linear except for the case when $\gamma, \lambda \to 0$ and $|\gamma|, \lambda \to \infty$ which will be discussed in the final part of this section, and for the case when $\lambda = 1$, then p_{ij} is independent of γ and so S_{ij} is also a linear function of

γ, viz.

$$(S_{ij})_{\lambda=1} = -2.52 - 0.13\gamma \quad (\gamma \neq 1). \tag{5.3.22}$$

Feuillebois (1980) did a numerical calculation of S_{ij} at large Péclet number for the case $\lambda = 1$, and found $S_{ij} = -2.7 + 0.1\gamma$. The small numerical differences are possibly a consequence of using different (and less accurate) data for the mobility function s in the near field, by Feuillebois.

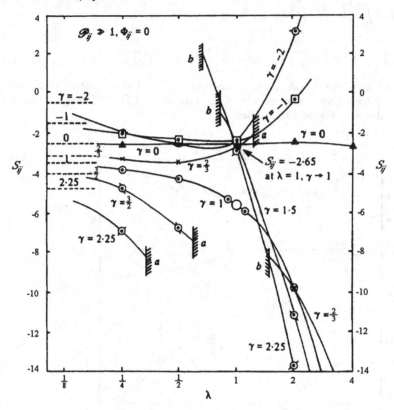

Fig. 6. The sedimentation coefficient at large Péclet number as a function of λ for various values of γ. (From Batchelor and Wen 1982)[12]

Some of the curves shown in these two figures have two sections with quite different behaviours which are separated by a gap corresponding to values of λ and γ which lie within the region in which some trajectories are of finite lengths.

The large circle around the point corresponding to $\lambda = 1, \gamma = 1$ in each of Figures 5 and 6 of Chapter 5 is a reminder (a) that the value at this point depends on how it is approached (the limiting value of S_{ij} is -2.65 when $\gamma \to 1$ is taken the last, the limiting value of S_{ij} is between -5.29 and -5.95 when $\lambda \to 1$ is taken the last); (b) that in a neighbourhood of this point the Péclet number is not large, and the limit value of S_{ij} when $\mathcal{P}_{ij} \to 0$ is -6.55. These results again indicate that it is evidently important to have accurate observations of the size variation in a monodisperse experiment.

The bewildering pattern of variations of S_{ij} shown in Figures 5 and 6 of Chapter 5 needs to be assimilated bit by bit. One striking feature of Figure 5 of Chapter 5 is the quite small variation of S_{ij} with λ at $\gamma = 0$. This reflects the fact that when the j-spheres are neutrally buoyant the main hydrodynamic effect of the presence of the j-spheres is to cause more dissipation in the flow field generated by each falling sphere i. When $\gamma \ll 1$ this additional dissipation is given by the Einstein formula (see the final part of this section) for the effective viscosity of a dispersion of spheres, and $S_{ij} \approx -2.5$; whereas as λ is not small the hydrodynamic interaction of the spheres i and j modifies that value by an amount which is evidently small. The occurrence of positive value of S_{ij} is also worthy of note, since experience with monodisperse systems might suggest that 'hindered settling' is normal at the absence of interparticle forces. The positive values of S_{ij} for $\gamma < 0$ and some values of λ greater than unity are no doubt a consequence of the strong counter flow associated with the rising j-particles.

3.2. Small Péclet Number

Here the pair-distribution function was found to be given by Eq. (5.2.19) correct to the leading order in $\mathcal{P}_{ij}$. We seek an expression for the mean velocity of particles of species i which is correct to the leading order in $\mathcal{P}_{ij}$, that is, correct to order $\mathcal{P}_{ij}^0$.

The direct contribution to S_{ij} due to gravity is given by Eq. (5.3.18), in which the values of $\mathbf{J}'$ and $\mathbf{J}''$ correct to order $\mathcal{P}_{ij}^0$ are found by substituting into Eq. (5.3.6) and Eq. (5.3.7) an expression for p_{ij} which contains only the leading term of Eq. (5.2.19) and so is spherically symmetric. The integrations with respect to the direction of $\mathbf{s}$ in $\mathbf{J}'$ and $\mathbf{J}''$ can be carried out giving[11,12]

$$S_{ij}^{(G)} = \left(\frac{1+\lambda}{2\lambda}\right)^3 \int_2^\infty (A_{11} + 2B_{11} - 3) \exp\left(-\frac{\Phi_{ij}}{kT}\right) s^2 \mathrm{d}s$$
$$+ \gamma \left(\frac{1+\lambda}{2}\right)^2 \int_2^\infty \left\{ (A_{12} + 2B_{12}) \exp\left(-\frac{\Phi_{ij}}{kT}\right) - \frac{3}{s} \right\} s^2 \mathrm{d}s$$
$$- \gamma(\lambda^2 + 3\lambda + 1). \tag{5.3.23}$$

The second direct contribution due to the interparticle force is given by Eq. (5.3.19). The magnitude of $S_{ij}^{(I)}$ is not determined by $\mathcal{P}_{ij}$ alone (Φ_{ij}/kT also being involved), and we therefore try to make it as accurate as possible by substituting the whole expression Eq. (5.2.19) for p_{ij} in Eq. (5.3.19). The leading term in Eq. (5.2.19) is spherically symmetric and makes no contribution to the integral in Eq. (5.3.19). After carrying out the integration with respect to the direction of $\mathbf{s}$ we find[11,12]

$$S_{ij}^{(I)} = \frac{1}{2}\lambda(\gamma\lambda^2 - 1) \int_2^\infty \left(\frac{2A_{12}}{1+\lambda} - A_{11}\right) \frac{\mathrm{d}\exp(-\Phi_{ij}/kT)}{\mathrm{d}s} Q(s) s^2 \mathrm{d}s. \tag{5.3.24}$$

The third direct contribution to S_{ij} is due to the Brownian thermodynamical effects, and is given by Eq. (5.3.20). Here we certainly need to substitute the whole of the expression Eq. (5.2.19) for p_{ij}. The spherically symmetric term of the leading order in Eq. (5.2.19) makes no contribution, and after carrying out the integration with respect to the direction of **s** we find[11,12]

$$S_{ij}^{(B)} = (\gamma\lambda^2 - 1)\left(\frac{1+\lambda}{2\lambda}\right)^3 \int_2^\infty \left\{ \frac{A_{11} - B_{11}}{s} + \frac{1}{2}\frac{dA_{11}}{ds} \right.$$

$$\left. - \frac{2(A_{12} - B_{12})}{(1+\lambda)s} - \frac{1}{1+\lambda}\frac{dA_{12}}{ds} \right\} \exp\left(-\frac{\Phi_{ij}}{kT}\right) Q(s)s^2 ds. \qquad (5.3.25)$$

As mentioned in section 5.2, $(\gamma\lambda^2 - 1)Q$ is a linear function of γ, because in this case of small Péclet number the effects of the forces applied to the particles i and j are independent perturbations of equilibrium systems. We may therefore write

$$(\gamma\lambda^2 - 1)Q(s,\gamma,\lambda) = Q'(s,\lambda) + \gamma Q''(s,\lambda). \qquad (5.3.26)$$

We may also introduce new sedimentation coefficients S_{ij}' and S_{ij}'' which do not depend on γ and are defined by the relation

$$6\pi\mu a_i\langle \mathbf{U}_i\rangle = \mathbf{F}_i^{(0)} + \sum_{j=1}^m \phi_j(S_{ij}'\mathbf{F}_i^{(0)} + \lambda^{-3}S_{ij}''\mathbf{F}_j^{(0)}), \qquad (5.3.27)$$

where $\mathbf{F}_i^{(0)}$ and $\mathbf{F}_j^{(0)}$ are the applied forces. The factor λ^{-3} has been inserted in

Table 1. Calculated values of S_{ij}', S_{ij}'' for $\mathcal{P}_{ij} \ll 1$, $\Phi_{ij} = 0$, and different values of λ. Note that $\Delta = \lambda^2 + 3\lambda + 1$. (From Batchelor and Wen 1982)[12].

λ	$S_{ij}'^{(G)}$	$S_{ij}'^{(B)}$	S_{ij}'	$S_{ij}''^{(G)} + \Delta$	$S_{ij}''^{(B)}$	$S_{ij}'' + \Delta$	$(S_{ij})_{\gamma=1} + \Delta$
0.125	− 2.34	− 0.03	− 2.37	0.01	0.01	0.02	− 2.35
0.25	− 2.24	− 0.07	−2.31	0.04	0.04	0.08	− 2.23
0.5	− 2.09	− 0.15	− 2.24	0.1	0.12	0.24	− 2.00
0.9	− 1.88	−0.25	− 2.13	0.25	0.24	0.49	−1.64
1.0	−1.83	−0.27	−2.10	0.28	0.27	0.55	−1.55
1.1	−1.78	−0.28	−2.06	0.32	0.30	0.62	−1.44
2	−1.45	−0.35	−1.80	0.48	0.47	0.95	−0.85
4	−1.04	−0.41	−1.45	0.64	0.68	1.32	−0.13
8	−0.69	−0.45	−1.14	0.75	0.88	1.63	0.49

Eq. (5.3.27) for convenience in our particular case in which $\mathbf{F}_i^{(0)}$ and $\mathbf{F}_j^{(0)}$ represent gravitational force proportional to the sphere volumes. Then from a comparison with Eq. (5.3.3), we see that

$$S_{ij} = S_{ij}' + \gamma S_{ij}''. \qquad (5.3.28)$$

The explicit expressions for $S_{ij}{}'$ and $S_{ij}{}''$ can be obtained from Eq. (5.3.23)–Eq. (5.3.25) by replacing $(\gamma\lambda^2-1)Q$ with $Q'+\gamma Q''$ and then identifying the coefficients of γ^0 and γ. Note that, although in this section the applied forces are parallel, the expression for $S_{ij}{}'$ and $S_{ij}{}''$ found in this way are valid for arbitrary directions of $\mathbf{F}_i^{(0)}$ and $\mathbf{F}_j^{(0)}$.

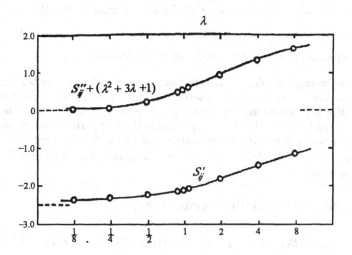

Fig. 7. The sedimentation coefficient as functions of λ at small Péclet number (From Batchelor and Wen 1982)[12]

We now consider the case $\Phi_{ij} = 0$, as the perturbation function in Eq. (5.2.19) was obtained in section 5.2. Then $S_{ij}^{(I)} = 0, S_{ij}^{(G)}$ and $S_{ij}^{(B)}$ can be obtained by replacing the factor $\exp(-\Phi_{ij}/kT)$ in Eq. (5.3.23), Eq. (5.3.25) by unity. Table 1 of Chapter 5 shows the results of the calculations of $S_{ij}^{'(G)}, S_{ij}^{'(B)}$ and $S_{ij}^{''(G)}, S_{ij}^{''(B)}$ for $\lambda = 1/8, 1/4, 1/2, 0.9, 1, 1.1, 2, 4, 8$. In the case of $S_{ij}^{''(G)}$ we have subtracted $-(\lambda^2 + 3\lambda + 1)$ from the calculated values since this is the known asymptotic form both for $\lambda \to 0$ and $\lambda \to \infty$ (see the final part of this section) and the subtraction reduces greatly the variation of the tabulated numbers over the whole range.

Figure 7 of Chapter 5 shows the calculated values of $S_{ij}{}'$ and $S_{ij}{}'' + \lambda^2 + 3\lambda + 1$ as functions of λ. It is evident that S_{ij}' approaches to -2.5 and that S_{ij}'' approaches to -1 as $\lambda \to 0$ (see the final part of this section). The theoretical prediction that $S_{ij}' \to 0$ and $S_{ij}'' + \lambda^2 + 3\lambda + 1 \to 0$ as $\lambda \to \infty$ are not in such clear agreement with the calculated values, but it is believed that because the calculations have not been taken to a sufficiently large value of λ. The convergence of the sedimentation coefficient in

Table 1 of Chapter 5 (or Figure 7 of Chapter 5) to their asymptotic values as $\lambda \to \infty$ is slow and does not begin to be apparent until λ exceeds 8.

A striking feature of the calculated values shown in Table 1 of Chapter 5 is the smallness of $|S_{ij}^{'(B)}|$ relative to $|S_{ij}^{'(G)}|$, except when $\lambda \gg 1$. The approximate relation

$$S_{ij} \approx S_{ij}^{'(G)} + \gamma S_{ij}^{''(G)} \tag{5.3.29}$$

which ignores the direct contribution from the Brownian thermodynamical effect would be sufficiently accurate for many practical purpose, except for $\lambda \gg 1$ and values of γ near zero when the dominance of the second term in Eq. (5.3.29) no longer exists.

3.3. The Effect of an Interparticle Force at Small Péclet Number

In this part we mainly consider the simplest case—the zero Péclet number case. It is different from the case investigated in section 5.1 since now $\Phi_{ij} \neq 0$. The pair-distribution function is no longer uniform, and is given by the Boltzmann distribution Eq. (5.2.15). The direct contribution to $\langle \Delta \mathbf{U}_i \rangle^{(I)}$ due to the interparticle force is zero, since the pair-distribution is spherically symmetric, and $\langle \Delta \mathbf{U}_i \rangle^{(B)}$ is also zero due to the same reason. Hence S_{ij} is determined only due to gravity i.e. $S_{ij}^{(G)}$, and from Eq. (5.3.21)[12] we have

$$S = \int_2^\infty \left\{ (A_{11} + 2B_{11} - 3 + A_{12} + 2B_{12})_{\lambda=1} \exp(-\Phi/kT) - \frac{3}{s} \right\} s^2 ds - 5$$

$$= -6.55 + \int_2^\infty (A_{11} + 2B_{11} - 3 + A_{12} + 2B_{12})_{\lambda=1} \{\exp(-\Phi/kT) - 1\} s^2 ds. \tag{5.3.30}$$

According to the results for the two-sphere mobility function s with $\lambda = 1$, (see Chapter 4), $A_{11} + 2B_{11} - 3 + A_{12} + 2B_{12}$ varies by only 6 percent over the range $2 < s < 2.2$, and as an approximation we may regard it as constant and equal to 1.32 over that range, and Φ is approximately zero when $s > 2.2$. We then have

$$S \approx -6.55 + 0.44\alpha, \tag{5.3.31}$$

where

$$\alpha = 3 \int_2^\infty (e^{-\Phi/k} - 1) s^2 ds = \frac{n}{\phi} \int_{s \geq 2} (p - 1) ds. \tag{5.3.32}$$

Here $\alpha\phi$ can be interpreted as the excess number of spheres (excess to the number for p=1) which are partners in close pairs. We have tested the accuracy of Eq. (5.3.31) by comparing it with the results obtained from Eq. (5.3.30). The agreement between the two expressions Eq. (5.3.30) and Eq. (5.3.31) is found to be good, the difference being less than 1% over all the calculated results. The relation Eq. (5.3.31) may therefore have value for experimental purposes.

Figure 8 of Chapter 5 shows the numerical evaluation of the integral Eq. (5.3.30) for the interparticle potential described in section 5.2. It gives the form of S as a function of ξ_0, the locations of the Coulomb potential barrier for the four different sphere radii, $a = 0.1, 0.5, 1.0, 2.0\mu$m. This figure indicates that the value of S depends quite strongly on ξ_0 at small values of ξ_0. Small values of ξ_0 imply a small double-layer thickness and a correspondingly large range of action of attractive force and a positive excess number of close pairs ($\alpha > 0$).

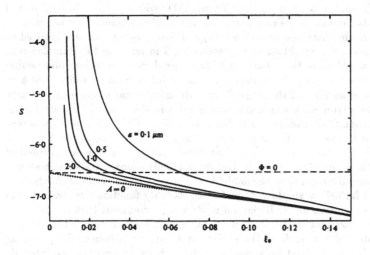

Fig. 8. The sedimentation coefficients for a dispersion of identical spheres of radius a as a function of ξ_0 specifying the location of the high Coulomb potential barrier surrounding each sphere. (From Batchelor and Wen 1982)[12]

Since close pairs fall more quickly than well-separated pairs the value of S is greater than -6.55. Large values of ξ_0 on the other hand imply the exclusion of pairs over a wider range of gap thicknesses and a deficiency of close pairs ($\alpha < 0$), and the associated values of S are less than -6.55. The value of ξ_0 at which $\alpha = 0$ depends on the sphere radius, and varies from 0.022 for $a = 2\mu$m to 0.065 for $a = 0.1\mu$m.

Incidentally, we may gain some idea of the effect of a variation in the composite Hamaker constant A by comparing the above results (for $A = 5.0 \times 10^{-21} J$) with those obtained for $A = 0$ (the dotted curve in Figure 8 of Chapter 5). In the latter case

$$\exp(-\Phi/kT) = \{ \begin{matrix} 0 & \text{when } 0 < \xi < \xi_0, \\ 1 & \text{when } \xi \geq \xi_0, \end{matrix}$$

and if we approximate the bracketed sum of mobility function s in Eq. (5.3.30) in the same manner as in Eq. (5.3.31) we find

$$S \approx -6.55 - 5.3 \left(\xi_0 + \frac{1}{2}\xi_0^2 \right), \qquad (5.3.33)$$

which is shown in Figure 8 of Chapter 5. The sedimentation coefficient S in this case is always smaller than that obtained in section 5.1 with $\Phi_{ij} = 0$ ($S = -6.55$).

Some previous authors have investigated the effect of interparticle forces on the sedimentation coefficient for a monodisperse system , although there are few quantitative results. The first were Goldstein and Zimm (1971) who adopted a force potential which represents the van der Waals attraction and Coulomb repulsion more realistically than the model discussed here. The range of their calculations was limit ed to serve their purpose, viz. to obtain an expression for S in terms of the Hamaker constant A which would allow the value of A to be inferred from a previous observation of the mean particle velocity. Their expression for S differs from that discussed here slightly because of an error of the principle in their calculations of the mean velocity of a particle. In a later work which also was in error in this part of the calculation, Reed and Anderson (1976) adopted a realistic form of the Coulomb repulsion potential but ignored the van der Waals potential. They found that values of S are less than that for $\Phi = 0$ which they recognized as a consequence of a greater fall speed of the excluded close pairs. Most recently, Dickinson (1980) adopted the same high discontinuous Coulomb barrier as in the model here but ignored the van der Waals force , and then used far-field asymptotic form for the mobility function s to evaluate the integral in the expression for S. Dickinson made a valid comment that the values of the mobility function s for small values of $r - 2a$, become less relevant as the distance of the high potential barrier from the surface of a particle is increased. A similar comment may be made about polydisperse systems, because many of the integrals weighted with a factor $\exp(-\Phi_{ij}/kT)$ involve also a pair-distribution function whose form near $s = 2$ is not easily found.

A number of observations of the settling speed of uniform colloidal spheres in a dilute liquid suspension have been made, usually under conditions such that the double layer thickness is a small fraction of the sphere radius (corresponding to a fairly high electrolytic strength of the liquid). The observed values of the sedimentation coefficient S have generally been in the range -5 to -6. Cheng and Schachman (1955) used polystyrene latex spheres of radius $0.13\mu m$ dispersed in a sodium chloride solution (0.098 mol dm^{-3}), and found $S = -5.1$[22]. In 1982, Buscall et al. also used polystyrene latex spheres of radius $1.55\mu m$ dispersed in a sodium chloride solution (10^{-3} mol dm^{-3}), and their measurements gave $S = -5.4 \pm 0.1$[23]. Goldstein and Zimm (1971) estimated that under the conditions of Cheng and Schachman's experiments the double layer thickness, or the Debye-Hückel screening length (κ^{-1}), was $1.0 \times 10^{-3}\mu m = 0.0077a$; and with a 100-times weaker electrolyte used in the experiments of Buscall et al. the double layer thickness would be 10 times larger and so equal to $0.0065a$ for these latter experiments.

We may try to relate these measurements on S to the calculated values shown in Figure 8 of chpater 5 by choosing a value of the dispersion $\xi_0 a$ so as to make the calculated value of S agree with the measured value. According to Figure 8 of Chapter 5, $S = -5.1$ for spheres of radius 0.13μm when $\xi_0 = 0.020$, that is, about 2.6 times the D.-H. length in Cheng and Schachman's experiments[22], and $S = -5.4$ for spheres of radius 1.55μm when $\xi_0 = 0.009$, that is, about 1.4 times the D.-H. length in the experiments of Buscall et al[23]. These are plausible values of ξ_0, and the fact that values of $\xi_0\kappa a$ are above unity is understandable, because the Coulomb potential remains appreciable until the sphere gap exceeds about two double-layer thicknesses. The height of the Coulomb barrier under the experimental conditions is of course also relevant to a choice of ξ_0, but information on this is not available. It seems reasonable to conclude that the positive differences between these measured values of S and -6.55 (Batchelor's theoretical value in 1972) could be a consequence of the van der Walls attractive forces causing as excess number of close pairs whose common fall speed exceeds the fall speed of an isolated sphere.

Another set of measurements was made by Kops-Werkhoven and Fijnaut (1981) who used stabilized silica spheres of radius 0.021μm dispersed in cyclohexane and found $S = -6 \pm 1$. This value of S corresponds to a value of ξ_0 of about 0.1 if the interparticle potential has the form assumed here. It does not seem possible to infer the position of the effective potential barrier surrounding these silica particles from the data given by Kops-Werkhoven and Fijnaut, although a value near 0.1 is plausible.

We now turn to the case of $\mathcal{P}_{ij} \ll 1$ but not zero. Unfortunately the exact perturbation solution Q of p_{ij} for the Eq. (5.2.22) with $\Phi_{ij} \neq 0$ at small Péclet number is not available. However in their 1982 paper Batchelor and Wen gave an argument that it can be approximated by the results obtained from Eq. (5.2.22) with $\Phi_{ij} = 0$ and $\mathcal{P}_{ij} \ll 1$ under the conditions $\lambda = 1$ while the model for Φ_{ij} mentioned in section 5.2 being used. Then the values of $Q(s)$ shown in Figure 5.2.4 for $\lambda = 1$ were therefore used to calculate $S_{ij}^{'(G)}, S_{ij}^{''(G)}, S_{ij}^{'(I)}, S_{ij}^{''(I)}, S_{ij}^{'(B)}$ and $S_{ij}^{''(B)}$. Note in this case all the three contributions to sedimentation coefficient are non-zero, and need to be computed. Here we would not give a detailed description of the results. For a detailed description of the results the reader is referred to the paper of Batchelor and Wen (1982)[12]. However, some main points are needed to be given here, that is, decreasing the double layer thickness ξ_0 will increase $S_{ij}^{'(I)}, S_{ij}^{'(B)}, S_{ij}^{'(G)}$ and the resultant S_{ij}', S_{ij}''. The opposite situations are also right when ξ_0 increases. The reason of the variations of the sedimentation coefficients with ξ_0 is the same as in the identical sphere case.

3.4. Two Species of Spheres with Very Different Radii or Densities

These cases have been mentioned above several times, and now we will obtain these results.

It was shown in section 5.2, when $\lambda \ll 1$ or $\lambda \gg 1$, $p_{ij} - 1$ is small, being of order λ^3 when $\lambda \ll 1$ and of order λ^{-3} when $\lambda \gg 1$, for any given values of the Péclet number. We may use this result now to obtain some asymptotic expressions for S_{ij} which likewisely are independent of $\mathcal{P}_{ij}$.

We consider first the direct contribution due to gravity given by Eq. (5.3.18). The tensor $\mathbf{J}'$ and $\mathbf{J}''$ in that equation can be rewritten as

$$\mathbf{J}' = \mathbf{I}\int_2^\infty (A_{11} + 2B_{11} - 3)s^2 ds$$

$$+\frac{3}{4\pi}\int_{s\geq 2}\left\{(A_{11}-1)\frac{ss}{s^2} + (B_{11}-1)\left(\mathbf{I} - \frac{ss}{s^2}\right)\right\}(p_{ij}-1)ds, \qquad (5.3.34)$$

$$\mathbf{J}'' = \mathbf{I}\frac{2\lambda}{1+\lambda}\int_2^\infty \left(A_{12} + 2B_{12} - \frac{3}{s}\right)s^2 ds$$

$$+\frac{3}{4\pi}\frac{2\lambda}{1+\lambda}\int_{s\geq 2}\left\{A_{12}\frac{ss}{s^2} + B_{12}\left(\mathbf{I} - \frac{ss}{s^2}\right)\right\}(p_{ij}-1)ds. \qquad (5.3.35)$$

and the orders of magnitude of all these integrals may be estimated .

The behaviours of $A_{11}, B_{11}, A_{12}, B_{12}$ when $\lambda \ll 1$ are shown in Eq. (4.2.29)–Eq. (4.2.32), whence we find

$$\int \lim_{\lambda\to 0}\left(\frac{A_{11}+2B_{11}-3}{\lambda^3}\right)s^2 ds = \int_2^\infty \left(-\frac{60}{s^2} + \frac{480}{s^4} - \frac{1600}{s^6}\right)ds = -20, \qquad (5.3.36)$$

and

$$A_{12} + 2B_{12} - \frac{3}{s} = O(\lambda^3). \qquad (5.3.37)$$

Using the above result for $p_{ij} - 1$, when $\lambda \ll 1$ it is shown

$$\mathbf{J}' \sim 20\lambda^3\mathbf{I}, \quad \mathbf{J}'' = O(\lambda^4), \qquad (5.3.38)$$

and

$$S_{ij}^{(G)} = -\frac{5}{2} - \gamma + O(\lambda). \qquad (5.3.39)$$

On the other hand, when $\lambda \gg 1$, we may use the result in Chapter 4, $A_{11} - 1$ and $B_{11} - 1$, and also $A_{11} + 2B_{11} - 3$, which are of order λ^{-1} when $\lambda \gg 1$. We also know from Eq. (4.2.20) that A_{12} and B_{12} are unchanged when λ is replaced by λ^{-1}, whence it follows from Eq. (5.3.37) that $A_{12} + 2B_{12} - 3s^{-1}$ is of order λ^{-3} when $\lambda \gg 1$. Hence when $\lambda \gg 1$

$$\mathbf{J}' = O(\lambda^{-1}), \quad \mathbf{J}'' = O(\lambda^{-3}), \qquad (5.3.40)$$

and the contribution to S_{ij} due to gravity is

$$S_{ij}^{(G)} = -\gamma(\lambda^2 + 3\lambda + 1) + O(\lambda^{-1}). \qquad (5.3.41)$$

These two expressions for $S_{ij}^{(G)}$, Eq. (5.3.39) and Eq. (5.3.41) can be given a physical interpretation. When a sphere of radius a_i is falling through a dispersion there are two direct consequences of its presence in the fluid. One is that there is a net volume flux within a spherical shell of radius $a_i + a_j$ surrounding each sphere j equal to the sum of $(4/3)\pi a_j^3 \mathbf{U}_j^{(0)}$ due to the motion of the rigid sphere itself and $(4/3)\pi a_j^3(3\lambda^{-1} + (3/2)\lambda^{-2})\mathbf{U}_j^{(0)}$ due to the motion of the fluid in the spherical shell surrounding the rigid sphere. The downward volume flux must be balanced by an equal upward flux in the remainder of the system, and so the mean velocity in the fluid accessible to the centre of the test sphere i is

$$-(1 + 3\lambda^{-1} + \frac{3}{2}\lambda^{-2})\phi_j \mathbf{U}_j^{(0)}, = -\gamma(\lambda^2 + 3\lambda + \frac{3}{2})\phi_j \mathbf{U}_i^{(0)}.$$

The other is that the motion of spheres j generates an environment for each sphere i in which the Laplacian of the fluid velocity at any point is non-zero. As explained in section 5.1, this affects the motion of the test sphere i and the change in the mean velocity of the test sphere i which takes all accessible positions with equal probability is

$$\frac{1}{2}\lambda^{-2}\phi_j \mathbf{U}_j^{(0)}, = \frac{1}{2}\gamma\phi_j \mathbf{U}_i^{(0)}.$$

Thus, the total change in the mean velocity of the test sphere i is

$$-\gamma(\lambda^2 + 3\lambda + 1)\phi_j \mathbf{U}_i^{(0)}. \tag{5.3.42}$$

When $\lambda \ll 1$ the expression Eq. (5.3.42) accounts for the second term on the right-hand-side of Eq. (5.3.39) and when $\lambda \gg 1$ it accounts for the whole of the explicit part of Eq. (5.3.41).

However, it is not quite true that a large sphere i does not disturb the environment. One of the influences of the presence of the sphere i for a neighbouring smaller sphere j is that the ambient velocity gradient at the position of the smaller sphere j is non-zero and that the rate of energy dissipation in the fluid by viscosity is increased by the presence of the smaller sphere j. All elements of the fluid surrounding the larger sphere i dissipate energy not as a fluid of viscosity μ, but as a fluid of viscosity μ containing, on average, n_j spheres of radius a_j per unit volume, and we know from the Einstein formula that the effective viscosity of such a mixture is $\mu(1 + (5/2)\phi_j)$ (see Chapter 7) correct to order of ϕ_j. The larger spheres i are therefore falling through a fluid medium whose effective viscosity is $\mu(1 + (5/2)\phi_j)$, and the corresponding fractional change in the fall speed of the larger sphere i is $-(5/2)\phi_j$. Note that it is not necessary that there be a large number of smaller spheres j in the neighbourhood of the larger sphere i; provided the probable number of smaller spheres j in unit volume outside the larger sphere i is uniform, it does not matter how small this number is. Thus the first term on the right-hand-side of Eq. (5.3.39) is also accounted for.

The contributions $S_{ij}^{(I)}$ and $S_{ij}^{(B)}$ are negligible in these cases. $S_{ij}^{(I)}$ is evidently small, and as already assumed, the interparticle force is negligible if the spheres gaps are comparable with the smaller one of a_i and a_j. For $S_{ij}^{(B)}$ we note that the expression within curly brackets in Eq. (5.3.25) is seen from Eq. (4.2.29)–Eq. (4.2.32) to be of order λ^3 when $\lambda \ll 1$. Hence, remembering the result concerning $p_{ij} - 1$, we find

$$S_{ij}^{(B)} = O(\lambda^4) \tag{5.3.43}$$

when $\lambda \ll 1$, at any given value of $\mathcal{P}_{ij}$. And when $\lambda \gg 1$ we use the fact that $A_{11} - 1$ and $B_{11} - 1$ behave as λ^{-1} (as shown in Chapter 4) to find

$$S_{ij}^{(B)} = O(\lambda^{-2}). \tag{5.3.44}$$

It appears therefore that, when $\lambda \ll 1$ and when $\lambda \gg 1$, the sedimentation coefficient is dominated by $S_{ij}^{(G)}$. At any Péclet number we have

$$S_{ij} = -\frac{5}{2} - \gamma + O(\lambda) \quad \text{when } \lambda \ll 1, \tag{5.3.45}$$

and

$$S_{ij} = -\gamma(\lambda^2 + 3\lambda + 1) + O(\lambda^{-1}) \quad \text{when } \lambda \gg 1. \tag{5.3.46}$$

The limit cases $\gamma \to 0$ or $|\gamma| \to \infty$ are quite straightforward. It was noted in section 5.2 that $L(s)$ and $M(s)$, and hence also p_{ij} for a given Péclet number, approach to finite limit s as $\gamma \to 0$ or $|\gamma| \to \infty$. The mobility function s do not depend on γ, and it is evident therefore from Eq. (5.3.5) that, at absence of any dependence of interparticle force effects on γ, S_{ij} is asymptoticall linear in γ as $|\gamma| \to \infty$ for a given value of $\mathcal{P}_{ij}$. At small Péclet number S_{ij} has already been seen to be a linear function of γ for all γ; and at large Péclet number we see from Eq. (5.3.21) that

$$\frac{S_{ij}}{\gamma} \sim \frac{1}{4}(1+\lambda)^2 \int_2^\infty \left\{ (A_{12} + 2B_{12})(p_{ij})_{|\gamma| \to \infty} - \frac{3}{s} \right\} s^2 \mathrm{d}s - (\lambda^2 + 3\lambda + 1), \tag{5.3.47}$$

as $|\gamma| \to \infty$.

4. Discussion

The asymptotic forms for the sedimentation coefficient S_{ij} for $\lambda \to 0$ and $\lambda \to \infty$, which has the symmetric relation like

$$S_{ij}(\lambda, \gamma) = S_{ji}(\lambda^{-1}, \gamma^{-1}) \tag{5.4.1}$$

are important, because they help to set limit s to the variation of S_{ij} with λ, γ and $\mathcal{P}_{ij}$. As $\lambda \to 0$ we have

$$S_{ij} \sim -2.5 - \gamma. \tag{5.4.2}$$

The theoretical result for the limit $\lambda \to \infty$ is

$$S_{ij} \sim -\gamma(\lambda^2 + 3\lambda + 1). \tag{5.4.3}$$

These asymptotic forms impose such strong constraints on the values of S_{ij} at small Péclet number, where we have the additional exact information that S_{ij} is a linear function of γ, and the simple empirical relation

$$S_{ij} = \frac{-2.5}{1 + 0.16\lambda} - \gamma \left(\lambda^2 + 3\lambda + \frac{\lambda^2}{1 + \lambda^3} \right) \tag{5.4.4}$$

is accurate to the first decimal place over almost the whole of the (γ, λ)-plane when $\mathcal{P}_{ij} \ll 1$. At large Péclet number the dependence of S_{ij} on γ and λ appears to be too complicated to allow approximate representation by simple algebraic expressions, owing to the strange difference of shape and position of the curves representing S_{ij}

as functions of γ or λ on the two sides of the region in which some trajectories are of finite length. There is also a difficulty of about the many-valuedness of S_{ij} which occurs at the central point $\gamma = 1, \lambda = 1$. The case of spheres of equal density ($\gamma = 1$) is perhaps of greatest practical interest, and the results for this case at both small and large values of the Péclet number are shown in Figure 9 of Chapter 5 The curve for $\mathcal{P}_{ij} \gg 1$ lies above that for $\mathcal{P}_{ij} \ll 1$ because close sphere pairs, which have a larger speed of fall, are more numerous at large Péclet number. The difference between the values of S_{ij} on two curves is rather small, except at values of λ above unity. At absence of evidence to the contrary, it would be reasonable to expect that S varies monotonically with $\mathcal{P}_{ij}$ for given λ when $\Phi_{ij} = 0$.

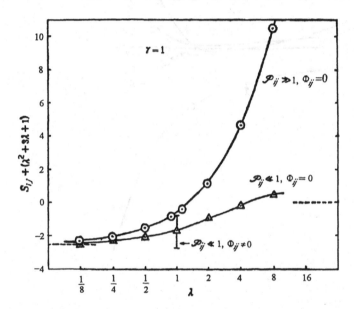

Fig. 9. Sedimentation coefficient S_{ij} as a function of λ with $\gamma = 1$ (From Batchelor and Wen 1982)[12]

For comparison with this range of variation of S_{ij} we also show in Figure 9 of Chapter 5 the range of variation resulting from the change of the position of the Coulomb potential barrier from $\xi_0 = 0.02$ to $\xi_0 = 0.20$ in the case of identical spheres of radius 1μm which exert forces on each other.

In 1989 Russel, Saville and Schowalter re-analysed Davis and Birdsell's experimental data[25]. Figure 10 of Chapter 5 gives their results[26]. From the figure we may find that the two data of the sedimentation coefficients in a bidisperse dispersion obtained by Davis and Birdsell for glass and acrylic spheres with $\mathcal{P}_{ij} = 10^7$, ,$\gamma = 0.11$, $\lambda = 1.0$ and $\gamma = 1.0$, $\lambda = 1/2$ agree with the theoretical predictions shown in Figure 6 of Chapter 5 calculated from sedimentation coefficient formula Eq. (5.3.21). Thus, the Batchelor and Wen's sedimentation theory for the case of a dilute polydisperse system is preliminarily confirmed.

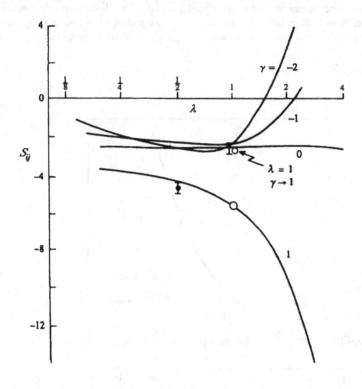

Fig. 10. The sedimentation coefficient as a function of the size ratio λ and the reduced density ratio γ for the case of large Péclet numbers. (From Russel, Saville and Schowalter 1989)[26].

——Predictions of Batchelor and Wen (1982)[12]. ● Data of Davis and Birdsell (1988)[25] for mixtures of glass beads ($\lambda = 0.52, \gamma = 1.0$) and glass and acrylic beads ($\lambda = 0.99, \gamma = 0.11$) with $\mathcal{P}_{ij} \approx 10^7$

CHAPTER 6

COAGULATION OF AEROSOL PARTICLES

1. Coagulation in a Dilute Unstable System

Coagulation between spheres (or between spheres and a collector like a sampler, or the thread of a spider web etc.), in a dilute unstable system is actually a process of particles which disappear from the dispersion.

Colloidal particles including aerosol and hydrosol particles which take a relative Brownian random motion, or a bulk motion (like simple shear-induced or a linear ambient flow -induced motion), or a differential settling under gravity , have relative motion between particles and then have a finite probability of colliding and sticking to one another; i.e. they may experience a Brownian coagulation , shear-induced coagulation , an linear ambient flow -induced coagulation and gravitational coagulation . The sticking efficiency of aerosol particles is a much more complicated function of their shape and surface conditions (roughness, absorbed vapors, etc.), the relative humidity of the air, the presence of foreign vapors in the air, the interparticle potential and other factors. Although little is known quantitatively about the sticking efficiency of aerosol particles, the fact that kinetic energy of the colliding particles is very small makes bounce-off unlikely. We shall therefore assume a sticking efficiency of unity in the sequel. The experimental results of coagulation of cloud droplet have proved this assumption.

When the dispersions undergo the coagulation process. the number density of the particles will decrease, and the average size of the particles will increase, the mean speed of fall of the particles will be accelerated. They will thus be separated from the medium more quickly than the stable system. Then the dispersions cannot remain dispersed and hence are referred to unstable systems.

Investigation of coagulation is an important subject. ·It has a close connections with the scavenging of atmospheric aerosols, the microphysics of clouds and precipitation, medical and health work, and some engineering, like chemical industry—the manufacturers of polymer latices who wish to insure stability of their dispersions, and also those who wish to flocculate particulate dispersions, and thereby separate finely divided solids from a liquid. In a word, coagulation of dispersion is such a common phenomenon in aerosol science and hydrosol science that it has attracted a lot of researchers to investigate this phenomenon since the beginning of this century.

The first attempt to estimate the rate of coagulation of a dispersion was made by Smoluchowski (1917). He proposed two assumptions, as the first step to make the study possible.

First, he suggested that all the cases for different values of the Péclet number could be divided into two regions, corresponding to two different asymptotic limit s, i.e. $\mathcal{P}_{ij} \to 0$ and $\mathcal{P}_{ij} \to \infty$. When $\mathcal{P}_{ij} \to 0$ the convection term can be neglected, and the interparticle potential term is assumed to be zero except a sticking force. The problem reduces to a complete probability type, the equation for pair-distribution function Eq. (4.4.27) reduces to the simplest diffusion equation, and is adopted to approach the problem. Then the coagulation for this case is named the Brownian coagulation. When $\mathcal{P}_{ij} \to \infty$ the Brownian motion term can be neglected completely. The problem reduces to a deterministic type, the trajectory analyses is adopted to approach the problem then the coagulation for these cases are named the gravitational coagulation or shear-induced coagulation , etc.

Secondly, the colloidal particles were assumed by Smoluchowski to move independently without any hydrodynamic interaction with each other even when the gap between spheres is very small, and without any interparticle force other than a sticking force at contact.

We now consider the case of the Brownian coagulation . Having determined the solution p_{ij} for the pure diffusion equation, the rate F_{ij} of coagulation can be obtained by integrating a flux of sphere j across the contact surface of $r = a_i + a_j$, viz.

$$F_{ij} = n_j \int_{r=a_i+a_j} \{-p_{ij}\mathbf{V}_{ij} + p_{ij}\mathbf{D}_{ij} \cdot \nabla_r(\Phi_{ij}/kT) + \mathbf{D}_{ij} \cdot \nabla p_{ij}\} \cdot \mathbf{n}dA(\mathbf{r}). \qquad (6.1.1)$$

The first and the second terms in the integrand are zero under the Smoluchowski 's assumptions, and the relative Brownian diffusivity of the two spheres would be constant when we neglect any effects of hydrodynamic interaction s on the diffusivity $\mathbf{D}_{ij}$ then $G \equiv H \equiv 1$, and

$$\mathbf{D}_{ij} = D_{ij}^{(0)}\mathbf{I}.$$

Substituting the solution p_{ij} determined from the pure diffusion equation with a constant diffusivity $D_{ij}^{(0)}$ into Eq. (6.1.1), we have the Smoluchowski 's Brownian coagulation rate, viz.

$$F_{ij} = 4\pi(a_i + a_j)D_{ij}^{(0)}n_j. \qquad (6.1.2)$$

Smoluchowski considered the dispersion as a monodisperse suspension, then $\lambda = 1, a_i = a_j = a, n_i = n_j = n_0, D_{ij}^{(0)} = D_{ii}^{(0)} = 2D_0$ where $D_0 = kT/6\pi\mu a$ and needs a Knudsen number correction $(1 + \alpha K_n)$ for aerosol particles of submicron size (Pruppacher and Klatt (1978)[3]). Thus Eq. (6.1.2) reduces to

$$F_{ij} = 16\pi a D_0 n_0. \qquad (6.1.3)$$

The variation of the number density n with respect to time t and the half-life time $t_{1/2}$ of the dispersion will be given by

$$n(t) = \frac{n_0}{1 + 8\pi a D_0 n_0 t}, \qquad (6.1.4a)$$

and

$$t_{1/2} = (8\pi a D_0 n_0)^{-1}. \qquad (6.1.4b)$$

However, the relative Brownian diffusivity cannot be identically equal to a constant as Smoluchowski assumed, when hydrodynamic interaction between the two spheres is included. In Chapter 4 we have shown that the longitudinal scalar function of the relative Brownian diffusivity G is proportional to ξ as $\xi \to 0$ (see Eq. (4.4.23)). Then G is zero at the contact surface $r = a_i + a_j$ through the action of the Stokes resistance due to the lubrication film, which will approach infinity as $\xi \to 0$. It is impossible under these circumstances to find a solution p_{ij} which can make the Brownian flux of sphere j ($G dp_{ij}/d\xi$) non-zero across the contact surface, and at the same time satisfies the absorbing wall condition, i.e. $p_{ij} = 0$ at $\xi = 0$. The finite energy of the Brownian motion of sphere j is not large enough for overcoming the Stokes resistance of the lubrication film between the two spheres as the gap between them approaches zero. For the Brownian coagulation to occur we need an attractive force which should be sufficiently large for overcoming the Stokes resistance. Derjaguin was the first who recognized this problem (1956)[27], and found eventually the necessarily attractive force—van der Waals force which can break down the barrier made by the lubrication film (Derjaguin and Muller 1967[28]). We shall introduce their work in the following part of this section.

1.1. The Rate of Coagulation as $\mathcal{P}_{ij} \to 0$

Under the condition $\mathcal{P}_{ij} = 0$ and $\Phi_{ij} \neq 0$, Eq. (4.4.27) becomes

$$\frac{\partial p_{ij}}{\partial t} = \nabla \cdot \{ p_{ij} \mathbf{D}_{ij} \cdot \nabla(\Phi_{ij}/kT) + \mathbf{D}_{ij} \cdot \nabla p_{ij} \}, \tag{6.1.5}$$

with the boundary conditions

$$p_{ij} = 0 \quad \text{at} \quad r = a_i + a_j, \quad p_{ij} \to 1 \quad \text{as} \quad r \to \infty. \tag{6.1.6}$$

The inner boundary condition now is an absorbing wall, reflecting the system is unstable. The derivative of p_{ij} with respect to time t should of course be non-zero in an unstable system. However, provided the dispersion is dilute, the rate of conversion of singlets into doublets is not too rapid (Wen and Batchelor 1983, 1985[13]), and the interparticle potential satisfies the requirements indicated by van de Ven and Mason (1977)[29] and by Melik and Fogler (1984) (note that van der Waals potential just satisfies these requirements.), a steady state can be approximately reached in the initial stage of the coagulation process. The divergence term of the relative velocity of the two spheres in Eq. (6.1.5) is thus equal to zero, we have

$$\nabla \cdot \{ p_{ij} \mathbf{D}_{ij} \cdot \nabla(\Phi_{ij}/kT) + \mathbf{D}_{ij} \cdot \nabla p_{ij} \} = 0. \tag{6.1.7}$$

The solution p_{ij} for Eq. (6.1.7) must be spherically symmetric, because the van der Waals potential is isotropic (see Eq. (4.3.20)) and the random Brownian motion is statistically isotropic. Note that the origin of the spherical polar coordinate system is put at the centre of the test sphere i. The Eq. (6.1.7) then reduces to an ordinary differential equation

$$\frac{1}{r^2}\frac{d}{dr}\left\{r^2\left[D_{ij}^{(0)}p_{ij}G\frac{d(\Phi_{ij}/kT)}{dr}+D_{ij}^{(0)}G\frac{dp_{ij}}{dr}\right]\right\}=0. \qquad (6.1.8)$$

Because the system is in a steady state, the flux of sphere j across any surface with radius $r > a_i + a_j$, is equal to the flux through the contact surface , thence $F_{ij} =$ constant. The solution of p_{ij} for Eq. (6.1.8) satisfying the boundary conditions Eq. (6.1.6) is

$$n_j p_{ij} = e^{-\Phi_{ij}/kT}\left[n_j - \frac{F_{ij}}{4\pi D_{ij}^{(0)}}\int_r^\infty \frac{\exp(\Phi_{ij}/kT)}{r^2G}dr\right], \qquad (6.1.9)$$

and the rate of coagulation F_{ij} is found to be

$$F_{ij} = 4\pi(a_i + a_j)C_\varphi D_{ij}^{(0)}n_j, \qquad (6.1.10)$$

where C_φ is given by

$$C_\varphi = \left\{2\int_2^\infty \frac{\exp(\Phi_{ij}/kT)}{s^2G(s)}ds\right\}^{-1}. \qquad (6.1.11)$$

Comparing Eq. (6.1.10) with Smoluchowski 's solution Eq. (6.1.2), there is a correction factor C_φ which represents the influence of the van der Waals force on Brownian coagulation , appearing in Eq. (6.1.10) (Derjaguin and Muller 1967)[28]. The correction factor C_φ due to the van der Waals force is very important. If $\Phi_{ij} = 0$, then $C_\varphi = 0, F_{ij} = 0$. Without the help of the attractive van der Waals potential, the Brownian coagulation cannot occur, no matter how large the Brownian diffusivity $D_{ij}^{(0)}$ may be. This is due to the asymptotic behaviour of the van der Waals force in the near field. We have indicated in Eq. (4.3.22) that the van der Waals potential Φ_{ij} varies as ξ^{-1}, and the attractive force varies as ξ^{-2}, as $\xi \to 0$. Thus the attractive potential at $\xi = 0$ is a singularity, and it has sufficiently large energy for overcoming the Stokes resistance of the lubrication film .

In his book "The mechanics of aerosol", Fuchs (1964)[2] underestimated the contributions of the van der Waals potential. According to his estimation, the molecular van der Waals force at most can increase the Brownian coagulation rate by 1–2 per cent. Obviously his conclusion was not correct. The point is that Fuchs continued using the Smoluchowski 's isolated particle assumption, hence he still adopted the constant relative Brownian diffusivity model, viz. $\mathbf{D}_{ij} = D_{ij}^{(0)}\mathbf{I}$. Under such condition, the Brownian coagulation still can occur without the help of the attractive van der Waals potential, so the attractive potential is no longer a necessary condition for the Brownian coagulation in his book[2].

We may understand why Fuchs did this mistake, just because his book was written before Derjaguin and Muller's work (1967)[28]. However, it is difficult for us to understand Pruppacher and Klett who still insisted using the Smoluchowski 's constant diffusivity assumption, and the corresponding results Eq. (6.1.3) and Eq. (6.1.4) in their 1978 book[3], since there had been a lot of investigations on the effects of hydrodynamic interaction on the relative Brownian diffusivity tensor $\mathbf{D}_{ij}$, like Deutch and Oppenheim (1971), Spielman (1970), Murphy and Aguirre (1972), Aguirre and Murphy (1973), and Batchelor (1976)[18] etc., before Pruppacher and Klett wrote their 1978 book[3]. And the conclusion of all the above investigations arrived at the same point as Derjaguin (1956,1967)[27,28] that the relative Brownian diffusivity tensor $\mathbf{D}_{ij}$ cannot be a constant due to the hydrodynamic interaction between the spheres under consideration.

In their 1978 book, Pruppacher and Klett stated: "The reciprocal dependence of the total particle concentration on time predicted by Eq. (6.1.4) (i.e. (12.41) in their book (1978)[3]) has been verified by extensive experimental studies over the last forty years (e.g. Whytlaw–Gray and Patterson (1932), Patterson and Cawood (1932), Whytlaw–Gray (1935), Artemov (1946), Devir (1963)) ⋯"[3].

It is noted (a) that all the experiments mentioned by Pruppacher and Klett (1978)[3] were made before 1967. It seemed to be impossible for these investigators to measure the interparticle potential and the correction factor C_φ, since the theory of the Brownian–interparticle potential coagulation was not established at that time, (b) the correction factor C_φ in Derjaguin and Muller's theory is independent of time, hence, the reciprocal dependence of the total particle concentration on time predicted by Eq. (6.1.4) should also be valid in Derjaguin and Muller's theory. Thus the reciprocal dependence of the particle concentration on time cannot be used to prove the validity of Smoluchowski's theory. At any rate, neglecting the effect of hydrodynamic interaction s on diffusivity is permissible only when the separation between the two spheres is large, whereas, neglecting the effect of hydrodynamic interaction s on the relative Brownian diffusivity at the intermediate separations especially as the gap between the two spheres approaching zero is unimaginable. More careful experimental work is needed to check the validity of Smoluchowski's theory.

Although we think that Derjaguin and Muller's theory is correct in principle, there was a deficiency in their calculations. They did not calculate accurately the mobility function , and hence, for the longitudinal scalar function of the diffusivity G, they only gave a very rough empirical expression with $\lambda = 1$ as follows

$$G = (1 + (4\xi)^{-1})^{-1}.$$

Comparing this expression for $G(s)$ with the one obtained in section 4.4, it is evident that their empirical formula overestimated the values of G, especially in the near field. When $\xi \to 0$, their expression gave $G \sim 4\xi$, whereas $G \sim 2\xi$ with $\lambda = 1$ according to our result shown in section 4.4, which is calculated from a more accurate mobility function (Jeffrey and Onishi 1984)[9]. Thus Derjaguin and Muller's work in 1967 overestimated the values of C_φ.

Using more accurate values of G obtained in section 4.4, and the van der Waals potential Eq. (4.3.20), we recalculated the values of C_φ with different A—the composite Hamaker constant , different λ—the size ratio at room temperature $T = 293^0\text{K}$. Table 1 of Chapter 6 gives the computed results.

Table 1. The values of the correct factor C_φ with $T = 293^\circ$K

A(erg) λ	10^{-14}	10^{-13}	10^{-12}	10^{-11}	10^{-10}
1	0.456	0.593	0.839	1.25	1.89
1/2	0.493	0.625	0.856	1.24	1.83
1/4	0.580	0.699	0.888	1.20	1.69
1/8	0.692	0.787	0.915	1.14	1.50

Obviously the new result of C_φ is smaller than Derjaguin and Muller's result with $\lambda = 1$. (They did not calculate the unequal-sized system.) From Table 1 of Chapter 6 we may also see that the values of C_φ decrease as λ decreases when the composite Hamaker constant A is large; and the values of C_φ increase as λ decreases when the composite Hamaker constant is small. This is due to the fact that the van der Waals attractive potential decreases as λ decreases, whereas the longitudinal scalar function of the diffusivity G increases as λ decreases.

1.2. The Rate of Coagulation at $\mathcal{P}_{ij} \to \infty$

As an example, we will mainly consider the gravitational coagulation , and only the interactions of particle pairs need to be considered, since the dispersion is a dilute polydisperse one.

According to Smoluchowski 's assumption, the Brownian motion can be neglected completely, and the trajectory analysis is adopted to calculate the rate of coagulation. Then the trajectories of sphere j with radius a_j, which is smaller than a_i—the radius of the test sphere i—can be calculated from the Lagrangian equation for sphere j. The goal is to find y_c defined as the initial horizontal offset of the centre of the lower sphere j from the vertical line through the centre of the upper test sphere i, thus a grazing trajectory can be found numerically. Note that the initial vertical separation z_0 for sphere j is taken to be large enough so that the two spheres fall independently at the beginning, thus the results will not depend on z_0. In practice it has been found that the computed results are generally insensitive to z_0 if $z_0 > 10^2 a_i$, for any size ratio λ. Figure 1 of Chapter 6 shows a schematic representation of a grazing trajectory of small sphere j, y_c is the critical, horizontal offset for the grazing trajectory . Then, all the spheres j enclosed inside the grazing trajectory will collide with the test sphere i, since there is not any Brownian motion . Thus the rate of coagulation F_{ij} can be calculated by

$$F_{ij} = \pi y_c^2 V_{ij}^{(0)} n_j. \tag{6.1.12}$$

From the Smoluchowski 's second assumption, we see that $y_c = a_i + a_j$, since sphere j will move independently without any interaction with the test sphere i and directly impact on it along a straight line. Then we have the Smoluchowski 's flux

$F_{ij}^{(0)}$, viz.

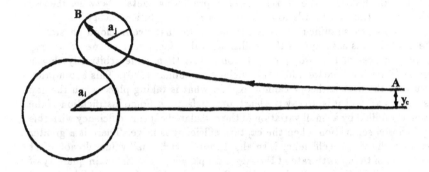

Fig. 1.   Schematic representation of the hydrodynamic interaction of a pair of spheres; y_c is the critical, horizontal offset for a grazing trajectory of the small sphere j

$$F_{ij}^{(0)} = F_{ij} = \pi(a_i + a_j)^2 V_{ij}^{(0)} n_j. \qquad (6.1.13)$$

It is customary to express the rate of coagulation at $\mathcal{P}_{ij} \to \infty$ in terms of the capture efficiency E_{ij} defined as the ratio of F_{ij} over the value that the particle flux would be if each sphere moved as if the others were absent (the so-called Smoluchowski 's flux, $F_{ij}^{(0)}$), viz.

$$E_{ij} = F_{ij}/F_{ij}^{(0)} = y_c^2/b(a_i + a_j)^2, \qquad (6.1.14)$$

where

$$F_{ij}^{(0)} = b\pi(a_i + a_j)^2 V_{ij}^{(0)} n_j, \qquad (6.1.15)$$

and b=1 for the case of gravity -induced coagulation, $b = 8/3\sqrt{3}$ for the case of a bulk axially symmetric pure straining motion -induced coagulation. Then we see from the impact model proposed by Smoluchowski , the capture efficiency E_{ij} is unity, i.e. $E_{ij} = 1$ in any cases.

Obviously Smoluchowski 's impact model overestimated the capture efficiency, since it neglected the hydrodynamic interaction s between the two spheres. When hydrodynamic interaction s are included, sphere j can move along a straight line only at large separations. When the separations are not large, the particles are forced by hydrodynamic interactions to flow around each other, and the y_c should be smaller than $a_i + a_j$, and $E_{ij} < 1$. Langmuir (1948) was the first who recognized this problem and according to his calculation found that E_{ij} even can be equal to zero if the size of sphere j, or more precisely, the Stokes number of sphere j is smaller than a critical value.

Another more serious problem is that the actual grazing trajectory does not exist when hydrodynamic interaction s are included, since the Stokes resistance of the lubrication film will approach to infinity as the gap between the two spheres approaches to zero. Then, the failure of the Stokes theory in predicting contact between the two spheres in a finite time now leads to difficulties in the coagulation theory.

For overcoming these difficulties it is customary, for instance in cloud physics, to make the arbitrariness assumption that collisions will in fact occur whenever the gap between the surfaces of the two spheres becomes less than some arbitrarily chosen small length (Davis and Sartor 1967 and Hocking and Jonas 1970). This assumption recognizes that the Stokes theory cannot explain what is taking place when the gap becomes very small, but it is, to say the least, unsatisfactory. Some justification of the assumption is fulfilled by a small variation of the calculated capture efficiency with this arbitrarily chosen separation when the capture efficiency is large. There is a greater variation when the capture efficiency is small, of course, such small values do not affect the calculations of the growth rate of the larger drops, which are the main quantity of interest in the application to cloud physics . The scavenging of atmospheric aerosol particles by the cloud droplets, however, has a low capture efficiency , and to calculate the extent of this scavenging process, and the resultant deposition of the pollutant on the ground, a more accurate assessment of the arbitrarily chosen length is required.

As an example, we consider the problem of coagulation between two liquid drop (Hokcking 1973)[30]. In his paper, Hocking pointed out that a complete discussion of the final stages of the approach of two liquid drops would be extremely complicated. When the gap is not too small, it is reasonable to assume that the drops are effectively rigid spheres and that the Stokes theory holds. As the gap decreases, a large number of effects could become important, such as the deformation of the drops, internal circulation within them, the compressibility of the air, rarefied gas effects and interparticle potential like the van der Waals potential. For the drops with radius in the range $10\mu m - 100\mu m$ Hocking has shown that the necessary modification to the Stokes theory when the gap becomes comparable with the mean free path λ_m of the air molecules is to use the Maxwell molecular slip flow approximation. On the other hand it was shown by Hocking and Jonas (1970) that drop deformation, internal circulation and electrical effects of magnitude typical of naturally occurring in clouds were all unimportant at this size gap, and the compressibility of the air was also negligible unless the gap becomes much smaller than the mean free path (Hocking 1973)[30],

As for the van der Waals attractive potential , Hocking(1973) indicated that if these forces were to be important, their magnitude must be comparable with the weight of the drops, that is[30],

$$A\lambda/8(1 + \lambda)\pi a_i^2 \rho_p g d^2 = O(1) \text{ or more,}$$

where A is the Hamaker constant ($A = 5 \times 10^{-13}$ erg for the present problem), $\rho_p = 1gm/cm^3, d = r - (a_i + a_j)$, and $\lambda = 1$. Then the van der Waals force become important when the gap is less than $3 \times 10^{-3}\mu m$ for drops of $100\mu m$ radius and less than $3 \times 10^{-2}\mu m$ for drops of $10\mu m$ radius. It follows that $10\mu m$ is the lower limit of the range of drop size for which the flow effects dominate over the effect of the van der Waals force .

The modification of the Stokes theory when the gap is comparable with the mean free path was obtained by Hocking (1973) and Davis (1972). In the calculations they used the Maxwell-slip flow approximation. The continuum equations were retained, but the boundary condition of no slip was replaced by the condition that the relative velocity at the boundary was proportional to the tangential stress there. The constant of proportionality was not exactly defined, but was of the same order as the mean free path.

The whole procedure used by Hocking (1973) for finding the resistance coefficients with the Maxwell-slip flow was the same as we have used in Chapter 3 only except that the boundary conditions was no longer a no-slip condition , but is a Maxwell-slip flow condition. When the relative longitudinal velocity of two spheres is $\mathcal{U}$, and the origin of the spherical polar. coordinate system is at the centre of sphere i, Hocking obtained that the resistance for sphere j is $-6\pi\mu a_j\mathcal{U}f_j$, and the resistance for sphere i is $-6\pi\mu a_i\mathcal{U}f_i$ where the resistance coefficient f_j was given by (Hocking 1973),[30]

$$f_j = \frac{4\lambda}{(1+\lambda)^3\xi\beta^2}[(1+\beta)\ln(1+\beta) - \beta], \qquad (6.1.16)$$

and

$$f_i = -\lambda f_j, \qquad (6.1.17)$$

where

$$\beta = 6\lambda_m/d, \text{ and } d = \xi\frac{(a_i + a_j)}{2} = r - (a_i + a_j).$$

When $\beta \gg 1$, Eq. (6.1.16) reduces to

$$f_j = \frac{\lambda}{3(1+\lambda)^2}\frac{1}{K_{ni}}\ln\beta, \qquad (6.1.18)$$

where K_{ni} is the Kundsen number for sphere i, i.e. $K_{ni} = \lambda_m/a_i$.

Thus, the force that opposes relative motion along the line connecting the centres is no longer inversely proportional to the gap, but rather it is inversely proportional to the mean free path and only logarithmically dependent on the gap. That means that the ξ^{-1} singularity of the Stokes resistance reduces to a logarithmical singularity $\log\xi^{-1}$. The logarithmical singularity is an integrable singularity, then as a consequence it permits that the two spheres contact for a finite time. The grazing trajectory can thus exit.

In the past, the influence of the van der Waals force on the gravitational capture efficiency of aerosol particles has been overlooked. Instead the generally accepted mechanism responsible for allowing the particles to overcome the Stokes resistance of the lubrication film is the effect of the Maxwell-slip flow model. Perhaps, Hocking was the first who recognized the significance of the contributions of the van der Waals forces. In his 1973 paper he indicated that the London–van der Waals forces were effective for promoting collisions between waterborne 1μm particles in a shear flow

(Curtis and Hocking 1970), and that 10μm is the upper limit of the range of drop size for which the effects of the van der Waals force dominate over the effects of Maxwell-slip flow , in gravitational coagulation . However his estimation for the aerosol (droplet suspended in air) still underestimated the role of the van der Waals force .

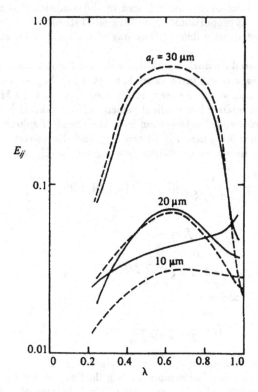

Fig. 2. The gravitational capture efficiency of water droplet in air: —, unretarded van der Waals force and no slip; − − −, Maxwell-slip flow model and no van der Waals force. (From Davis 1984[31])

Using a trajectory analysis Davis (1984)[31] has made a comparison between the effects of the two different factors. The purpose of this comparison was to see the relative importance of Maxwell-slip flow and of van der Waals force in promoting the capture efficiency between sedimenting aerosol particles. In Figure 2 of Chapter 6, the relation of E_{ij} versus λ is given for $a_i = 10, 20$ and 30μm with $\mu = 1.7 \times 10^{-4}$P, $\rho_0 = 1.3 \times 10^{-3}$gm/m^3, $\rho_i = \rho_j = 1$gm/cm^3, and $A = 5 \times 10^{-13}$ erg (Davis 1984)[31]. The solid lines are based on the consideration of the unretarded van der Waals force with no slip at the drop surfaces. The dashed lines represent the Maxwell-slip model with no interparticle forces. From Figure 2 of Chapter 6 it is apparent that the discrete-molecule effect gives a larger capture efficiency when a_i is greater than 20μm, but that the van der Waals force are the dominant mechanism for particles smaller than 20μm. Thus the van der Waals force have a greater effect in increasing collisions of ordinary droplets in clouds than was earlier realized. Only for drops of radius larger than 20μm, the discrete-molecule effect can play an important role. Therefore, the van der Waals attractive forces are a very important factor for overcoming the Stokes resistance of lubrication film both for the Brownian coagulation of particles of submicron size when $\mathcal{P}_{ij} \ll 1$, and for the gravitational coagulation of particles of radius as large as twenty microns when $\mathcal{P}_{ij} \gg 1$.

In aerosol the density of the particles is greater than that of the surrounding air, and often the inertia of the particles significantly influences their motion (large Stokes number , Stk,) even when the inertia of the surrounding air is negligible.

When the particles are sufficiently massive, their inertia must be included while determining their trajectories, and the hydrodynamic forces no longer balance the applied forces, then the mobility function in Chapter 4 cannot be used again to determine the trajectories of the particles, instead, we need to use the resistance function (Jeffrey and Onishi 1984) which were also mentioned in Chapter 4 briefly to determine the trajectories of the particles[9]. We note that the Stokes number is proportional to the particle density, whereas the Reynolds number is proportional to the fluid density. Therefore, only when the particle density is much larger than the fluid density may the particle inertia be significant while the fluid inertia remains negligible. This is true only in aerosol dispersions but not in hydrosol dispersions in general.

When $Stk=0$, the particles follow the trajectories computed by the mobility function s; as Stk increases from zero, the particles deviate from these trajectories—they tend to move in straight lines rather than flow around each other. Thus we may expect the capture efficiency E_{ij} to increase as the Stokes number increases, i.e. as the radius a_i increases.

On the other hand, for the ordinary aerosol particles they are of micron size or sub-micron size. The radius of most aerosol particles are smaller than 10μm. We see that for a typical aerosol system the Stokes number is of order unity when a_i is approximately 10μm, the values of the Stokes number are smaller than unity for most of the aerosol particles. Therefore, neglecting the particle inertia (putting $Stk=0$, and use the mobility function) is permissible in calculating the capture efficiency of the most aerosol particles.

Figure 3 of Chapter 6 gives a comparison of the gravitational capture efficiencies of two cases—one is neglecting the particle inertia—$Stk=0$, the other is including the particle inertia $Stk \neq 0$. Davis 1984)[30]. These are calculated for a typical aerosol dispersion in which $\mu = 1.7 \times 10^{-13}$P. $\rho_0 = 1.3 \times 10^{-3}$gm/cm^3, $\rho_i = \rho_j = 1.0$gm/cm^3, $A = 5 \times 10^{-13}$ erg and $kT = 4 \times 10^{-14}$ erg.

Again the solid lines in Figure 3 of Chapter 6 are obtained using the unretarded van der Walls potential with $Stk \neq 0$, and the dashed lines include the retardation effects with $Stk \neq 0$. For a comparison, the dotted line is the capture efficiency for $\lambda = 1/2$ with unretarded van der Waals potential and $Stk=0$. The case $\lambda = 1/2$ has been chosen because the effect of particle inertia is the largest when $\lambda = 1/2$. In

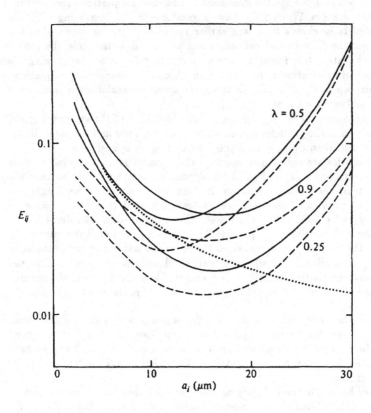

Fig. 3. The gravitational capture efficiency as a function of the radius of the larger sphere (a_i) for a typical aerosol; $\mu = 1.7 \times 10^{-4}$P, $\rho_0 = 1.3 \times 10^{-3}$gm/cm^3, $\rho_i = \rho_j = 1.0$gm/cm^3 and $A = 5 \times 10^{-13}$erg, $kT = 4 \times 10^{-14}$erg. — unretarded van der Waals force with $Stk \neq 0$, ——— retarded van der Waals force with $Stk \neq 0$, $\cdots$ unretarded van der Waals force with $Stk = 0$.

this case, the dependence on Stk as a function of λ has a maximum when $\lambda = 3/7$ with $\beta(\equiv \rho_j/\rho_i) = \gamma = 1$. When $\lambda \to 0$ the mass of the smaller particle is negligible, and $Stk \to 0$; when $\lambda \to 1$ the relative velocity of the two spheres approaches zero, and again $\to 0$. The dotted line in Figure 3 of Chapter 6 coincides with the solid line ($\lambda = 1/2$) when $a_i < 10\mu$m , and it shows that for the most aerosol particles we may use the zero-Stokes number approximation for calculating their capture efficiency . Of course, this approximation cannot be used again when the size of the aerosol particles under consideration is larger than 10μm. Figure 3 of Chapter 6 shows the capture efficiency would then be severely underestimated when $a_i \geq 10\mu$m, if we neglect the particle inertia as its radius is larger than 10μm. Within the range of radius larger than 10μm, we must include the effect of particle inertia which is very important in cloud physics , and is named 'inertia capture —a special term in cloud physics .

2. The Rate of Coagulation in a Dilute Unstable System at Large Péclet Number

So far, we have reviewed briefly the improvements made in recent years to the Smoluchowski 's assumption that the interactions between spheres could be neglected completely. On the other hand the method of trajectory analysis proposed by Smoluchowski for the case of $\mathcal{P}_{ij} \rightarrow \infty$ remained unchanged without any modification until the eighties.

In 1983 and 1985, Wen and Batchelor presented a novel method to approach this problem[13], which first yielded an exact asymptotic expression for the rate of formation of doublets . Instead of trajectory analysis, we suggested that the method of equation for pair-distribution function, which has been used in the case of $\mathcal{P}_{ij} \ll 1$ for years since Smoluchowski (1917), can also apply to the case of $\mathcal{P}_{ij} \rightarrow \infty$.

The small rigid spherical particles are dispersed in air (or liquid) with statistical homogeneity. We assume that the particles are sufficiently large so that the Brownian motion is negligible (large Péclet number and large Q_{ij} number), but not so large that the inertia of both the particles and the surrounding fluid is important (small Reynolds number and small Stokes number). For a typical aerosol system, these restrictions require that the particle radii lie between 2μm and 10μm, whereas for a typical hydrosol system, these restrictions imply that the particle radii lie between 2μm and 30μm. The primary concern is with the case of sedimenting spheres of non-uniform size and density, but the method is applicable to other types of the forced motion of the spheres. For example, we shall give detailed results in particular for the case of a bulk axisymmetric pure straining motion. This is for comparison with those already obtained by Zeichner and Schowalter (1977)[32] based on trajectory analysis . Figure 4 of Chapter 6 illustrates the two cases. (Wen and Batchelor 1984)[13]

The method involves a consideration of the pair-distribution function $n_j p_{ij}(\mathbf{r})$, which is governed by Eq. (4.4.27) in its steady–state form with the boundary conditions Eq. (6.1.6). In the above section we showed that the dispersion under consideration can approximately reach a steady–state at the initial stage of the coagulation process. Then, when $\mathcal{P}_{ij} \gg 1$ and $Q_{ij} \gg 1$ there is a thin boundary layer of p_{ij} near $r = a_i + a_j$ outside which the steady–state form of Eq. (4.4.27) reduces to

$$\nabla \cdot (p_{ij}\mathbf{V}_{ij}) = 0. \qquad (6.2.1)$$

In the case of sedimenting spheres, $\mathbf{V}_{ij}$ is the relative velocity of an i and a j spheres which are far from other spheres and the solution of Eq. (6.2.1) with the outer boundary condition $p_{ij} \rightarrow 1$ as $r \rightarrow \infty$ was found in section 5.2. (see equation

(5.2.7)), viz.

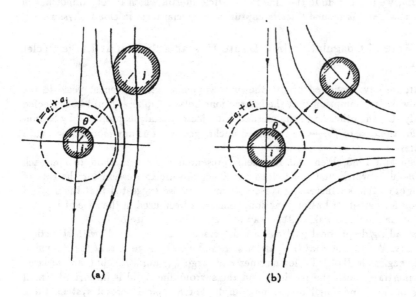

(a)   (b)

Fig. 4. Definition sketch showing the externally imposed relative motion of two spheres in two cases,(a) spheres of different sizes and densities falling under gravity and (b) spheres of a uniform size in fluid undergoing an axisymmetric compressional pure straining motion (From Wen and Batchelor 1984)[13]

$$p_{ij}(\mathbf{r}) = q(s) = \exp\left\{\int_s^\infty \left(\frac{L-M}{\frac{1}{2}sL} + \frac{1}{L}\frac{dL}{ds}\right)\right\}ds, \qquad (6.2.2)$$

where L and M are the longitudinal and transverse scalar function of $\mathbf{V}_{ij}$, (see Chapter 4). We are interested in the behaviour of this integral near $\xi = 0$ where $q(s)$ is singular, and we note for later purposes that if $M(s)$ were constant, with value M^* say, at small value of ξ we should have

$$q \sim \xi^{-\alpha}, \text{ where } \alpha = 1 - \frac{M^*}{L_1}, \text{ as } \xi \to 0. \qquad (6.2.3)$$

The function M does tend to a constant, with value M_0, as $\xi \to 0$, but the approach is extremely slow and through most of a boundary layer of small thickness δ (made dimensionless by dividing over the length $(a_i + a_j)/2$) M is approximately constant with the different values $M(\delta)$. Thus the power α in the asymptotic form Eq. (6.2.3) can be regarded as a slowly-varying function of the thickness of the boundary layer in which the pair-distribution function p_{ij} is being calculated. It appears that in all cases $0 < \alpha < 1$. Statements similar in all respects to those made in this paragraph may be made about the case of a dispersion subjected to a steady axisymmetric pure straining motion .

Since the convection-dominated form of the pair-distribution function is singular at $r = a_i + a_j$, a more convenient dependent variable φ for investigation of the boundary layer is

$$\varphi(\mathbf{r}) = p_{ij}(\mathbf{r})/q(r), \qquad (6.2.4)$$

which satisfies the conditions

$$\varphi = 0 \ \text{ at } \ \xi = 0, \quad \varphi \to 1 \ \text{ as } \ \xi \to \infty.$$

2.1. The Boundary Layer Region

The boundary layer near $r = a_i + a_j$ is by definition a region in which either or both of the last terms within the curly brackets in Eq. (4.4.27) are important despite the largeness of the Péclet number and Q_{ij} number. To see which of these two terms is the more important we must examine the asymptotic form of $\mathbf{D}_{ij}$ and Φ_{ij}. The longitudinal scalar function of the relative diffusivity G is given by

$$G(r) \sim \frac{(1+\lambda)^2}{2\lambda}\xi, \qquad \text{as } \ \xi \to 0. \qquad (6.2.5)$$

The interparticle force potential Φ_{ij} is in general the sum of contributions representing the van der Waals attraction and electrostatic repulsion , but at very small separations of the two spheres in an unstable system, the van der Waals attraction is dominant and is represented by

$$\Phi_{ij}(r) \sim -\frac{A\lambda}{3(1+\lambda)^2\xi} \ \text{ as } \ \xi \to 0. \qquad (6.2.6)$$

This asymptotic form is acceptably accurate for surface-to-surface separations less than 0.1μm; at larger separations the effect of phase lag of the polarizing electromagnetic waves between the two spheres is significant. Bearing in mind that only the derivative of p_{ij} in the direction of $\mathbf{r}$ has large magnitude within the boundary layer , it is evident that the term in Eq. (4.4.27) representing the van der Waals attractive forces dominates that due to the Brownian diffusion .

The equation describing the steady–state boundary layer when $\mathcal{P}_{ij}$ and Q_{ij} are both large is thus

$$\nabla \cdot \{p_{ij}\mathbf{V}_{ij} - p_{ij}\mathbf{D}_{ij} \cdot \nabla(\Phi_{ij}/kT)\} = 0. \qquad (6.2.7)$$

When the asymptotic form for Φ_{ij} and $\mathbf{D}_{ij}$ are used and a new variable φ defined by Eq. (6.2.4) is introduced the boundary layer equation for the pair-distribution function becomes (Wen and Batchelor 1983, 1984, 1985)[13]

$$\frac{\mathbf{V}_{ij} \cdot \nabla\varphi}{V_{ij}^{(0)}} = \frac{1}{6Q_{ij}q}\frac{\partial(q\varphi/\xi)}{\partial\xi}. \qquad (6.2.8)$$

The thickness of the boundary layer is evidently of order $Q_{ij}^{-1/2}$.

A similarity solution of Eq. (6.2.8) can be found for φ with a method which exploits the known linearity of $\mathbf{r} \cdot \mathbf{V}_{ij}$ in ξ near $\xi = 0$ and which is analogous to that used in problems of mass transfer at a large Péclet number (Batchelor 1979)[33]. In both cases to be considered the velocity $\mathbf{V}_{ij}$ is axisymmetic about the vertical direction, and its spherical polar components are of an approximate form

$$V_\xi / V_{ij}^{(0)} = -\xi \beta_1 f_1(\theta), \qquad V_\theta / V_{ij}^{(0)} = \beta_2 f_2(\theta) \text{ as } \xi \to 0, \qquad (6.2.9)$$

where $\theta = 0$ is the axis of symmetry and the functions $f_1(\theta)$ and $f_2(\theta)$ are known for the two cases of a bulk pure straining motion and sedimenting spheres. The relation Eq. (4.3.2) shows that $f_1 = \cos\theta, f_2 = \sin\theta$ in the latter case. Both V_ξ/ξ and V_θ approach constants as $\xi \to 0$ for given θ, but in the case of V_θ the approach is extremely slow (owing to a dependence on $(\log \xi^{-1})$). And it is necessary to choose an effective value of the constant β_2 which is representative of the values of $V_\theta / V_{ij}^{(0)} f_2$ in the boundary layer . That is to say, in the calculations we regard β_2 as a slowly-varying function of the boundary layer thickness and so of Q_{ij}. It is supposed that β_1, β_2 are both positive.

The velocity is not solenoidal, but it can be made so by multiplying by a suitably chosen function $h(\theta)$, whence we may introduce a stream function ψ defined by

$$\psi = Q_{ij}^{1/2} \xi \beta_2 f_2 h, \qquad h = \frac{1}{f_2} \exp\left(\int \frac{2\beta_1 f_1}{\beta_2 f_2} d\theta \right), \qquad (6.2.10)$$

and

$$\frac{Q_{ij}^{1/2} h V_\xi}{V_{ij}^{(0)}} = -\frac{1}{2} \frac{\partial \psi}{\partial \theta}, \qquad \frac{Q_{ij}^{1/2} h V_\theta}{V_{ij}^{(0)}} = \frac{\partial \psi}{\partial \xi}$$

within the boundary layer where $\xi \ll 1$. We now transform Eq. (6.2.8) to (ψ, θ) as independent variables, giving

$$\frac{3}{\beta_2 f_2 h^2} \frac{\partial \varphi}{\partial \theta} = \frac{1}{q} \frac{\partial}{\partial \psi}\left(\frac{q\varphi}{\psi}\right).$$

With a further replacement of the variable θ by τ, where

$$\tau = \int_0^\theta \frac{1}{3}\beta_2 f_2 h^2 d\theta, \qquad (6.2.11)$$

we have

$$\frac{\partial \varphi}{\partial \tau} = \frac{1}{q} \frac{\partial}{\partial \psi}\left(\frac{q\varphi}{\psi}\right). \qquad (6.2.12)$$

The solution of Eq. (6.2.12) is applicable when q is proportional to a power of ξ. Now it was noted earlier (see Eq. (6.2.3) and the remark following it) that the form of $q(\xi)$ near $\xi = 0$ is approximately

$$K\xi^{-\alpha}, \qquad (6.2.13)$$

where α may be regarded as a slowly–varying function of the boundary-layer thickness δ. Inserting this form for q, the equation for φ becomes,

$$\frac{\partial \varphi}{\partial \tau} = \psi^\alpha \frac{\partial}{\partial \psi} \left(\frac{\varphi}{\psi^{\alpha+1}} \right), \tag{6.2.14}$$

with boundary conditions on φ

$$\varphi = 0 \text{ at } \psi = 0, \quad \varphi \to 1 \text{ as } \psi \to \infty.$$

The solution can depend only on the similarity variable ψ^2/τ, whence we find (Wen and Batchelor 1983)[13]

$$\varphi = \left(\frac{\psi^2}{\psi^2 + 2\tau} \right)^{\frac{1}{2}(\alpha+1)}. \tag{6.2.15}$$

2.2. The Expression for the Capture Efficiency

The flux of j-particles across the contact surface $r = a_i + a_j$ enclosing an i-particle may now be evaluated from Eq. (6.1.1). The Brownian diffusion term is negligible and $\mathbf{n}.\mathbf{V}_{ij} = 0$ at $r = a_i + a_j$, so

$$F_{ij} = n_j \int_0^\pi \left\{ q\varphi \frac{\mathbf{r}.\mathbf{D}_{ij}.\mathbf{r}}{r^2 kT} \frac{d\Phi_{ij}}{dr} r^2 \right\}_{r \downarrow a_i + a_j} 2\pi \sin\theta d\theta.$$

Then with equations Eq. (6.2.5), Eq. (6.2.6), Eq. (6.2.13) and Eq. (6.2.15) we find

$$F_{ij} = \frac{n_j A}{9\mu} \frac{(1+\lambda)^2}{\lambda} K Q_{ij}^{\frac{1}{2}(\alpha+1)} \int_0^\pi \left(\frac{\beta_2^2 f_2^2 h^2}{2\tau} \right)^{\frac{1}{2}(\alpha+1)} \sin\theta d\theta, \tag{6.2.16}$$

provided the variable τ is defined over the whole range $0 \leq \theta \leq \pi$. (If it is not, as a consequence, f_2 changes sign at an interior point of this range, more than one integration is needed to cover the surface of the sphere.) Hence our result for the capture efficiency from Eq. (6.1.15), Eq. (6.1.16) is

$$E_{ij} = \frac{K}{3b} Q_{ij}^{\frac{1}{2}(\alpha-1)} \int_0^\pi \left(\frac{\beta_2^2 f_2^2 h^2}{2\tau} \right)^{\frac{1}{2}(\alpha+1)} \sin\theta d\theta, \tag{6.2.17}$$

in which α and β_2 as well as h and τ, are slowly–varying functions of the boundary-layer thickness, and thus of Q_{ij}.

It will be observed that since the non-dimensional boundary-layer thickness δ is proportional to $Q_{ij}^{-1/2}$ we may write

$$K Q_{ij}^{\frac{1}{2}(\alpha-1)} = (\delta Q_{ij}^{\frac{1}{2}})^{(\alpha-1)} \delta q^*(\delta), \tag{6.2.18}$$

where $q^*(\xi)$ is the power function Eq. (6.2.13) that represents $q(\xi)$ approximately. This allows Eq. (6.2.17) to be written as

$$E_{ij} = c\delta q^*(\delta), \tag{6.2.19}$$

where c is a constant (more accurately, a slowly–varying function of Q_{ij}) of order unity given by

$$c = \frac{(\delta Q_{ij}^{\frac{1}{2}})^{\alpha-1}\beta_2^{\alpha+1}}{3b} \int_0^\pi \left(\frac{f_2^2 h^2}{2\tau}\right)^{\frac{1}{2}(\alpha+1)} \sin\theta d\theta. \tag{6.2.20}$$

There seems to be a possible slight gain in accuracy while replacing $q^*(\delta)$ by the true form of $q^*(\delta)$, giving finally (Wen and Batchelor 1983) [13].

$$E_{ij} = c\delta q(\delta). \tag{6.2.21}$$

The form of Eq. (6.2.21) indicates that, as one might expect, a certain fraction of all the j-spheres in the boundary layer are captured. The power of Q_{ij} in Eq. (6.2.17), which is the most important part of the expression and which is negative, is determined by the dependence of two quantities on Q_{ij} , one being the boundary-layer thickness and another the number density of j-spheres at the outer edge of the boundary layer . Evidently, Smoluchowski 's 'impact' model of the capture process contains little of the real physics.

For definiteness we shall define the boundary-layer thickness in every case as being the value of ξ at which the term ψ^2 and 2τ in the denominator of Eq. (6.2.15) representing effects of van der Waals forces and convection respectively are equal. The actual boundary-layer thickness varies with θ, but it is a single measure that we need and so we choose $\theta = 0$ as the angular coordinate of the point at which particles are entering the boundary layer normally and define,

$$\delta = Q_{ij}^{-\frac{1}{2}} \lim_{\theta\to 0} \left(\frac{2\tau}{\beta_2^2 f_2^2 h^2}\right)^{\frac{1}{2}}, \tag{6.2.22}$$

this location being chosen because the flux reaches the maximum there.

2.3. Results for a Dispersion of Sedimenting Spheres

Since we have shown that $f_1 = \cos\theta, f_2 = \sin\theta$ in this case, it follows that the function $h(\theta)$ defined by Eq. (6.2.10) is given by

$$h(\theta) = (\sin\theta)^{\frac{2\beta_1}{\beta_2}-1}, \tag{6.2.23}$$

and that the variable τ given by Eq. (6.2.11) is

$$\tau = \frac{1}{3}\beta_2 \int_0^\theta (\sin\theta)^{\frac{4\beta_1}{\beta_2}-1} d\theta. \tag{6.2.24}$$

This integral should be evaluated numerically in general, but for the limit ed purpose of calculating the boundary-layer thickness from the definition Eq. (6.2.22) we have the analytical result

$$\delta = (6\beta_1 Q_{ij})^{-\frac{1}{2}}. \tag{6.2.25}$$

Numerical data for q as a function of s for a number of different combinations of λ and γ are available in this case (Batchelor and Wen 1982)[12]. The explicit expression Eq. (6.2.20) for the 'constant' c is found, using Eq. (6.2.25) and recalling that $b = 1$ for sedimenting spheres, it becomes

$$c = \frac{1}{3}(6\beta_1)^{\frac{1}{2}(\alpha-1)}\beta_2^{\alpha+1} \int_0^\pi \left\{ \frac{(\sin\theta)^{\frac{4\beta_1}{\beta_2}}}{2\tau} \right\}^{\frac{1}{2}(\alpha+1)} \sin\theta d\theta. \tag{6.2.26}$$

All quantities appearing in expression Eq. (6.2.26) for c are now known, since $\beta_1 = L_1 = (dL/ds)_{s=2}, \beta_2 = M(\delta)$ and $\alpha = 1 - M(\delta)/L_1$. Therefore, E_{ij} can be found finally from Eq. (6.2.21).

Figure 5 of Chapter 6 shows all calculated results. The four solid lines show the variation of E_{ij} with respect to Q_{ij} for spheres of the same density and four different size ratios, $\lambda = 0.9, 0.5, 0.25, 0.125$ (the case $\gamma = 1, \lambda = 1$ being singular because two identical spheres have zero relative velocity, see section 5.2 for further comment on this case). When $\lambda = 1$ the pair-distribution function is independent of γ, and the long-dashed line represents the value of E_{ij} for this case.

In 1984 Davis calculated the gravitational capture efficiency under the same condition as Wen and Batchelor's, but using the traditional trajectory analysis . [31] Thus it provided an opportunity to check the validity of the new method proposed by Wen and Batchelor.[13] Figure 6 of Chapter 6 shows a comparison of the two methods. The solid lines and the dashed lines in Figure 6 of Chapter 6 are the results calculated by the trajectory analysis (Davis 1984)[31], while the dotted lines in this figure are the results calculated by the new method. These asymptotic results, obtained in a very different manner, are in close agreement with Davis' results, provided $Q_{ij} > 10^3$, showing the validity of the new method. Incidentally, from the boundary-layer theory we may see further the reason why the van der Waals force are a more important factor than the effects of mean free path in overcoming the Stokes resistance of the lubrication film for the general aerosol particles. This can be understood explicitly by considering the parameter $\delta_{m'}$

$$\delta_m = \frac{(6L_1 Q_{ij})^{\frac{1}{2}}}{2\lambda_m/(a_i + a_j)},$$

which is the ratio of the boundary-layer thickness to the mean free path scaled by the

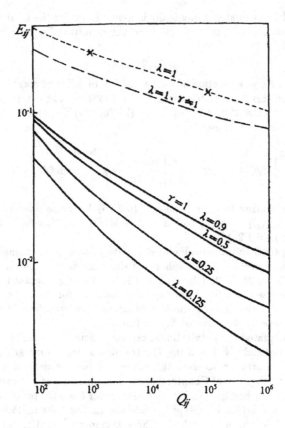

Fig. 5. The calculated asymptotic relation between the capture efficiency E_{ij} and the dimensionless number Q_{ij} measuring the relative magnitude of effects of the externally imposed motion and the van der Waals attraction. (From Wen and Batchelor 1984 and 1985)[13]

average radius of the two spheres. When δ_m is large the interparticle forces become important at a certain value of the gap size for which Maxwell-slip effects are minor; for small δ_m the situation is reversed. Moreover, since Q_{ij} is proportional to a_i^4, δ_m is then inversely proportional to the first power of a_i, we may expect that the van der Waals force will be the most important for small particles (i.e. most of aerosol particles).

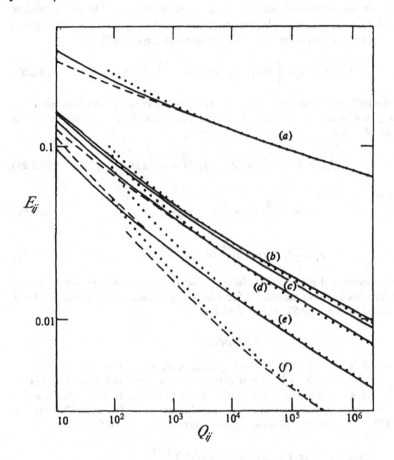

Fig. 6. The gravitational capture efficiency E_{ij} as a function of Q_{ij} for $(a)\lambda = 1, \gamma \neq 1$ (arbitrary); $(b)\gamma = 1, \lambda = 0.9$; $(c)\gamma = 1, \lambda = 3/4$; $(d)\gamma = 1, \lambda = 1/2$; $(e)\gamma = 1, \lambda = 1/4$; $(f)\gamma = 1, \lambda = 1/8$. The dashed lines are the asymptotic results; the solid lines are from the complete numerical solution; the dotted lines are reproduced from Wen and Batchelor (1984).[13] (From Davis 1984)[31]

2.4. Results for a Dispersion in Steady Axisymmetric Pure Straining Motion

We consider only the case of spheres of uniform size here, and their density is not relevant. The principal rates of strain of the bulk motion are chosen as $-E, 1/2E, 1/2E$, where $E > 0$ (see Figure 4 of Chapter 6), in conformity with our convention that particles are entering the boundary layer normally at $\theta = 0$. Thus it is an axisymmetric compressional motion under consideration. The representative relative speed $V_{ij}^{(0)}$ appearing in the dimensionless groups $\mathcal{P}_{ij}$ and Q_{ij} is defined as $(a_i + a_j)E/2$. The expression for $\mathbf{V}_{ij}$ is (Batchelor and Green 1972)[34]

$$\mathbf{V}_{ij} = \mathbf{E} \cdot \mathbf{r} - \left\{ A(s)\frac{\mathbf{rr}}{r^2} + B(s)\left(\mathbf{I} - \frac{\mathbf{rr}}{r^2}\right)\right\} \cdot \mathbf{E} \cdot \mathbf{r}, \qquad (6.2.27)$$

where $\mathbf{E}$ is the bulk rate of the strain tensor , and the behaviour of the function $1 - A$ and B near $\xi = 0$ is similar to that of the functions L and M. We see from this expression for $\mathbf{V}_{ij}$ that

$$f_1(\theta) = \frac{1}{4} + \frac{3}{4}\cos 2\theta, \qquad f_2(\theta) = \frac{3}{4}\sin 2\theta. \qquad (6.2.28)$$

It follows that

$$h(\theta) = \frac{4}{3}(\tan\theta)^{\frac{\beta_1}{3\beta_2}}(\sin 2\theta)^{\frac{\beta_1}{\beta_2}-1}, \qquad (6.2.29)$$

and that

$$\tau = \frac{4}{9}\beta_2 \int_0^\theta (\tan\theta)^{\frac{2\beta_1}{3\beta_2}}(\sin 2\theta)^{\frac{2\beta_1}{\beta_2}-1}d\theta, \qquad (6.2.30)$$

which must be evaluated numerically. In this case f_2 changes sign at $\theta = \pi/2$, and this is therefore the limiting value of θ for which τ is defined. The boundary layer thickness defined by Eq. (6.2.22) is found to be

$$\delta = (6\beta_1 Q_{ij})^{-\frac{1}{2}}$$

again, as indeed it should be. The integration with respect to θ in expression Eq. (6.2.26) for the constant c here can extend to $\theta = \pi/2$, and since the flux to the portion of the surface of the sphere $\pi/2 \leq \theta \leq \pi$ is clearly half the whole, so we may simply double the value of the integral over the range $0 \leq \theta \leq \pi/2$. Then giving b the value $8/3\sqrt{3}$ appropriate to axisymmetric pure straining, we find

$$c = \frac{\sqrt{3}\beta_2^{\alpha+1}}{4(6\beta_1)^{\frac{1}{2}(\alpha-1)}} \int_0^{\frac{\pi}{2}} \left\{\frac{(\tan\theta)^{\frac{2\beta_1}{3\beta_2}}(\sin 2\theta)^{\frac{2\beta_1}{\beta_2}}}{2\tau}\right\}^{\frac{1}{2}(\alpha+1)} \sin\theta d\theta, \qquad (6.2.31)$$

This complicated expression requires numerical evaluation.

Values of the function $q(s)$ for a bulk pure straining motion have been calculated by Batchelor and Green (1972)[34], with the aid of data for the two scalar functions $A(s)$ and $B(s)$ in Eq. (6.2.27) obtained by Lin, Lee and Sather (1970).

The constant β_1 in Eq. (6.2.9) is here given by

$$\beta_1 = 2\lim_{\xi \to 0} \frac{1-A}{\xi} = 8.154. \qquad (6.2.32)$$

The two parameters α and β_2 which are regarded as slowly–varying functions of Q_{ij} were determined with the same procedure as for the case of sedimenting spheres. The expression $q(s)$ which is the analogue of Eq. (6.2.2) is (Batchelor and Green 1972)[34]

$$q(s) = \frac{1}{1-A} \exp\left\{ \int_r^\infty \frac{2(B-A)}{r(1-A)} dr \right\},$$ (6.2.33)

and since the slowly–varying functions $B(\xi)$ is approximately constant with value $B(\delta)$ over most of the boundary layer it follows that

$$\alpha = 1 - \frac{3B(\delta)}{\beta_1}.$$ (6.2.34)

Similarly for the parameter β_2 define by Eq. (6.2.9) we have

$$\beta_2 = 2[1 - B(\delta)].$$ (6.2.35)

The functions $h(\theta)$ and $\tau(\theta)$ in Eq. (6.2.29), Eq. (6.2.30) can be evaluated and finally the 'constant' c given by Eq. (6.2.31) can be found. The capture efficiency $E_{ij} = c\delta q(\delta)$ can thus be calculated.

The calculated relation between E_{ij} and Q_{ij} is shown in Figure 5 of Chapter 6. A bulk pure straining motion is evidently more effective in promoting coagulation than differential settling due to gravity . (The short dashed line represents the values of E_{ij} for this case, and is higher than any curves which represent the values of gravitational capture efficiency E_{ij}). This appears to be a consequence of the larger values of the convection-dominated pair-distribution function near the inner boundary generated by a bulk pure straining motion .

Calculation of the relative particle trajectories of two equal-sized spheres in a bulk axisymmetric pure straining motion and the associated values of the capture efficiency for two values of Q_{ij} viz. 10^3 and 10^5 respectively have been reported by Zeichner and Schowalter (1977)[32]. The motion considered by Zeichner and Schowalter was an axisymmetric extensional flow , with one positive principal rate of strain, by contrast here we have considered an axisymmetric compressional motion . However the two sets of results can be compared because it can be shown by Wen and Batchelor's method, that the total flux F_{ij} is the same for the two cases. The proof amounts to a reworking of the above analysis for the extensional motion with $\theta = 0$ and $\tau = 0$ now at the equator (where particles are approaching the inner boundary normally), the result being the same expression Eq. (6.2.31) for c. The fact that the total flux F_{ij} is unchanged by reversal of the convective motion is remarkable, because the distribution of the flux density over the inner boundary is quite different in two cases, but is not a surprise since the same kind of result has been obtained in several problems of mass transfer from a body in a given flow field (see Batchelor 1979[33] in particular for analogous results for flow with a large Péclet number). The comparison with Zeichner and Schowalter's result shown in Figure 6 of Chapter 6 suggests that the asymptotic results obtained by Wen and Batchelor's new method are again acceptably accurate both at $Q_{ij} = 10^3$ and $Q_{ij} = 10^5$. (The cross symbols in Figure 6 of Chapter

6 represent the values of E_{ij} calculated by Zeichner and Schowalter[32] using the traditional trajectory analysis).

2.5. The Problem of $M(\delta)$ Approximation

The comparisons with Davis (1984)[31] and Zeichner & Schowalter's (1977)[32] results show preliminarily the validity of Wen and Batchelor's new method. However the transverse scalar function of the relative velocity $M(\xi)$ assumed by Wen and Batchelor (1983, 1984, 1985)[13] to be approximately a constant $M(\delta)$ within the boundary layer , is actually a function of ξ. According to Eq. (4.3.13) it should be (Jeffrey and Onisi 1984)[9]

$$M(\xi) = \frac{M_0(\log \xi^{-1})^2 + M_1 \log \xi^{-1} + M_3}{(\log \xi^{-1})^2 + e_1 \log \xi^{-1} + e_2} + O[\xi(\log \xi^{-1})^3]. \qquad (6.2.36)$$

It follows from Eq. (6.2.36) that the asymptotic form of the outer solution of the pair-distribution function near $\xi = 0$ is no longer a negative power of ξ like Eq. (6.2.3) and Eq. (6.2.13), but is (Wen and Lin 1987)[35]

$$q = \frac{q_0}{\xi^\alpha(\log \xi^{-1} - e_3)^{\beta_1}(\log \xi^{-1} - e_4)^{\beta_2}} \quad \text{as} \quad \xi \to 0, \qquad (6.2.37)$$

where

$$\left.\begin{aligned}
\alpha &= 1 - \frac{M_0}{L_1}, \\[2mm]
e_3 &= \frac{-e_1 + \sqrt{e_1^2 - 4e_2}}{2}, e_4 = \frac{-e_1 - \sqrt{e_1^2 - 4e_2}}{2}, \\[2mm]
\beta_1 &= \frac{1}{L_1}\left(\frac{2M_3 - M_1 e_1 + M_0 e_1^2 - 2M_0 e_2}{2(e_1^2 - 4e_2)} + \frac{M_1 - M_0 e_1}{2}\right), \\[2mm]
\beta_2 &= \frac{1}{L_1}\left(-\frac{2M_3 - M_1 e_1 + M_0 e_1^2 - 2M_0 e_2}{2(e_1^2 - 4e_2)} + \frac{M_1 - M_0 e_1}{2}\right),
\end{aligned}\right\} \qquad (6.2.38)$$

with L_1 being the same as before, i.e. the coefficient in the asymptotic form of $L(\xi)$ near $\xi = 0$, viz. $L = L_1\xi + O(\xi^2 \log \xi^{-1})$.

Thus the equation for φ Eq. (6.2.8) becomes (Wen and Lin 1987)[35],

$$-L_1(\eta \cos \theta + \frac{1}{\eta})\frac{\partial \varphi}{\partial \eta}$$

$$+\frac{\sin \theta}{2}\frac{M_0(\log(\delta\eta)^{-1})^2 + M_1 \log(\delta\eta)^{-1} + M_3}{(\log(\delta\eta)^{-1})^2 + e_1 \log(\delta\eta)^{-1} + e_2}\frac{\partial \varphi}{\partial \theta}$$

$$= \frac{L_1\varphi}{\eta^2}[-(\alpha + 1) + \frac{\beta_1}{\log(\delta\eta)^{-1} - e_3} + \frac{\beta_2}{\log(\delta\eta)^{-1} - e_4}], \qquad (6.2.39)$$

where the enlarged radial coordinate in the boundary layer η is defined as $\eta = \xi/\delta$.

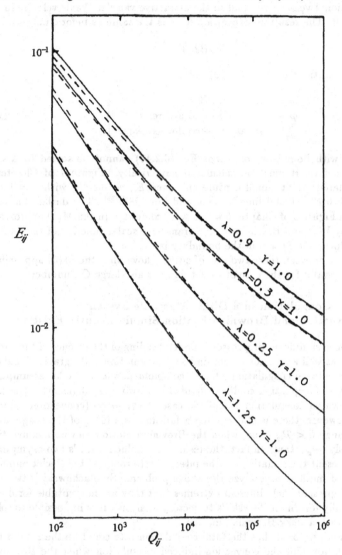

Fig. 7. The gravitational capture efficiency E_{ij} as a function of Q_{ij}. The solid lines are calculated from the actual variation of M with Eq. (6.6.36) in boundary layer . The dashed lines are reproduced in the $M(\delta)$ approximation suggested by Wen and Batchelor (From Wen and Lin 1987)[35]

Here the boundary-layer thickness δ is defined as the location at which the longitudinal relative gravitational velocity is equal to the attractive van der Waals velocity in the direction of $\theta = 0$. The resulting expression for δ is the same as before, i.e.

$$\delta = (6L_1 Q_{ij})^{-1/2}.$$

The boundary conditions for Eq. (6.2.39) are

$$\left. \begin{array}{lll} \varphi = 0 & \text{at} & \eta = 0, \\ \varphi \to 1 & \text{as} & \eta \to \infty \text{ upstream,} \\ \frac{\partial \varphi}{\partial \eta} \to 0 & \text{as} & \eta \to \infty \text{ downstream.} \end{array} \right\} \qquad (6.2.40)$$

Eq. (6.2.39) with boundary conditions Eq. (6.2.40) cannot be solved by a similar method as before, it must be calculated numerically. Figure 7 of Chapter 6 shows the calculated gravitational capture efficiency E_{ij} versus Q_{ij} with $\gamma = 1$ and $\lambda = 0.9, 0.5, 0.25, 0.125$. (solid lines). (Wen and Lin 1987)[35]. The dashed lines are reproduced from Figure 5 of Chapter 6 which were computed in the $M(\delta)$ approximation suggested by Wen and Batchelor to the transverse scalar function of the relative gravitational velocity $M(\xi)$ within the boundary layer .

The agreement between these two sets of curves shows that the $M(\delta)$ approximation is sufficiently valid for the case of Péclet number and large Q_{ij} number.

3. The Rate of Coagulation in a Dilute Unstable System When Convection and Brownian Motion Simultaneously Exist

There have been significant progresses of understanding of the essence of the Brownian coagulation as well as the convection-induced coagulation—like gravitational coagulation , shear-induced coagulation, etc., since Smoluchowski made his assumption that all the cases of coagulation could be divided into two main divisions —the case of $\mathcal{P}_{ij} \to 0$ (Brownian coagulation), and the case of $\mathcal{P}_{ij} \to \infty$ (convection-induced coagulation). However, there is still missing a full understanding of the coagulation process in the range $0 < \mathcal{P}_{ij} < \infty$, when the Brownian motion and bulk convection are simultaneously important. In fact, the results of Smoluchowski 's two asymptotic limit cannot represent the situations of the intermediate range of the Péclet number. We need a unified method to analyze the above problem. Smoluchowski 's two different methods represent such different extremes that they are not suitable for doing this. And especially for Smoluchowski 's trajectory analysis , it is impossible to solve a problem even with a weak Brownian motion .

In 1959, Mason proposed that the total coagulation rate could be the sum of the Brownian coagulation and the convection-induced coagulation when the Brownian and convection motion are both at present. Swift and Friedlander (1964) concluded on the basis of their experiments that the two processes were linearly independent over a wide range of conditions. Then it would be proper for us to sum the two coagulation rate as the total coagulation rate when the two processes are simultaneously important. However, investigations made in recent years by van de Ven and

Mason (1977)[29], Feke and Schowalter (1983)[36], Melik and Fogler (1984)[39], Wang and Wen (1989, 1990)[37,14], Wen and Zhang (1990)[38], Wen ,Zhang and Lin (1991)[15] have arrived at the same conclusion that is contrary to the above assumption.

All the research work mentioned above used the method of the complete equation for the pair-distribution function . Indeed, this is the only unified method which we need to solve the coagulation problem in the whole range $0 \leq \mathcal{P}_{ij} \leq \infty$. The results of section 6.2 already indicated that the results of the capture efficiency obtained in the method of the boundary-layer equation for the pair-distribution function as $\mathcal{P}_{ij} \to \infty$ agree with those obtained from the trajectory analysis . We now use this method to solve some coagulation problems with the Péclet number being finite.

3.1. The Rate of Coagulation at $\mathcal{P}_{ij} \gg 1$ but $\neq \infty$

In this paragraph, we mainly consider the case of strong gravitation–weak Brownian coagulation rate. This work was done recently by Wen and Zhang (1990)[38], and Wen, Zhang and Lin(1991)[15]. In order to find the effect of weak Brownian motion on the gravitational coagulation, we need to solve the Eq. (4.4.27) in its steady–state form, with boundary condition Eq. (4.4.29) and Eq. (4.4.31), since the dispersion is still a dilute statistically homogeneous polydisperse system. The size of the particles is still in the range 2μm $-$ 10μm for aerosol system to ensure the Péclet number and Q_{ij} number being sufficiently large, and the Reynolds number and the Stokes number being sufficiently small.

It has been proved by Feke and Schowalter (1983)[36] in their calculation of weak Brownian/strong shear-induced coagulation that this kind of problems is a regular perturbation problem, since a regular perturbation expansion in p_{ij} suffices not only to match the outer boundary condition, but also to match the inner boundary condition. It is also true for the case of weak Brownian/Strong gravity -induced coagulation. The Brownian diffusion term is a small quantity compared to the van der Waals force within the boundary layer . Therefore it must be a regular perturbation problem for the case of weak Brownian/Strong gravity -induced coagulations.

Expanding the solution p_{ij} in inverse Péclet number , we have

$$p_{ij} = p_{ij}^{(0)} + \mathcal{P}_{ij}^{-1} p_{ij}^{(1)} + O(\mathcal{P}_{ij}^{-2}). \qquad (6.3.1)$$

Substituting Eq. (6.3.1) into the dimensionless steady–state form of Eq. (4.4.27) the perturbation equations for $p_{ij}^{(0)}$ and $p_{ij}^{(1)}$ are obtained,

$$\tilde{\nabla} \cdot \left\{ \tilde{\mathbf{V}}_{ij} - \frac{1}{Q_{ij}} \tilde{\mathbf{D}}_{ij} \cdot \tilde{\nabla} \tilde{\Phi}_{ij} \right\} p_{ij}^{(0)} = 0, \qquad (6.3.2)$$

$$\tilde{\nabla} \cdot \left\{ \tilde{\mathbf{V}}_{ij} - \frac{1}{Q_{ij}} \tilde{\mathbf{D}}_{ij} \cdot \tilde{\nabla} \tilde{\Phi}_{ij} \right\} p_{ij}^{(1)} = \tilde{\nabla} \cdot \{ \tilde{\mathbf{D}}_{ij} \cdot \tilde{\nabla} p_{ij}^{(0)} \}, \qquad (6.3.3)$$

with boundary conditions

$$\left.\begin{array}{ll} p_{ij}^{(0)} = p_{ij}^{(1)} = 0 & \text{at } s = 2, \\ p_{ij}^{(0)} \to 1, p_{ij}^{(1)} \to 0 & \text{as } s \to \infty \text{ upstream,} \\ \frac{\partial p_{ij}^{(0)}}{\partial s} = \frac{\partial p_{ij}^{(1)}}{\partial s} = 0 & \text{as } s \to \infty \text{ dowmstream.} \end{array}\right\} \qquad (6.3.4)$$

The tilde refers to dimensionless quantities.

When $\mathcal{P}_{ij} \gg 1, Q_{ij}$ must also be much larger than unity, since Q_{ij} is proportional to $\mathcal{P}_{ij}$ with the constant of proportionality kT/A being of order unity (see equation (5.2.1)), then the solution $p_{ij}^{(0)}$ of Eq. (6.3.2) is actually $q\varphi$, where φ is the analytical solution Eq. (6.2.15) derived by Wen and Batchelor. In their computation Feke and Schowalter also included the case of $Q_{ij} = O(1)$, even $Q_{ij} < O(1)$ besides the case of $Q_{ij} \gg 1$. Of course, Feke and Schowalter recognized that the conditions where $Q_{ij} = O(1)$, and $\mathcal{P}_{ij} \gg 1$, may not be realized in practice, since one expects $kT/A = O(1)$. They did their calculation in this way because they hoped to facilitate a comparison of the results to earlier work in which Q_{ij} is finite but the Brownian motion is ignored.[36]

We will calculate $p_{ij}^{(0)}, p_{ij}^{(1)}$ by making use of Feke and Schowalter method which has been improved by introduction of the analytical solution φ Eq. (6.2.15) to $p_{ij}^{(0)}$. We rewrite Eq. (6.3.2) and Eq. (6.3.3) as

$$F_1 \frac{\partial p_{ij}^{(0)}}{\partial s} + F_2 \frac{\partial p_{ij}^{(0)}}{\partial \theta} = F_3 p_{ij}^{(0)}, \qquad (6.3.5)$$

$$F_1 \frac{\partial p_{ij}^{(1)}}{\partial s} + F_2 \frac{\partial p_{ij}^{(1)}}{\partial \theta} = F_3 p_{ij}^{(1)} + \tilde{\nabla} \cdot (\tilde{D}_{ij} \cdot \tilde{\nabla} p_{ij}^{(0)}). \qquad (6.3.6)$$

Now the problem is axisymmetric about the direction **g**, hence the azimuthal angle is irrelevant, only the polar angle θ is referred. In Eq. (6.3.5), Eq. (6.3.6), and

$$\begin{aligned} F_1 &= [\tilde{V}_{ij} - \frac{1}{Q_{ij}}\tilde{D}_{ij} \cdot \tilde{\nabla}(\tilde{\Phi}_{ij})] \cdot \mathbf{e}_s \\ &= -L(s)\cos\theta - \frac{G(s)}{Q_{ij}}\frac{d\tilde{\Phi}_{ij}}{ds}, \end{aligned} \qquad (6.3.7)$$

$$F_2 = \frac{\tilde{V}_{ij} \cdot \mathbf{e}_\theta}{s} = \frac{M(s)\sin\theta}{s}, \qquad (6.3.8)$$

$$\begin{aligned} F_3 &= -\tilde{\nabla} \cdot [\tilde{V}_{ij} - \frac{1}{Q_{ij}}\tilde{D}_{ij} \cdot \tilde{\nabla}(\tilde{\Phi}_{ij})] \\ &= \frac{2}{s}[L(s)\cos\theta + \frac{G(s)}{Q_{ij}}\frac{d\tilde{\Phi}_{ij}}{ds}] + \frac{dL(s)}{ds}\cos\theta \\ &+ \frac{1}{Q_{ij}}[\frac{dG(s)}{ds}\frac{d\tilde{\Phi}_{ij}}{ds} + G(s)\frac{d^2\tilde{\Phi}_{ij}}{ds^2}] - \frac{2M(s)}{s}\cos\theta, \end{aligned} \qquad (6.3.9)$$

and

$$\begin{aligned} \tilde{\nabla} \cdot (\tilde{D}_{ij} \cdot \tilde{\nabla}p_{ij}^{(0)}) &= \frac{2}{s}G(s)\frac{\partial p_{ij}^{(0)}}{\partial \theta} \\ &+ \frac{dG(s)}{ds}\frac{\partial p_{ij}^{(0)}}{\partial \theta} + G(s)\frac{\partial^2 p_{ij}^{(0)}}{\partial s^2} \\ &+ \frac{H(s)}{s}\frac{\partial^2 p_{ij}^{(0)}}{\partial \theta^2} + \frac{H(s)}{s^2\tan\theta}\frac{\partial p_{ij}^{(0)}}{\partial \theta}. \end{aligned} \qquad (6.3.10)$$

Rearranging Eq. (6.3.5), and Eq. (6.3.6) into a coupled system of ordinary differential equations, one obtains

$$\frac{ds}{F_1} = \frac{d\theta}{F_2} = \frac{dp_{ij}^{(0)}}{F_3 p_{ij}^{(0)}} = \frac{dp_{ij}^{(1)}}{F_3 p_{ij}^{(1)} + \tilde{\nabla} \cdot (\tilde{D}_{ij} \cdot \tilde{\nabla} p_{ij}^{(0)})}. \tag{6.3.11}$$

The first set of equations in Eq. (6.3.11) represents relative sphere trajectories in absence of the Brownian motion , and since the functions F_i are known, $p_{ij}^{(0)}$ can be calculated along the trajectories. However $p_{ij}^{(0)}$ needs not to be calculated numerically in the present case. In contrary the solutions for $p_{ij}^{(1)}$ are of course more difficult to obtain because of the presence $\tilde{\nabla} \cdot (\tilde{D}_{ij} \cdot \tilde{\nabla} p_{ij}^{(0)})$ in the case of Feke and Schowalter.[36] It needs to be calculated on a trajectory bundle in close proximity to a primary trajectory, along which the term $\tilde{\nabla} \cdot (\tilde{D}_{ij} \cdot \tilde{\nabla} p_{ij}^{(0)})$ could be estimated.

The calculations of $\tilde{\nabla} \cdot (\tilde{D}_{ij} \cdot \tilde{\nabla} p_{ij}^{(0)})$ are now simplified owing to the introduction of φ to $p_{ij}^{(0)}$ in the present case. Then

$$\begin{aligned}
\tilde{\nabla} \cdot (\tilde{D}_{ij} \cdot \tilde{\nabla} p_{ij}^{(0)}) &= \tfrac{2}{s} G \tfrac{dq}{ds} + \tfrac{dG}{ds} \tfrac{dq}{ds} \\
&+ G \tfrac{d^2 q}{ds^2} \qquad \text{when} \quad s - 2 > 10\delta
\end{aligned} \tag{6.3.12}$$

and

$$\begin{aligned}
&\tilde{\nabla} \cdot (\tilde{D}_{ij} \cdot \tilde{\nabla} p_{ij}^{(0)}) \\
&= \left[\frac{2}{s} G + \frac{dG}{ds} \right] \left[\frac{dq}{ds} \varphi + q \frac{\partial \varphi}{\partial s} \right] \\
&+ G \left[\varphi \frac{d^2 q}{ds^2} + 2 \frac{dq}{ds} \frac{\partial \varphi}{\partial s} + q \frac{\partial^2 \varphi}{\partial s^2} \right] \\
&+ \frac{H}{s^2} q \frac{\partial^2 \varphi}{\partial \theta^2} + \frac{H}{s^2 \tan \theta} q \frac{\partial \varphi}{\partial \theta} \qquad \text{when} \quad s - 2 > 10\delta
\end{aligned} \tag{6.3.13}$$

where δ is the boundary-layer thickness defined in Eq. (6.2.25).

Because of all the functions on the right-hand-side of Eq. (6.3.12) and Eq. (6.3.13) being known, $\tilde{\nabla} \cdot (\tilde{D}_{ij} \cdot \tilde{\nabla} p_{ij}^{(0)})$ can then be evaluated along only the primary trajectories. The whole bundle of trajectories is no longer needed.

By the definition of the rate of coagulation F_{ij} Eq. (6.1.1), and of the capture efficiency E_{ij} Eq. (6.1.14), we may see that

$$E_{ij} = E_{ij}^{(0)} + \mathcal{P}_{ij}^{-1} E_{ij}^{(1)} + O(\mathcal{P}_{ij}^{-2}), \tag{6.3.14}$$

where

$$E_{ij}^{(0)} = c\delta q(\delta), \tag{6.3.15}$$

(see Eq. (6.2.21)) and

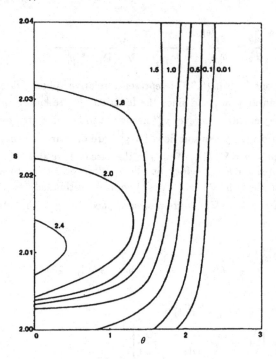

Fig. 8. Contours of constant $p_{ij}^{(0)}$ in relative gravitational convection with $\gamma = 1, \lambda = 1/2, Q_{ij} = 10^4$. (From Wen, Zhang and Lin 1991)[15]

$$E_{ij}^{(1)} = \frac{s^2}{4} \int_A \left\{ -p_{ij}^{(1)} \tilde{\mathbf{V}}_{ij} + \frac{1}{Q_{ij}} p_{ij}^{(1)} \tilde{\mathbf{D}} \cdot \tilde{\nabla} \tilde{\Phi}_{ij} \right. $$
$$\left. + \quad \tilde{\mathbf{D}}_{ij} \cdot \tilde{\nabla} p_{ij}^{(0)} \right\} \cdot \mathbf{n} dA, \tag{6.3.16}$$

where s is the radius of a spherical surface A enclosing the test sphere i and A is concentered with the test sphere i .

Curves of constant $p_{ij}^{(0)} (\equiv q\varphi)$ are shown in Figure 8 of Chapter 6 for the case of $\gamma = 1, \lambda = 1/2$, $Q_{ij} = 10^4$. As one would expect, a maximum exits for p_{ij}, and occurs at $\theta = 0$ and $s \approx 2.01$. In the wake of the test sphere i $(\theta \approx \pi)$ $p_{ij}^{(0)}$ drops significantly below unity. Again, as one would expect from intuition and from Figure 8 of Chapter 6, diffusion effects are minimal at large particle separations because of the homogeneity of the pair-distribution function .

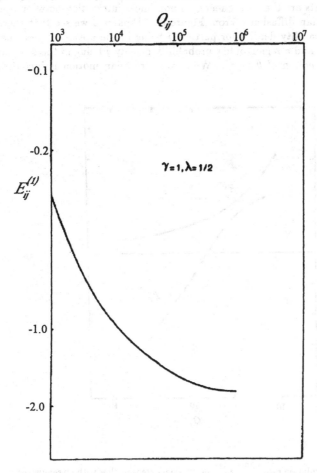

Fig. 9. Predicted the Brownian correct factor $E_{ij}^{(1)}$ for the gravitational capture efficiency with $\gamma = 1, \lambda = 1/2$, and no repulsive potential. (From Wen, Zhang and Lin 1991)[15]

Numerical results for E_{ij} in Eq. (6.3.14) are shown in Figure 9 of Chapter 6. It is found from Figure 9 of Chapter 6 that the Bownian motion can decrease the gravitational capture efficiency , i.e. $E_{ij}^{(1)} < 0$. This result agrees qualitatively with those obtained by Feke and Schowalter (1983)[36] when they calculated the effects of weak Brownian diffusion on the uniaxial extensional flow -induced coagulation rate and on the simple shear-induced coagulation rate provided $Q_{ij} \gg O(1)$. Figure 10 and 11 of Chapter 6 show this result.

The above results are the consequences of non-linear interaction between the convection and Brownian diffusion . From Figure 8 of chpater 6 we see that there is a region of high probability density for particle j being in the upstream area (i.e. the direction of $\theta \approx 0$), and a region of low probability density j being in the downstream area, (i.e. the direction of $\theta \approx \pi$). When the Brownian motion is included, the

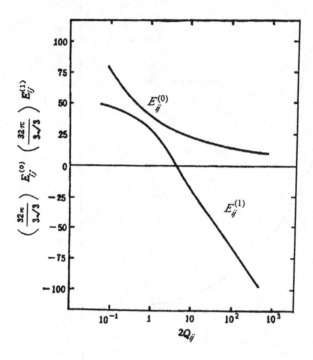

Fig. 10. Predicted coagulation rate $E_{ij}^{(0)}$, $E_{ij}^{(1)}$ for uniaxial extension flow with no repulsive potential. (From Feke and Schowalter 1983)[36]

Brownian diffusion will thus transfer particle j from the upstream area along the stream lies (see Figure 4 of Chapter 6 (a)) into the downstream area, from where the convection forces it to move downstream without being caught by the particle i. It thus deceases the coagulation rate. Clearly, this is a process which interacts closely with the strength of the basic flow . In all cases shown above it is immediately seen that the idea of additivity, suggested by Swift and Friedlander (1964) does not apply. The additivity hypothesis proposed by Swift and Friedlander is not valid in that regime.

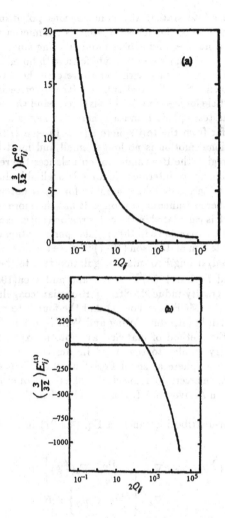

Fig. 11. Predicted coagulation rates for simple shear flow with no repulsive potential. (a) $E_{ij}^{(0)}$ (b) $E_{ij}^{(1)}$. (From Feke and Schowalter 1983)[36]

3.2. The Rate of Coagulation at $\mathcal{P}_{ij} \ll 1$ but $\neq 0$

In this paragraph, we will mainly consider the case of weak gravity -induced/Strong

Brownian coagulation . This problem was solved by Melik & Fogler (1984)[38] and Wang and Wen (1989,1990) [37,14].

The dispersion is still a dilute statistically homogeneous polydisperse of small rigid spherical particles. The size of the aerosol system is submicron size to ensure the Péclet number , Reynolds number and Stokes number being small.

We will still use the Eq. (4.4.27) in its steady–state form with boundary conditions Eq. (4.4.29) and Eq. (4.4.31). In contrast with the above case, here the problem is a singular perturbation problem. Near the test sphere i (the inner region) Brownian motion balances the interparticle force —van der Waals force—and the relative gravity -induced motion between the test sphere i and a sphere j is negligible when Péclet number is small. However, far from the test sphere i ($r > (a_i + a_j)/2\mathcal{P}_{ij}$, the outer region) the relative gravitational motion is no longer small and must be taken into account. Then in the outer region the Brownian motion balances the relative gravity -induced motion, and the influence of interparticle force is negligible due to its rapid decay. Thus, an expansion in terms of Péclet number for p_{ij} is not valid for large distances of sphere j from sphere i (inner expansion). It has therefore to be matched by a separate expansion which is calculated in the outer region (outer expansion). The method of matched asomptotic expansion s in the singular perturbation theory is then used. Using this method van de Ven and Mason (1977)[29] were the first who calculated the case of weak shear-induced/strong Brownian coagulation rate to the second term of order $\mathcal{P}_{ij}^{1/2}$, and Melik and Fogler (1984)[38] and Wang and Wen(1989,1990)[39,14] calculated the case of weak gravity-induced/ Strong Brownian coagulation rate up to the second term of order $\mathcal{P}_{ij}$ (Melik and Fogler) and the fourth term of order $\mathcal{P}_{ij}^2$ (Wang and Wen). We will mainly introduce Wang and Wen's work in this section.

The basic procedure of the method of matched asomptotic expansion s used in Wang and Wen's work is very similar to that used in Acrivos and Taylor's work (1962)[40] on mass transfer to a sphere at small Péclet number. It is of interest to note that the two problems have a certain connection. The result of coagulation rate under consideration agrees with Acrivos and Taylor's result, when the radius of sphere j approaches to zero.

The equation for the pair-distribution function Eq. (4.4.27) in its dimensionless steady-state form is

$$
\mathcal{P}_{ij}\nabla_s \cdot \{\frac{\mathbf{V}_{ij}}{V_{ij}^{(0)}}\} \;-\; \nabla_s \cdot \left\{ p_{ij}\frac{\mathbf{D}_{ij}}{D_{ij}^{(0)}} \cdot \nabla_s(\frac{\Phi_{ij}}{kT}) \right\}
$$
$$
-\; \nabla_s \cdot \left\{ \frac{\mathbf{D}_{ij}}{D_{ij}^{(0)}} \cdot \nabla_s p_{ij} \right\} = 0
\tag{6.3.17}
$$

with the boundary conditions

$$
p_{ij} = 0 \text{ at } s = 2, \quad p_{ij} \rightarrow 1 \text{ as } s \rightarrow \infty.
\tag{6.3.18}
$$

The scalar functions L, M, G, H in $\mathbf{V}_{ij}$ and $\mathbf{D}_{ij}$ have been obtained in Chapter 4. Because of the decomposition of the solution into an "inner" and an "outer" expansions, whose necessity has been mentioned above, only the far field asymptotic forms for them are needed.

We substitute the far field asymptotic expressions for the mobility functions (Jeffrey and Onishi 1984)[9] into the scalar functions L, M, G, H given by Batchelor (1982).[11] The following far field asymptotic forms are obtained, viz.

$$L(s) = 1 + \frac{L_1}{s} + O(s^{-3}), \qquad (6.3.19)$$

$$M(s) = 1 + \frac{M_1}{s} + O(s^{-3}), \qquad (6.3.20)$$

$$G(s) = 1 + \frac{G_1}{s} + O(s^{-3}), \qquad (6.3.21)$$

$$H(s) = 1 + \frac{H_1}{s} + O(s^{-3}), \qquad (6.3.22)$$

where

$$L_1 = \frac{3(1 - \lambda^3 \gamma)}{(\lambda^2 \gamma - 1)(1 + \lambda)}, \text{ and } M_1 = \frac{L_1}{2}, \qquad (6.3.23)$$

$$G_1 = \frac{-6\lambda}{(1 + \lambda)^2}, \text{ and } H_1 = \frac{G_1}{2}. \qquad (6.3.24)$$

Note that the definitions of the coefficients L_1, M_1, G_1, H_1, are different from those mentioned in the previous sections. Here they are coefficients of L, M, G, H in the far field asymptotic form whereas they are coefficients of L, M, G, H in the near field asymptotic form discussed in the previous sections.

The divergence term of $\mathbf{V}_{ij}/V_{ij}^{(0)}$ is given by Eq. (4.3.9), and the far field asymptotic form for the scalar function $W(s)$ in Eq. (4.3.9) is given by Eq. (4.3.11).

We now turn to the problem of the interparticle potential Φ_{ij}. Only the case of rapid flocculation is considered here. The interparticle potential is thus dominated by the attractive van der Waals potential. The expression for Φ_{ij} is given by Eq. (4.3.21) (Hamaker 1937)[17]. It is easy to show that the far field asymptotic form for the van der Waals potential of Eq. (4.3.21) is[14]

$$\Phi_{ij} = -\frac{1024}{9} \frac{A\lambda^3}{(1 + \lambda)^6} \frac{1}{s^6} + O(s^{-8}). \qquad (6.3.25)$$

Of course the Hamaker's expression for the van der Waals potential does not consider the effects of retardation. van de Ven and Mason (1977)[29] said: "... the interaction energy at large particle separations is determined by the van der Waals attraction. At such separations the forces are retarded and V_{int} (i.e. Φ_{ij} in this book) is proportional to s^{-2}...". To include the retardation effects in the outer region is certainly necessary. However, the fairly slow decay of the van der Waals potential proposed by van de Ven and Mason (1977)[29], and then by Melik and Fogler (1984)[38] in their outer region analysis cannot be correct, since the retardation effects are to weaken but not to strengthen the van der Waals force. The decay of the retarded van der Waals potential must be more rapid than s^{-6}. For the case of equalized system,

Feke and Schowalter (1983) cited a far field asymptotic expression for the retarded van der Waals potential as follows[36],

$$\Phi_{ij} = -\frac{16A}{9s^6}\left(\frac{2.45}{p} - \frac{2.04}{p^2}\right) \qquad s \gg 1. \qquad (6.3.26)$$

In this expression $p = 2\pi(s-2)/\lambda_L$, where λ_L is the dimensionless London wavelength scaled on the radius of the spheres. Just as we expect, Eq. (6.3.26) does show a more rapid s^{-7} decay as $s \to \infty$.

Although decay as s^{-2} is not correct, we shall see later that it does not affect Melik and Fogler's two-term expansion result (1984) [38]. However, it will certainly affect the third and the fourth terms of the expansion, which are calculated in this section.

In the case of rapid flocculation, the repulsive potential V_R is approximated by a thin double layer potential (hydrosol dispersion, see section 5.2). For an unequal-sized system Melik and Fogler (1984)[38] gave the following form

$$V_R = \pm\frac{1}{2}\epsilon_0 a_i \psi_0^2 \frac{\lambda}{1+\lambda}\ln[\pm\exp(-\tau(s-2))], \qquad (6.3.27)$$

where ϵ_0 is the dielectric strength, ψ_0 the surface potential, and the parameter τ is the dimensionless reciprocal of the Debye-Hückel double layer thickness scaled on the reciprocal of the average radius $(a_i + a_j)/2$. $\tau \gg 1$, since Eq. (6.3.27) is valid only for thin double layer s.

From Eq. (6.3.27) we may see that the far field asymptotic form for the thin double layer repulsive potential is

$$V_R = O(e^{-\tau s}) \quad \text{as} \quad s \to \infty. \qquad (6.3.28)$$

Obviously, the decay of this type of repulsive potential is much more rapid than the negative power decay of the attractive van der Waals potential.

3.3. Inner Expansion and Outer Expansion for p_{ij}

We shall calculate a four-term inner expansion and a three-term outer expansion for p_{ij}. We choose a spherical polar coordinate system such that its origin is at the centre of the test sphere i and the direction of the polar axis coincides with $V_{ij}^{(0)}$. Thus the problem is axisymmetric about the polar axis, and θ is the polar angle.

According to the method of the matched asomptotic expansion s, we construct "inner" and "outer" expansions. The inner expansion is assumed to be of the form

$$p_{ij} = \sum_{n=1}^{\infty} p_{ij}^{(n)}(s, \theta)t_n(\epsilon) \quad \text{with} \quad t_1(\epsilon) = 1, \qquad (6.3.29)$$

where the perturbation parameter $\epsilon \equiv \mathcal{P}_{ij}$, $t_n(\epsilon)$ $(n = 1, 2, \cdots)$ and $\hat{t}_n(\epsilon)$ $(n = 1, 2, \cdots)$ in the outer expansion Eq. (6.3.31) are gauge functions s (van Dyke 1975), which are not necessarily simple powers of ϵ and for the moment are restricted only by requirements

$$\lim_{\epsilon \to 0} \frac{t_{n+1}(\epsilon)}{t_n(\epsilon)} = 0, \text{ and } \lim_{\epsilon \to 0} \frac{\widehat{t}_{n+1}(\epsilon)}{\widehat{t}_n(\epsilon)} = 0.$$

The inner equation is the same as Eq. (6.3.17). The set of equations for $p_{ij}^{(n)}$ $(n = 1, 2, \cdots)$ can be obtained by substitution of Eq. (6.3.29) into Eq. (6.3.17). The boundary conditions imposed on $p_{ij}^{(n)}$ are

$$p_{ij}^{(n)} = 0 \ (n \geq 1) \quad \text{at} \quad s = 2.$$

These boundary conditions are insufficient to uniquely determine $p_{ij}^{(n)}$. However, additional conditions at $s \to \infty$ are furnished by matching the inner and the outer expansion in their common domains of validity.

The outer expansion for p_{ij} is assumed to be of the form,

$$\widehat{p}_{ij} = \sum_{n=1}^{\infty} \widehat{p}_{ij}^{(n)}(\rho, \theta)\widehat{t}_n(\epsilon) \quad \text{with} \quad \widehat{t}_1(\epsilon) = 1. \tag{6.3.31}$$

The contracted radial coordinate ρ with $\rho = \epsilon s$, is introduced in the outer region so that the perturbation parameter ϵ can be scaled out in the outer equation, then we have

$$\nabla_\rho \cdot \left\{ \widehat{p}_{ij} \frac{\mathbf{D}_{ij}}{D_{ij}^{(0)}} \cdot \nabla_\rho \left(\frac{\Phi_{ij}}{kT} \right) + \frac{\mathbf{D}_{ij}}{D_{ij}^{(0)}} \cdot \nabla_\rho \widehat{p}_{ij} - \frac{\mathbf{V}_{ij}}{V_{ij}^{(0)}} \widehat{p}_{ij} \right\} = 0. \tag{6.3.32}$$

The set of equations for $\widehat{p}_{ij}^{(n)}$ $(n = 1, 2, \cdots)$ can also be obtained by substituting the expansion Eq. (6.3.31) into Eq. (6.3.32). The boundary conditions are

$$\widehat{p}_{ij}^{(1)} \to 1, \quad \widehat{p}_{ij}^{(n)} = 0 \ (n \geq 2) \quad \text{as} \quad \rho \to \infty. \tag{6.3.33}$$

These boundary conditions are also insufficient. However, additional conditions at $\rho \to 0$ are imposed by the requirements that the outer and inner expansion s be matched.

To see the similarity of the problem to the mass transfer problem, we now look for the equation for the first outer expansion term $\widehat{p}_{ij}^{(1)}$.

Substituting Eq. (6.3.31) into Eq. (6.3.32) and taking the leading term yields the equation for the first term of the outer expansion. Putting the far field asymptotic forms for $\mathbf{V}_{ij}, \mathbf{D}_{ij}$, and Φ_{ij}, Eq. (6.3.19)—Eq. (6.3.22) and Eq. (6.3.26) in the resulting equation and again taking the leading term one obtains

$$\nabla_\rho^2 \widehat{p}_{ij}^{(1)} - \frac{\mathbf{V}_{ij}^{(0)}}{V_{ij}^{(0)}} \cdot \nabla_\rho \widehat{p}_{ij}^{(1)} = 0. \tag{6.3.34}$$

The van der Waals attractive potential term disappears in the equation due to its rapid decay as s^{-6}. The Eq. (6.3.34) with constant diffusivity and a uniform stream field is very similar to the convective–diffusion equation in the mass transfer problem. Hence it is not surprising that the problem of coagulation at small Péclet number can be tackled by the method used in the mass transfer problem at small Péclet number . The method of Acrivos and Taylor (1962)[40] is thus used here.

Following the procedure of Acrivos and Taylor (1962)[40], it can be shown that for the inner expansion the solutions are

$$t_2(\epsilon) = \epsilon, \quad t_3(\epsilon) = \epsilon^2 \ln \epsilon, \quad t_4(\epsilon) = \epsilon^2, \tag{6.3.35}$$

and

$$p_{ij}^{(1)} = e^{-\frac{\Phi_{ij}}{kT}} \left\{ 1 - 2C_\varphi \int_s^\infty \frac{e^{\frac{\Phi_{ij}}{kT}}}{s^2 G(s)} ds \right\}. \tag{6.3.36}$$

$$p_{ij}^{(2)} = C_\varphi p_{ij}^{(1)} + Q(s) P_1(\cos \theta), \tag{6.3.37}$$

$$p_{ij}^{(3)} = \frac{5H_1 - 4M_1}{3} C_\varphi p_{ij}^{(1)}, \tag{6.3.38}$$

$$p_{ij}^{(4)} = R_0^{(4)} + R_1^{(4)} P_1(\cos \theta) + R_2^{(4)} P_2(\cos \theta), \tag{6.3.39}$$

where $P_l(\cos \theta)(l = 0, 1, 2)$ are the Legendre polynomials. The solution $p_{ij}^{(1)}$ given by Eq. (6.3.36) is the same as the zero- Péclet-number solution Eq. (6.1.9). This is a pure diffusion result. The scalar function $Q(s)$ in the particular solution for the second inner expansion term $p_{ij}^{(2)}$ (see Eq. (6.3.37)) satisfies the following ordinary differential equation

$$\frac{d}{ds} \left\{ s^2 G \left[Q \frac{d}{ds} \left(\frac{\Phi_{ij}}{kT} \right) + \frac{dQ}{ds} \right] \right\} - 2HQ$$
$$= \frac{2C_\varphi L}{G} + s^2 p_{ij}^{(1)} \left[W - L \frac{d}{ds} \left(\frac{\Phi_{ij}}{kT} \right) \right], \tag{6.3.40}$$

with the boundary conditions

$$Q = 0 \quad \text{at} \quad s = 2, \quad Q \to C_\varphi \quad \text{as} \quad s \to \infty. \tag{6.3.41}$$

The complementary solution $R_0^{(4)}$ for the fourth inner expansion term $p_{ij}^{(4)}$ is given by

$$\begin{aligned} R_0^{(4)} =\ & A_0^{(4)} p_{ij}^{(1)} + \frac{1}{3} e^{-\frac{\Phi_{ij}}{kT}} \int_2^s e^{\frac{\Phi_{ij}}{kT}} \left\{ \frac{LQ}{G} \right. \\ & + C_\varphi \left[(s-2) + (4M_1 - 5H_1) \log \frac{s}{2} \right] \frac{d}{ds} \frac{\Phi_{ij}}{kT} \\ & \left. + C_\varphi \left[1 + \frac{4M_1 - 5H_1}{s} \right] \right\} ds - \frac{1}{3} C_\varphi \left[(s-2) \right. \\ & \left. + (4M_1 - 5H_1) \log \frac{s}{2} \right], \end{aligned} \tag{6.3.42}$$

where

$$\begin{aligned} A_0^{(4)} =\ & C_\varphi \left(\frac{2M_1 - 3H_1}{2} + \frac{5H_1 - 4M_1}{3} C_E - \frac{4M_1 - 3H_1}{3} \log 2 + C_\varphi - \frac{2}{3} \right) \\ & - \frac{1}{3} \int_2^\infty e^{\frac{\Phi_{ij}}{kT}} \left\{ \frac{LQ}{G} + C_\varphi \left[(s-2) + (4M_1 - 5H_1) \log \frac{s}{2} \right] \frac{d}{ds} \left(\frac{\Phi_{ij}}{kT} \right) \right. \\ & \left. + C_\varphi \left[1 + \frac{4M_1 - 5H_1}{s} \right] \right\} ds. \end{aligned} \tag{6.3.43}$$

Here C_E is the Euler constant $C_E = 0.577216$. The precise forms of the scalar functions $R_1^{(4)}$ and $R_2^{(4)}$ in the particular solutions of the fourth term $p_{ij}^{(4)}$ of the inner expansion are not required here, since only the coagulation rate is considered in this section.

In the outer region, the first three terms of the outer expansion are found to be

$$\hat{t}_2(\epsilon) = \epsilon, \quad \hat{t}_3(\epsilon) = \epsilon^2, \tag{6.3.44}$$

and

$$\hat{p}_{ij}^{(1)} = 1, \tag{6.3.45}$$

$$\hat{p}_{ij}^{(2)} = -\frac{2C_\varphi}{\rho} e^{-\frac{1}{2}\rho(1-\cos\theta)}, \tag{6.3.46}$$

$$\hat{p}_{ij}^{(3)} = e^{\frac{1}{2}\rho\cos\theta}\{e^{-\frac{1}{2}\rho}\sum_{l=0}^{2}\pi B_l^{(3)} P_l(\cos\theta)\sum_{m=0}^{l}\frac{(l+m)!}{(l-m)!m!}\frac{1}{\rho^{(m+1)}}\}$$

$$+\sum_{l=0}^{2} u_l^{(3)}, \tag{6.3.47}$$

where the integral constants $B_l^{(3)}(l = 0, 1, 2)$ are

$$B_0^{(3)} = \frac{C_\varphi}{\pi}\left(H_1 - 2C_\varphi - \frac{5H_1 - 4M_1}{3}C_E\right), \tag{6.3.48}$$

$$B_1^{(3)} = -\frac{2C_\varphi}{\pi}(M_1 - H_1)(C_E - 1), \tag{6.3.49}$$

$$B_2^{(3)} = \frac{C_\varphi}{3\pi}(H_1 - 3M_1)(3 - C_E), \tag{6.3.50}$$

and the particular solutions $u_l^{(3)}(l = 0, 1, 2)$ are

$$u_0^{(3)} = C_\varphi P_0(\cos\theta)\left\{\frac{2H_1}{\rho^2}e^{-\frac{1}{2}\rho}\right.$$

$$\left. - \frac{5H_1 - 4M_1}{3}\left[\frac{e^{\frac{1}{2}\rho}}{\rho}\int_\rho^\infty \frac{e^{-x}}{x}dx + \frac{\log\rho}{\rho}e^{-\frac{1}{2}\rho}\right]\right\}, \tag{6.3.51}$$

$$u_1^{(3)} = 2C_\varphi(M_1 - H_1)P_1(\cos\theta)\left[\frac{1}{\rho}\left(1 - \frac{2}{\rho}\right)e^{\frac{1}{2}\rho}\int_\rho^\infty \frac{e^{-x}}{x}dx\right.$$

$$\left. -\frac{1}{\rho}\left(1 + \frac{2}{\rho}\right)e^{-\frac{1}{2}\rho}\log\rho - \frac{2}{\rho^2}e^{-\frac{1}{2}\rho}\right], \tag{6.3.52}$$

$$u_2^{(3)} = -\frac{1}{3}C_\varphi(H_1 - 2M_1)P_2(\cos\theta)\left[\frac{1}{\rho}\left(1 - \frac{6}{\rho}\right.\right.$$

$$\left. + \frac{12}{\rho^2}\right)e^{\frac{1}{2}\rho}\int_\rho^\infty \frac{e^{-x}}{x}dx + \frac{1}{\rho}\left(1 + \frac{6}{\rho} + \frac{12}{\rho^2}\right)e^{-\frac{1}{2}\rho}\log\rho \tag{6.3.53}$$

$$\left. + \frac{6}{\rho^2}\left(1 + \frac{6}{\rho}\right)e^{-\frac{1}{2}\rho}\right].$$

3.4. A Four-Term Expansion for the Dimensionless Coagulation Rate

For the case of $\mathcal{P}_{ij} \ll 1$, it is usual to express the rate of the doublet formation in terms of the dimensionless coagulation rate Nusselt number $\mathcal{N}_{ij}$ defined as the ratio of F_{ij} (see Eq. (6.1.1)) to the value of F_{ij} at $\mathcal{P}_{ij} = 0$, viz.

$$\mathcal{N}_{ij} = \frac{F_{ij}}{4\pi(a_i+a_j)C_\varphi D_{ij}^{(0)} n_j}$$

$$= \frac{1}{C_\varphi} \int_0^\pi \left\{ G \left[p_{ij} \frac{d}{ds} \left(\frac{\Phi_{ij}}{kT} \right) + \frac{\partial p_{ij}}{\partial s} \right] \right\}_{s=2} \sin\theta d\theta. \tag{6.3.54}$$

Expanding p_{ij} in terms of Legendre polynomials $P_l(\cos\theta)$, it is evident that by virtue of the orthogonality of $P_l(\cos\theta)$, only those $P_0(\cos\theta)$ terms in p_{ij}—say p_{ij}^0 —contribute to the flux integral in Eq. (6.3.54). Thus we have

$$\mathcal{N}_{ij} = \frac{2}{C_\varphi} \lim_{s\to 2} \left\{ G[p_{ij}^0 \frac{d}{ds} \left(\frac{\Phi_{ij}}{kT} \right) + \frac{dp_{ij}^0}{ds} \right\}. \tag{6.3.55}$$

Substituting the inner expansion Eq. (6.3.29) with Eq. (6.3.35)–Eq. (6.3.39) into Eq. (6.3.55), one can obtain a four-term expansion $\mathcal{N}_{ij}$, namely

$$\mathcal{N}_{ij} = 1 + C_\varphi \mathcal{P}_{ij} + \frac{2\lambda^4\gamma - 3\lambda^3\gamma + 3\lambda - 2}{(\lambda^2\gamma - 1)(1+\lambda)^2} C_\varphi \mathcal{P}_{ij}^2 \log \mathcal{P}_{ij}$$
$$+ A_0^{(4)} \mathcal{P}_{ij}^2 + o(\mathcal{P}_{ij}^2), \tag{6.3.56}$$

where the coefficient of the fourth term $A_0^{(4)}$ is given by Eq. (6.3.43). The first two terms of the four-term expansion Eq. (6.3.56) agree with the results of Melik and Fogler (1984)[38]. Their incorrect s^{-2} decay of the van der Waals potential does not affect the two terms of the expansion, since the s^{-2} decay makes the interparticle potential in the equations for the first two terms of the outer expansion disappear. However the s^{-2} decay will certainly affect the third and the fourth term which are calculated in this section, since in that case there will be an inner expansion term appearing as a inhomogeneous term in the corresponding equations for the outer expansion , whereas the more rapid s^{-6} decay will still make them disappear.

In order to gain a better understanding of the coupled Brownian and gravity - induced coagulation process, the dimensionless coagulation rate $\mathcal{N}_{ij}$ as a function of the Péclet number has been calculated for a typical aerosol dispersion in which $\gamma = 1, \lambda = 1/2, A = 5 \times 10^{-13}$erg., and $kT = 4 \times 10^{-14}$erg., (then $A/kT = 12.5$) (Davis 1984)[31]. From Eq. (6.1.11) and Eq. (6.3.43), it can be shown that $C_\varphi = 0.7724$, and $A_0^{(4)} = -0.09135$. Figure 12 of Chapter 6 shows the computed results for $0 \leq \mathcal{P}_{ij} \leq 1$. Curve 1 in Figure 12 of Chapter 6 corresponds to the result of Melik and Fogler 1984. The term $C_\varphi \mathcal{P}_{ij}$ is the leading term of the effects of gravity -induced motion on the Brownian coagulation rate. It is always a positive term. In the outer region, the leading term of the relative gravitational velocity of the two spheres is $\mathbf{V}_{ij}^{(0)}$, which

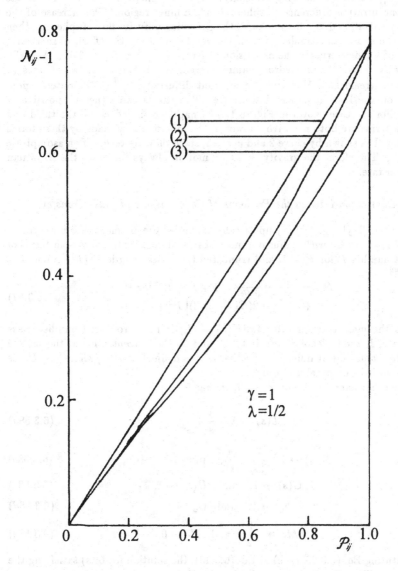

Fig. 12. The dimensionless coagulation rate at small Péclet number s (1) $\mathcal{N}_{ij} = 1 + C_\varphi \mathcal{P}_{ij}$ (2) the first three terms of expansion Eq. (6.3.56), (3) the four-term expansion Eq. (6.3.56) (From Wang and Wen 1990)[14]

is a uniform stream field. The uniform stream field transfers sphere j from infinity to $s \sim O(\mathcal{P}_{ij}^{-1})$ without changing the concentration of sphere j. It thus increases the overall concentration difference of sphere j in the inner region. The increase of the overall concentration difference enhances the Brownian diffusive flux of sphere j , then enhances the Brownian coagulation rate. Curves 2 and 3 show the effects of successive additions of further terms in the expansion of Nusselt number Eq. (6.3.56). Actually they show the effects of the hydrodynamic interaction s between the two spheres on the uniform stream field $\mathbf{V}_{ij}^{(0)}$ and the constant diffusivity $D_{ij}^{(0)}$ in the outer region, and then on the Brownian coagulation rate. The effects can either be positive or negative. For the case shown in Figure 12 of Chapter 6 the effects of the third and the fourth terms are both negative. However, the modifications made by these terms must be small as shown by curve 2 and curve 3, since they are terms of an asomptotic expansion . Therefore, the gravity -induced motion always increases the Brownian coagulation rate.

3.5. The Connection between the Problems of Coagulation and Mass Transfer

Acriros and Taylor (1962) asymptotically expanded the dimensionless mass transfer rate N for a sphere with radius a immersed in a uniform external flow in terms of the Péclet number P for $P \ll 1$, and truncated the series to order $O(P^2)$,then N is given by[40]

$$N = \frac{F}{4\pi a K (C_1 - C_0)} = 1 + P + 2P^2 \log P$$
$$+ \left[\frac{121}{240} + 2(C_E + \log 2) \right] P^2 + o(P^2). \tag{6.3.57}$$

Here F is the mass transfer rate, $4\pi a K(C_1 - C_0)$ is the zero-Péclet-number mass transfer rate, K the molecular diffusivity, C_1 and C_0 the concentration at the surface of the sphere and that at infinity respectively. P is defined as $aU_\infty / 2K$, where U_∞ is the velocity of the external uniform flow .

On the other hand, as $\lambda \to 0, a_j \to 0$, we have

$$L(s) \to 1 - \frac{3}{s} + \frac{4}{s^3}, \tag{6.3.58a}$$

$$M(s) \to 1 - \frac{3}{2s} - \frac{4}{s^3}, \quad \text{and} \quad W(s) \to 0, \tag{6.3.58b}$$

$$G(s) \to 1, \quad \text{and} \quad H(s) \to 1, \tag{6.3.58c}$$

$$\Phi_{ij} \to 1, \quad \text{and} \quad C_\varphi \to 1, \tag{6.3.58d}$$

$$M_1 \to -\frac{3}{2}, \quad \text{and} \quad H_1 \to 0. \tag{6.3.58e}$$

Substituting Eq. (6.3.28 $a - e$) in Eq. (6.3.40), the solution for $Q(s)$ satisfying the boundary condition Eq. (6.3.41) can be found to be

$$Q(s) = -1 + \frac{3}{s} - \frac{3}{s^2} + \frac{2}{s^3}. \tag{6.3.59}$$

Then substitution of Eq. (6.3.58 $a - e$) and Eq. (6.3.59) in Eq. (6.3.43) yields

$$A_0^{(4)} = \frac{121}{240} + 2(C_E + \log 2). \qquad (6.3.60)$$

With the results Eq. (6.3.58 d) and Eq. (6.3.60), the four-term expansion for the dimensionless coagulation rate Eq. (6.3.56) reduces to

$$\mathcal{N}_{ij} = \frac{F_{ij}}{4\pi a_i D_{ij}^{(0)} n_j} = 1 + \mathcal{P}_{ij} + 2\mathcal{P}_{ij}^2 \log \mathcal{P}_{ij}$$
$$\qquad\qquad (6.3.61)$$
$$+ \left[\frac{121}{240} + 2(C_E + \log 2) \right] \mathcal{P}_{ij}^2 + o(\mathcal{P}_{ij}^2).$$

The Péclet number $\mathcal{P}_{ij}$ now reduces to $a_i V_i^{(0)} / 2 D_{ij}^{(0)}$, where $V_i^{(0)}$ is the Stokes terminal velocity of the test sphere i under gravity. Comparing Eq. (6.3.57) with Eq. (6.3.61), it appears that Eq. (6.3.61) agrees with Eq. (6.3.57), if we make the following assumption that as $\lambda \to 0, a_j \to 0$, then $a_i \to a, D_{ij}^{(0)} \to K, n_j \to (C_1 - C_0), V_i^{(0)} \to U_\infty$.

The fact that the coagulation rate in the small limit of the radius of sphere j agrees with the mass transfer rate is remarkable, but is not surprising. In fact, it is easy to understand from the view point of the physical model for coagulation. When $\lambda \to 0, a_j \to 0$, the effect of sphere j on the flow field due to the settling of sphere i disappears, and a sphere j moves in the same way as a fluid point. Thus the flow field tends to that produced by a sphere immersed in a given uniform flow. As $\lambda \to 0, \mathbf{D}_{ij}$ tends to $D_{ij}^{(0)} \mathbf{I}, \Phi_{ij} \to 0(A \neq 0$ is required). Thus the coagulation model formally reduces to the mass transfer model.

4. Discussion

In the above paragraphs, we have used the far field asymptotic form for the van der Waals potential Eq. (6.3.25) without considering the retardation effects and the repulsive potential. However, it is not difficult to show that the form of the four-term expansion for Nusselt number $\mathcal{N}_{ij}$ Eq. (6.3.56) would be unchanged even when we include the effects of retardation and the repulsive potential.

According to the DLVO theory (Derjaguin and Landau 1941, Verwey and Overbeek 1948), the total interparticle potential can be obtained by summing the attractive and repulsive potential. From Eq. (6.3.28) we have seen that for the case of rapid flocculation, the decay of the thin double layer potential is much more rapid then s^{-6}, as $s \to \infty$. Hence the repulsive potential term cannot appear in the outer equations, and the form of the resulting expansion for $\mathcal{N}_{ij}$ should be the same as before. Of course, the values of C_φ and $A_0^{(4)}$ should be changed when the repulsive potential is considered.

The far field asymptotic form of the retarded van der Waals potential for an unequal-sized system seems to be not available. The expression Eq. (6.3.26) is valid only for equal-sized systems. Perhaps the right-hand-side of Eq. (6.3.26) should be multiplied by a numerical coefficient which depends on λ, and the dimensionless London wavelength λ_L should be scaled by the average radius $(a_i + a_j)/2$ when unequal-sized system is considered. In the case of the near field asymptotic expression, the numerical coefficient is $4\lambda/(1 + \lambda)^2$ (Davis 1984)[31], and the dimensionless London wavelength is just scaled by $(a_i + a_j)/2$ (Melik and Fogler 1984). If the above supposition is right, then the decay of the retarded attractive potential would still be s^{-7} as $s \to \infty$, and would still make it disappear in the outer equations. The same is true even if the above supposition is not right. At any rate, the decay of the retarded van der Waals potential should be more rapid than s^{-6}, since retardation effect is to weaken the van der Waals potential. Thus the form of the four-term expansion for $\mathcal{N}_{ij}$ Eq. (6.3.56) would be unchanged even if we include retardation effects. The only things which changed are the values of C_φ and $A_0^{(4)}$. They should be smaller than those obtained from Eq. (4.3.21) and Eq. (6.3.25), since both retardation and repulsive potential contribute negative effects on the van der Waals attractive potential . Incidentally, the incorrect s^{-2} decay used by Melik and Fogler (1984)[38] should also affect the value of C_φ (i.e. W_{Br}^{-1} in their paper). The values of their C_φ would possibly be larger than the actual values of C_φ due to the fairly slow s^{-2} decay of the retarded attractive potential .

Finally we shall return to the problem of additivity assumption made by Swift and Friedlander (1964). From Eq. (6.1.10), we see that the Brownian flux $(F_{ij})_{Br}$ is equal to $4\pi(a_i + a_j)C_\varphi D_{ij}^{(0)} n_j$. From Eq. (6.1.1) and Eq. (6.1.15) we see that gravitational flux $(F_{ij})_{Gr}$ is equal to $\pi(a_i + a_j)^2 E_{ij} V_{ij}^{(0)} n_j$. According to additivity assumption , the total coagulation rate in its dimensionless form, the Nusselt number $\mathcal{N}_{ij}$, should be

$$\mathcal{N}_{ij} = \frac{(F_{ij})_{Br} + (F_{ij})_{Gr}}{4\pi(a_i + a_j)C_\varphi D_{ij}^{(0)} n_j} = 1 + \frac{E_{ij}}{2C_\varphi}\mathcal{P}_{ij}. \qquad (6.4.1)$$

Comparing Eq. (6.3.56) with Eq. (6.4.1) we note that (a) the expansion Eq. (6.4.1) due to the additivity assumption ends at the second term, whereas the expansion Eq. (6.3.56) deduced by the method of the complete equation for p_{ij} which are solved by the singular perturbation technique can have more terms; and (b) the coefficients of the second term of the two expansions are different, the additivity is justified only when $E_{ij} = 2C_\varphi^2$. The comparisons made by Melik and Fogler (1984)[38] indicated that under certain conditions the additivity assumption gives results which are in an approximate agreement with the singular perturbation analysis, even though, this is only a coincidence, and there is no physical basis for the additivity being justified.

On the other hand for the case of weak shear-induced/strong Brownian coagulation, van de Ven and Mason (1977) gave the following result using a singular perturbation technique to solve the complete equation for the pair-distribution function , viz.[29]

$$\mathcal{N}_{ij} = 1 + c_1 \mathcal{P}_{ij}^{1/2} + o(\mathcal{P}_{ij}^{1/2}), \qquad (6.4.2)$$

whereas by the additivity assumption they obtained

$$\mathcal{N}_{ij} = 1 + c_2 \mathcal{P}_{ij}, \qquad (6.4.3)$$

where c_1 and c_2 are coefficients of the second terms. A striking feature of Eq. (6.4.3)

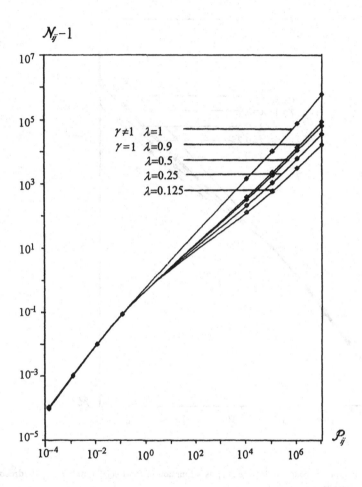

Fig. 13. The dimensionless coagulation rate $\mathcal{N}_{ij}$ as a function of the Péclet number $\mathcal{P}_{ij}$ (aerosol) (Qiao and Wen,1996[44])

is that it is different in form from Eq. (6.4.2). Therefore, van de Ven and Mason concluded that there is no theoretical foundation for the additivity assumption . For the whole range of the Péclet number $0 < \mathcal{P}_{ij} < \infty$, the complete equation method

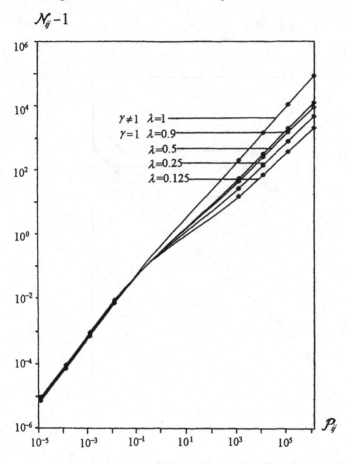

Fig. 14. The dimensionless coagulation rate $\mathcal{N}_{ij}$ as a function of the Péclet number $\mathcal{P}_{ij}$ (hydrosol) (Qiao and Wen,1996[44])

for the pair-distribution function should be the unique method to solve the problems of coagulation, when both convection and Brownian motion simultaneously exit.

On the basis of the Eq. (6.3.14)and Eq. (6.3.56) and using interpolation method, we obtain the pictures of the dimensionless coagulation rate $\mathcal{N}_{ij}$ as a function of the parameter $\mathcal{P}_{ij}$ over the whole range of the Péclet number including small, intermediate, and large Péclet number for the case of the coupled gravitational and Brownian coagulation.

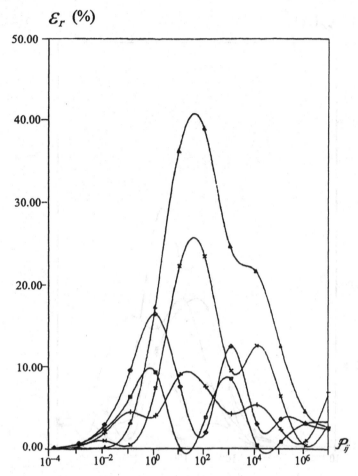

Fig. 15. The error distribution of the dimensionless coagulation rate made by the additivity assmption(aerosol) (Qiao and Wen,1996[44])

Although the first term of the asomptotic expansion for the capture efficiency E_{ij} in the range of large Péclet number is a constant term (see Eq. (6.3.14)), the first term of the corresponding expansion for dimensionless coagulation rate $\mathcal{N}_{ij}$ is by definition directly proportional to $\mathcal{P}_{ij}$,viz. (Qiao and Wen,1996[44])

$$\mathcal{N}_{ij} = \frac{\pi(a_i + a_j)^2 V_{ij}^{(0)} E_{ij} n_j}{4\pi(a_i + a_j) C_\varphi D_{ij}^{(0)} n_j} = \frac{E_{ij}^{(0)}}{2C_\varphi}\mathcal{P}_{ij} + \frac{E_{ij}^{(1)}}{2C_\varphi} + O(\mathcal{P}_{ij}^{-1}). \qquad (6.4.4)$$

Thus, we interpolate Eq. (6.4.4) and Eq. (6.3.56) and obtain the pictures of $\mathcal{N}_{ij}(\mathcal{P}_{ij})$

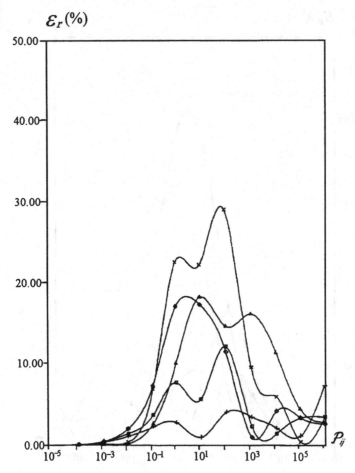

Fig. 16. The error distribution of the dimensionless coagulation rate made by the additivity assmp-tion(hydrosol) (Qiao and Wen,1996[44])

curves (Figure 13 of Chapter 6(aerosols) and Figure 14 of Chapter 6 (hydrosols) Qiao and Wen 1996[44]). The pictures also show the dependence of $\mathcal{N}_{ij}(\mathcal{P}_{ij})$ curves on λ and γ(see Figure 13 of Chapter 6 and Figure 14 of Chapter 6). From Figure 13 and 14 of Chapter 6 we may see that $\mathcal{N}_{ij}$ will decrease as λ decreases when $\mathcal{P}_{ij} \gg 1$. This is because the small j-particles are more easily subjected to the influence of convection, and will tend to be convected around the i-particle and downstream without collection. Then the coagulation rate will decrease as the radius of j-particle decreases. On the other hand, $\mathcal{N}_{ij}$ will increase as λ decreases when $\mathcal{P}_{ij} \ll 1$. This is because in this case the longitudinal scalar function G of the relative Brownian diffusivity tensor is dominant, and G will increase as the radius of j-particle decreases. Thus, the results in the case of $\mathcal{P}_{ij} \ll 1$ turn out contrary to the case of $\mathcal{P}_{ij} \gg 1$. Then, the $\mathcal{N}_{ij}(\mathcal{P}_{ij})$ curves show a turning region in the range of intermediate Péclet number , especially for the case of small λ (Figure 13 and 14 of Chapter 6). Besides, we made a further examination about the validity of the additivity assumption. The numerical results show that the effects of weak Brownian diffusion on gravity -induced coagulation are all minus for the case of $Q_{ij} = 10^3 - 10^6$, and for various λ and γ, and then demonstrate further that the additivity assumption has no theoretical foundation. However, the numerical errors of the dimensionless coagulation rate $\mathcal{N}_{ij}$ made by the additivity assumption are small (at most several per cent) for the case of $\mathcal{P}_{ij} \ll 1$ with the repulsion number being zero (see Figure 15 of Chapter 6(aerosols) and Figure 16 of chpater 6(hydrosols)[44]). This indicates that the additivity assumption has some practical value in this regime. On the other hand, for the case of large Péclet number , especially for the case of intermediate Péclet number , the numerical errors of the dimensionless coagulation rate $\mathcal{N}_{ij}$ made by the additivity assumption càn be as large as 40% (aerosol) (see Figure 15 of Chapter 6, Qiao and Wen 1996[44]) or 30% (hydrosol)(see Figure 16 of Chapter 6, Qiao and Wen 1996[44]) even with the repulsion number being zero. This shows that the additivity assumption has no practical value in this regimes.

CHAPTER 7
SOME OTHER PROBLEMS IN THE DYNAMICS OF AEROSOLS

1. Mass/Heat Transfer from a Particle Suspended in Flow Fields

In Chapter 1 and Chapter 2, it is indicated that the rate of mass transfer of an aerosol particle at small Péclet number can be approximated by the zero-Péclet number mass transfer rate Q_0. Then, the rate of mass transfer (e.g. the condensation rate or evaporation rate of a droplet) is proportional to its radius a. Thus,

$$Q_0 = 4\pi a D_m(C_1 - C_0), \tag{7.1.1}$$

where D_m is the molecular diffusivity of some diffusible quantity (in Eq. (2.4.14) it is replaced by the vapour molecular diffusivity D_0), C_1 is the concentration of that diffusible quantity at the particle surface, and C_0 is the concentration of that diffusible quantity far from the particle.

When the Péclet number is not exactly equal to zero, it is shown by Eq. (2.4.13) that some modifications are needed. And generally speaking, the correction $f(Pe)$ is larger than unity, that is, the convection will increase the rate of mass transfer . To evaluate the values of $f(Pe)$, there have been a lot of literatures since 1960s, and Batchelor (1979) gave a comprehensive summation and made some new developments.[33]

The concentration of the diffusible quantity in the fluid, to be denoted by C, satisfies the convective and diffusion equation

$$\frac{\partial C}{\partial t} + \mathbf{u} \cdot \nabla C = D_m \nabla^2 C, \tag{2.4.7}$$

where $\mathbf{u}$ is the local fluid velocity relative to the axes such that the particle has zero translational velocity. The fluid velocity $\mathbf{u}$ will be regarded as determined by specification of the ambient flow field, i.e. by the velocity $\mathbf{U}$ that the fluid would have in the absence of the particle and to which $\mathbf{u}$ tends as $r(= |\mathbf{x}|) \to \infty$, concretely,

$$\mathbf{u} \to \mathbf{U} \qquad \text{as} \qquad r(= |\mathbf{x}|) \to \infty. \tag{7.1.2}$$

Two alternative forms of the ambient flow field are suggested in practical transfer problems. If the density of the particle is different from that of the fluid, the particle is subject to gravitational and inertial forces which give it a translational motion relative to the fluid. Provided then the variation of the ambient fluid velocity over a distance of one particle dimension is small compared to the particle velocity relative to the fluid, the flow relative to the particle is actually that due to a particle held in a uniform stream. Thus here U is U_0, and is independent of x. If on the other hand the particle has the same density as the fluid, it has no translation velocity relative to the fluid. Provided the linear dimensions of the particle are small compared to distances over which the velocity gradient in the ambient flow field changes significantly, the flow near the particle is then actually that due to a force-free particle immersed in fluid in which the ambient velocity varies linearly with position. Thus here

$$U_i = G_{ij}x_j, \qquad\qquad\qquad (7.1.3)$$

where G is the second-rank velocity gradient tensor. Generally speaking, an aerosol particle is not a force-free particle. Its density is much larger than the air density. Thus, the relative flow near an aerosol particle is actually that due to a particle held in a uniform stream whose velocity is equal to and opposite to its gravitational terminal velocity U_0. If on the other hand the linear ambient velocity $G_{ij}x_j$ is much larger than its terminal velocity U_0, the aerosol particle can be regarded as a force-free particle, and the flow near it is effectively that due to a particle held in a linear ambient flow . Most previous theoretical work on the problem of mass or heat transfer from a particle in a moving fluid concerns the case of uniform ambient velocity. The new developments consider the many-sided case of linear variation of the ambient, velocity.[33]

Since G is a second-rank tensor with eight independent components (when $\nabla \cdot U = 0$), a large number of quite different ambient flow fields are described by this linear variation of U, some of which are of particular interest. The gradient tensor G can as usual be written as

$$G = E + \Omega, \qquad\qquad\qquad (7.1.4)$$

where the symmetrical part E represents a pure straining motion and the anti-symmetrical part Ω represents a rigid-body rotation with angular velocity $\frac{1}{2}\omega$ as

$$\Omega_{ij} = -\frac{1}{2}\epsilon_{ijk}\omega_k,$$

where ω being the vorticity of the ambient flow . The straining motion can be specified by three scalar parameters to determine the orientation of the three orthogonal principal axes of $E(p_1, p_2, p_3$, say; their precise meanings need not be spelt out) and by the three principal rates of strain, E_1, E_2, E_3, only two of which are independent because $E_1 + E_2 + E_3 = 0$. The angular volocity $\frac{1}{2}\omega$ can now be specified by its three components in the directions of the principal axes of the rate-of-strain tensor E, to be denoted by $\Omega_1, \Omega_2, \Omega_3$. A representative magnitude of the rate-of-strain tensor , E, say, will be used to define the Péclet number , and the remaining quantities needed to specify the ambient flow can then be taken as three orientation parameters, one ratio of the principal rates of strain and three angular velocity components.

The orientation of the ambient flow is relevant only when the particle itself has an orientational shape. In the case of a spherical particle, the particle orientation is irrelevant at all Péclet number s. Consequently in these cases there is no dependence of the transfer rate on the orientation parameters p_1, p_2, p_3, and only four parameters remain, aside from the Péclet number . Three of these four remaining parameters disappear for the important class of pure straining motion s (when $\Omega_1 = \Omega_2 = \Omega_3 = 0$), leaving one defining parameter unspecified. And in the case of two-dimensional ambient flow , we have

$$E_3 = 0 \quad (\text{and } E_2 = -E_1), \quad \Omega_1 = 0, \ \Omega_2 = 0,$$

again leaving one defining parameter (viz. Ω_3/E), with the particular case of simple shearing flow corresponding to $\Omega_3/E = \pm 1/\sqrt{2}$.

Solving Eq. (2.4.7) with the boundary conditions

$$\left. \begin{array}{ll} C = C_1 & \text{at the particle surface} \quad A, \\ C = C_0 & \text{as} \quad r \to \infty, \end{array} \right\} \tag{7.1.5}$$

we may obtain the mass transfer rate Q, since it is a flux integral at the particle surface, i,e.

$$Q = - \int_A D_m \nabla C \cdot \mathbf{n} dS, \tag{7.1.6}$$

where $\mathbf{n}$ is the unit outward normal to the surface element dS. The dimensionless mass transfer rate Nu (the Nusselt number) is defined as Q divided by Q_0 (the transfer rate at zero-Péclet number), then we have

$$Nu = \frac{Q}{Q_0} = f(Pe). \tag{2.4.13}$$

Later, we will find the asymptotic solution for $f(Pe)$ at small Péclet number.

For the case of a spherical aerosol particle the solution for Eq. (2.4.7) in a stationary fluid at zero-Péclet number is as follows,

$$C - C_0 = \frac{Q_0}{4\pi D_m r}. \tag{7.1.7}$$

When Pe is larger than zero and smaller than unity, the effect of convection is to modify this distribution at large values of r and to change the transfer rate. Concentration gradients become weaker with increasing distance from the particle, and the convection term $\mathbf{U} \cdot \nabla C$ becomes more important as $r/a \to \infty$ no matter how small the Péclet number may be. On the assumption that spatial differentiation changes the magnitude of a quantity by a factor r^{-1}, we estimate the ratio of the convection and diffusion terms as of order rU/D_m, where U is the magnitude of the ambient velocity at distance r from the particle. Thus at position near the particle, where rU/D_m is the same order of magnitude as the Péclet number Pe, diffusion effects are dominant, whereas at large distances such that $rU/D_m \gg 1$ the local distribution of C is dominated by convection effects. The method of matched asymptotic expansion s can then be used to approach the problem. However, here we adopt a simple intuitive procedure which is adequate for the purpose of obtaining the leading term in the expression for the change in the transfer rate due to convection.

We denote by r_c (which is defined to the order of magnitude only) the large value of r at which the convection and diffusion terms in the governing equation are comparable in magnitude; thus rU/D_m is of order unity at $r = r_c$ (giving $r_c/a = O(Pe^{-1})$ for a particle in translation motion and $r_c/a = O(Pe^{-1/2})$ for a linear ambient velocity field). The mass transfer from the particle surface to $r = r_c$ is brought about primarily by diffusion, and

$$C - C_0' \approx \frac{Q}{4\pi D_m r_c}$$

at $r = r_c$ in a case where the total rate of transfer is Q. The mass transfer in the range $r_c < r < \infty$ on the other hand takes place primarily by the much more efficient process of convection, and the associated change in concentration is negligible. The effect of convection on the total rate of transfer is thus equivalent to an increase of the overall concentration difference, by a fraction

$$\frac{Q}{4\pi D_m r_c (C_1 - C_0')}, \approx \frac{Q}{4\pi D_m r_c (C_1 - C_0)} \sim \frac{a}{r_c}. \tag{7.1.8}$$

And since in the linear pure diffusion problem the rate of transfer is proportional to

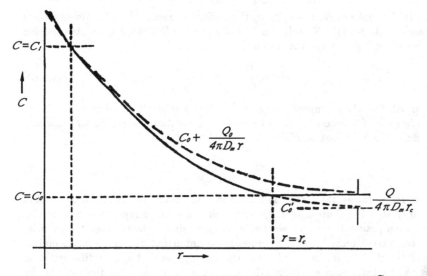

Fig. 1. The unbroken line shows schematically the distribution of concentration C at small Péclet number . For $r < r_c$, where diffusion effects are dominant, the curve is of the form $C_0 + \frac{Q}{4\pi D_m r} +$ const. For $r > r_c$, where convection effects are dominant, C is approximately constant. The broken curve shows the concentration distribution at the zero-Péclet number with the same inner and outer boundary conditions, (From Batchelor 1979[33]).

the overall concentration drop, we see that the fractional increase in the transfer rate due to convection when $Pe \ll 1$ is

$$\frac{\Delta Q}{Q_0} \sim \frac{a}{r_c}. \tag{7.1.9}$$

Figure 1 of Chapter 7 illustrates this simplified description of the concentration distribution (Batchelor 1979)[33]

The above relation Eq. (7.1.9) gives the dependence of ΔQ on the Péclet number (as Pe for a particle in translational motion and as $Pe^{1/2}$ for a linear ambient velocity field) but for numerical information we must match the inner and outer asymptotic regions to be better than only order of magnitude. In the inner region the pure diffusion equation is applicable, with the boundary condition $C = C_1$ at the particle surface A. In the outer region where convection is dominant the fluid velocity $\mathbf{u}$ is approximately equal to the ambient value $\mathbf{U}$ and the governing equation is

$$\mathbf{U} \cdot \nabla C = D_m \nabla^2 C, \tag{7.1.10}$$

with the boundary condition $C \to C_0$ as $r \to \infty$ and an inner boundary condition representing the fact that C is spherically symmetrical and determined by diffusion alone there. The concentration distribution in the outer region is thus the same as if the particle were replaced by a point source of strength Q in the given ambient flow field. In the inner part of this outer region $C - C_0$ is approximately equal to $Q/4\pi D_m r$, and an improved approximation at small values of r, assuming analytic dependence on r, will be

$$C - C_0 \approx \frac{Q}{4\pi D_m r} + \Delta C, \tag{7.1.11}$$

where the second term ΔC is independent of r but may depend on the direction of $\mathbf{x}$ (this departure from spherical symmetry being a consequence of the effect of convection). This is the outer boundary condition for the distribution of concentration in the inner region.

The additional transfer rate from the particle due to convection is now obtainable as a consequence of the fact that in the inner region C satisfies the diffusion equation and an outer boundary condition of the form Eq. (7.1.11) at some large value of r, say $r = R$. We may write C as the sum of two solutions of $\nabla^2 C = 0$, C' and C'' say, where

$$C' = C_1 \quad \text{for} \quad \mathbf{x} \text{ on } A \quad \text{and} \quad C' = C_0 + \frac{Q}{4\pi D_m r} + <\Delta C> \quad \text{at} \quad r = R,$$

and

$$C'' = 0 \quad \text{for} \quad \mathbf{x} \text{ on } A \quad \text{and} \quad C'' = \Delta C - <\Delta C> \quad \text{at} \quad r = R,$$

where $<\Delta C>$ denotes the mean of ΔC over all directions of $\mathbf{x}$. Near $r = R$ the harmonic function C'' can be expressed as a series of spherical harmonics from which those of degree 0 and -1 are excluded, because their means over all directions of $\mathbf{x}$ are non-zero, and so C'' makes no contribution to the transfer across any enclosed surface. Also we know that if the outer boundary condition for C' were $C' = C_0 + Q/4\pi D_m R$ at

$r = R$ the rate of transfer from the particle surface would be Q_0. The rate of transfer from the particle surface associated with C' evidently differs from Q_0 in consequence of a change in the overall concentration difference (between the particle surface and infinity) by the amount $- < \Delta C >$. The additional rate of transfer due to convection is therefore given by

$$\frac{Nu - Nu^0}{Nu^0} = \frac{\Delta Q}{Q} = -\frac{< \Delta C >}{C_1 - C_0}. \qquad (7.1.12)$$

Determination of ΔQ is thus reduced to the problem of finding the term ΔC in the expression Eq. (7.1.11) for the concentration at small values of r in the case of a point source in the given ambient flow field.

1.1. A Particle in Steady Translational Motion through the Fluid

Eq. (7.1.10) is now reduced to

$$\mathbf{U}_0 \cdot \nabla C = D_m \nabla^2 C. \qquad (7.1.13)$$

We require the solution of Eq. (7.1.13) corresponding to a point source of constant strength Q at the origin. By superimposing the distributions of concentration for a sequence of instantaneous point sources in a uniform stream we find

$$C(\mathbf{x}) = C_0 + Q \int_0^\infty \exp\left\{ \frac{-(\mathbf{x} - t\mathbf{U}_0)^2}{4 D_m t} \right\} \frac{dt}{(4\pi D_m t)^{3/2}}.$$

The integration of the above expression can be carried out, giving

$$C(\mathbf{x}) = C_0 + \frac{Q}{4\pi D_m r} \exp\left(\frac{\mathbf{x} \cdot \mathbf{U}_0 - r U_0}{2 D_m} \right). \qquad (7.1.14)$$

Thus, at small values of r,

$$C - C_o \approx \frac{Q}{4\pi D_m r} + \frac{Q U_0}{8\pi D_m^2}(\cos\theta - 1), \qquad (7.1.15)$$

where θ is the angle between the position vector $\mathbf{x}$ and the free-stream velocity $\mathbf{U}_0$.

The second term on right-hand-side of Eq. (7.1.15) can be identified with the quantity ΔC in the general argument from which we may obtain its mean value $< \Delta C >$, i.e.

$$< \Delta C >= \frac{-Q U_0}{8\pi D_m^2}. \qquad (7.1.16)$$

Thus we see that the fractional increase in the transfer rate due to the effect of convection is, to leading order,

$$\frac{Nu - Nu^0}{Nu^0} = \frac{\Delta Q}{Q_0} = \frac{Q_0 U_0}{8\pi D_m^2 (C_1 - C_0)} = \frac{1}{2} Nu^0 Pe, \qquad (7.1.17)$$

where $Pe = aU_0/D_m$ and the Nusselt number Nu^0 at $Pe = 0$ is unity, when $Q_0 = 4\pi a D_m(C_1 - C_0)$.

1.2. A Point Source in Fluid with Steady Linear Ambient Flow Field

Substituting Eq. (7.1.3) for **U** into Eq. (7.1.10) one obtains

$$\frac{\partial C}{\partial t} + G_{ij}x_j\frac{\partial C}{\partial x_i} = D_m\frac{\partial^2 C}{\partial x_i \partial x_i}, \tag{7.1.18}$$

of which the transform is

$$\frac{\partial \widehat{C}}{\partial t} - G_{ij}k_j\frac{\partial \widehat{C}}{\partial k_i} = -D_m k^2 \widehat{C}, \tag{7.1.19}$$

where the Fourier transform $\widehat{C}$ is defined by

$$\widehat{C}(\mathbf{k},t) = \int e^{-i\mathbf{k}\cdot\mathbf{x}}\{C(\mathbf{x},t) - C_0\}d\mathbf{x}. \tag{7.1.20}$$

Instead of trying to solve Eq. (7.1.19) directly it is simpler to note that the solution must be of the form

$$\widehat{C}(\mathbf{k},t) = Q\exp(-D_m k_i k_j B_{ij}), \tag{7.1.21}$$

and to choose the symmetric tensor **B** as a function of t so as to satisfy the equation. Now the expression Eq. (7.1.21) satisfies the Eq. (7.1.19) provided

$$k_i k_j \frac{\mathrm{d}B_{ij}}{\mathrm{d}t} = k^2 + 2k_i k_l G_{ij}B_{jl}$$

which requires

$$\frac{\mathrm{d}B_{ij}}{\mathrm{d}t} = \delta_{ij} + G_{il}B_{jl} + G_{jl}B_{il}. \tag{7.1.22}$$

All the components of **B** can be found as functions of t from this equation and the boundary condition

$$B_{ij} \sim \delta_{ij}t \quad \text{as} \quad t \to 0, \tag{7.1.23}$$

but we shall do this explicitly only for certain linear ambient flow fields.

The Fourier transform of the concentration distribution for a source maintained at the origin of steady strength Q is now found by integrating Eq. (7.1.21)

$$\widehat{C}(\mathbf{k}) = Q\int_0^\infty \exp(-D_m k_i k_j B_{ij})\mathrm{d}t. \tag{7.1.24}$$

The distribution in physical space is the Fourier transform of Eq. (7.1.24), viz.

$$C(\mathbf{x}) = C_0 + \frac{Q}{(4\pi D_m)^{3/2}}\int \exp\left(\frac{-x_i x_j b_{ij}}{4D_m D^2}\right)\frac{\mathrm{d}t}{D^{\frac{1}{2}}}, \tag{7.1.25}$$

where D denotes the determinant of the matrix $\mathbf{B}$ and b_{ij} is the co-factor of the matrix element B_{ij}. This expression for C can be written as

$$C(\mathbf{x}) = C_0 + \frac{Q}{4\pi D_m r} + J(\mathbf{x}), \tag{7.1.26}$$

where

$$J = \frac{Q}{(4\pi D_m)^{3/2}} \int_0^\infty \left\{ \frac{1}{D^{1/2}} \exp(\frac{r^2}{4D_m t} - \frac{x_i x_j b_{ij}}{4D_m D} - \frac{1}{t^{3/2}} \right\} \exp\left(\frac{-r^2}{4D_m t}\right) dt. \tag{7.1.27}$$

It follows from Eq. (7.1.23) that

$$tb_{ij}/D \sim \delta_{ij} \quad \text{as} \quad t \to 0,$$

with an error t; hence, as $r \to 0$, J approaches to a constant, which can be identified with ΔC in Eq. (7.1.11). We have

$$\Delta C = \lim_{r \to 0} J = \frac{Q}{(4\pi D_m)^{3/2}} \int_0^\infty (D^{-\frac{1}{2}} - t^{-\frac{3}{2}}) dt. \tag{7.1.28}$$

The fractional increase in the transfer rate due to convection follows from Eq. (7.1.12) and Eq. (7.1.28), and since Q differs from Q_0 by a small quantity only, we have, to leading order,

$$\frac{Nu - Nu^0}{Nu^0} = \frac{\Delta Q}{Q} = \frac{Q_0}{(C_1 - C_0)(4\pi D_m)^{3/2}} \int_0^\infty (t^{-3/2} - D^{-1/2}) dt$$

$$= \frac{Nu^0 a}{(4\pi D_m)^{1/2}} \int_0^\infty \frac{1 - (t^3/D)^{-\frac{1}{2}}}{t^{\frac{3}{2}}} dt. \tag{7.1.29}$$

We now look for B_{ij} and D for some particular ambient velocity distributions.

1.2.1. The additional transfer rate due to a steady simple shearing motion

For a steady simple shearing motion with velocity components $(\gamma x_2, 0, 0)$ the only non-zero components of $\mathbf{G}$ and $\mathbf{E}$ are

$$G_{12} = \gamma, \ E_{12} = E_{21} = \frac{1}{2}\gamma.$$

The Eq. (7.1.22) and boundary condition Eq. (7.1.23) are satisfied by

$$B_{11} = t(1 + \frac{1}{3}\gamma^2 t^2), \ B_{22} = t, \ B_{33} = t, \ B_{12} = \frac{1}{2}\gamma t^2, \ B_{23} = B_{31} = 0.$$

Hence

$$D = (B_{11}B_{22} - B_{12}^2)B_{33} = t^3 \left(1 + \frac{1}{12}\gamma^2 t^2\right), \tag{7.1.30}$$

and so from Eq. (7.1.29)

$$\frac{Nu - Nu^0}{Nu^0} = Nu^0 \left(\frac{a^2\gamma}{4\pi D_m}\right)^{1/2} \int_0^\infty \frac{1 - (1 + \frac{1}{12}q^2)^{-\frac{1}{2}}}{q^{\frac{3}{2}}} dq. \tag{7.1.31}$$

The integral in Eq. (7.1.31) can be carried out, giving

$$\frac{Nu - Nu^0}{Nu^0} = 0.257 Nu^0 \left(\frac{a^2\gamma}{D_m}\right)^{1/2} \quad \text{or} \quad 0.305 Nu^0 \frac{a^2 E^{\frac{1}{2}}}{D_m}, \tag{7.1.32}$$

where $E = (E_{ij}E_{ij})^{1/2} = \gamma/\sqrt{2}$.

1.2.2. The additional transfer rate due to an ambient steady pure straining motion

In this case $\mathbf{G} = \mathbf{E}$. We then cchoose the axes which coincide with the principal axes of $\mathbf{E}$, and the three principal rates of strain will be denoted by E_1, E_2, E_3. The Eq. (7.1.22) and boundary condition Eq. (7.1.23) are here satisfied by

$$B_{11} = \frac{\exp(2E_1 t) - 1}{2E_1}, \quad B_{22} = \frac{\exp(2E_2 t) - 1}{2E_2}, \quad B_{33} = \frac{\exp(2E_3 t) - 1}{2E_3}$$

$$B_{ij} = 0 \qquad (i \neq j) \tag{7.1.33}$$

and so

$$D = B_{11}B_{22}B_{33} = \frac{\sinh(E_1 t)\sinh(E_2 t)\sinh(E_3 t)}{E_1 E_2 E_3}. \tag{7.1.34}$$

Hence we find from Eq. (7.1.29)

$$\frac{Nu - Nu^0}{Nu^0} = Nu^0 \left(\frac{a^2 E}{4\pi D_m}\right)^{1/2} \int_0^\infty [1$$

$$- \left\{\frac{q^2 E_1 E_2 E_3 / E_3}{\sinh(E_1 q/E)\sinh(E_2 q/E)\sinh(E_3 q/E)}\right\}^{\frac{1}{2}} \frac{dq}{q^{\frac{3}{2}}}, \tag{7.1.35}$$

where $E^2 = E_1^2 + E_2^2 + E_3^2$. Note that the additional transfer due to convection is unchanged by reversing the signs of E_1, E_2 and E_3, which is reminiscent of a general theorem due to Brenner(1967). The integral in Eq. (7.1.35) can be evaluated numerically for particular values of the ratios of the principal rates of strain. The following two cases are the simplest ones:

(i) two-dimensional pure straining motion , for which

$$E_1 = -E_2, \; E_3 = 0 \quad \text{and} \quad E = \sqrt{2}|E_1|,$$

the result being

$$\frac{Nu - Nu^0}{Nu^0} = 0.428 Nu^0 \left(\frac{a^2|E_1|}{D_m}\right)^{1/2} \quad \text{or} \quad 0.360 Nu^0 \frac{a^2 E^{\frac{1}{2}}}{D_m}, \tag{7.1.36}$$

(ii) axisymmetric pure straining motion , for which

$$E_1 = E_2 = -\frac{1}{2}E_3 \quad \text{and} \quad E = (\frac{3}{2})^{\frac{1}{2}}|E_3|,$$

the result being

$$\frac{Nu - Nu^0}{Nu^0} = 0.399 Nu^0 \left(\frac{a^2|E_3|}{D_m}\right)^{1/2} \quad \text{or} \quad 0.360 Nu^0 \frac{a^2 E^{\frac{1}{2}}}{D_m}. \qquad (7.1.37)$$

The transfer rate for an ambient pure straining motion can be regarded as a function of E and one parameter which determines the type of straining motion and which we may choose as $(E_1 - E_2)/E$. The whole set of geometrical different pure straining motion s are covered if we begin with axisymmetric compression in the x_3 direction (for which $E_1 > 0, E_2 > 0$ and $(E_1 - E_2)/E = 0$) and decrease E_2 with E_1 fixed until we reach $E_2 = 0$ and $(E_1 - E_2)/E = 1/\sqrt{2}$, corresponding to a two-dimensional motion in (x_3, x_1) plane, and then decrease E_2 further with E_1 fixed until we reach $E_2 = -\frac{1}{2}E_1$ and $(E_1 - E_2)/E = \sqrt{\frac{3}{2}}$, corresponding to axisymmetric extension in the x_1 direction. But bearing in mind that the transfer rate is invariant to flow reversal, we see that it is sufficient to consider values of $(E_1 - E_2)/E$ in the range from 0 to $1/\sqrt{2}$ and that the value of Nu is stationary at the two end points of this range. The fact that the transfer rate is the same multiple of $(a^2 E/D_m)^{\frac{1}{2}}$ at the two end points of this range indicates that the relation

$$\frac{Nu - Nu^0}{Nu^0} = 0.36 Nu^0 \left(\frac{a^2 E}{D_m}\right)^{1/2} \qquad (7.1.38)$$

is likely to give accurate results for any pure straining motion .

A third case for which a solution of Eq. (7.1.22) is obtainable in a simple closed form is a two-dimensional motion, which we may represent by

$$G_{12} = E_{12} - \Omega, \ G_{21} = E_{12} + \Omega,$$

all other components of **G** being zero. The solution of Eq. (7.1.22) is here

$$B_{11}(E_{12} + \Omega) - \Omega t = B_{22}(E_{12} - \Omega) + \Omega t = \frac{\frac{1}{2}E_{12}\sinh\{2(E_{12}^2 - \Omega^2)^{\frac{1}{2}}t\}}{(E_{12}^2 - \Omega^2)^{\frac{1}{2}}},$$

$$B_{12} = -\frac{\frac{1}{2}E_{12}}{E_{12}^2 - \Omega^2} + \frac{\frac{1}{2}\cosh\{2(E_{12}^2 - \Omega^2)^{1/2}t\}}{E_{12}^2 - \Omega^2},$$

$$B_{33} = t, \qquad B_{23} = B_{32} = 0.$$

Hence

$$D = \frac{E_{12}^2 t \sinh^2\{E_{12}^2 - \Omega^2)^{1/2}t\}}{(E_{12}^2 - \Omega^2)^2} - \frac{\Omega^2 t^3}{E_{12}^2 - \Omega^2}, \qquad (7.1.39)$$

which reduces to Eq. (7.1.30) in the particular case $E_{12} = -\Omega = \frac{1}{2}\gamma$ corresponding to a simple shearing motion and to Eq. (7.1.34) in the case $\Omega = 0$ corresponding pure straining motion with principal rates of strain $E_{12}, -E_{12}, 0$. The transfer rate follows from Eq. (7.1.29), although numerical integration seems to be necessary. It is hardly worth undertaking since we already have the result for the cases $|\Omega/E_{12}| = 0$ and will discuss in a moment the case $|\Omega/E_{12}| \geq 1$.

An approximate expression for $(Nu - Nu^0)/Nu^0$ which holds for an even wider class of linear ambient velocity distribution than the pure straining motion or the two-dimensional motion would of course be useful. Such an approximation can be found by writing B_{ij} as a power series in t:

$$B_{ij}(t) = t\delta_{ij} + t^2 B_{ij}^{(2)} + t^3 B_{ij}^{(3)} + \cdots\cdots.$$

Using Eq. (7.1.22) to determine the first few coefficients, we find

$$B_{ij}^{(2)} = \frac{1}{2}(G_{ji} + G_{ij}) = E_{ij}$$

$$B_{ij}^{(3)} = \frac{2}{3}E_{il}E_{jl} + \frac{1}{3}(E_{il}\Omega_{jl} + E_{jl}\Omega_{il}).$$

If now we choose the axes of reference to coincide with the principal axes of the rate-of-strain tensor $\mathbf{E}$, the non-diagonal components of $\mathbf{B}^{(2)}$ and $\mathbf{B}^{(3)}$ are zero and so, correct to terms of order t^2,

$$\left(\frac{D}{t^3}\right)^{1/2} = \left(\frac{B_{11}B_{22}B_{33}}{t^3}\right)^{1/2} = 1 + \frac{1}{12}E^2 t^2, \tag{7.1.40}$$

where $E = E_{ij}E_{ij}$ as before. The value of $(Nu - Nu^0)/Nu^0$ corresponding to the approximation for D is seen in Eq. (7.1.29) as

$$\frac{Nu - Nu^0}{Nu^0} = \frac{\Gamma(\frac{1}{4})\Gamma(\frac{3}{4})}{4(12\pi^2)^{\frac{1}{4}}} Nu^0 \left(\frac{a^2 E}{D_m}\right)^{\frac{1}{2}} = 0.337 Nu^0 \left(\frac{a^2 E}{D_m}\right)^{\frac{1}{2}}. \tag{7.1.41}$$

Further terms in the series describe the way in which D/t^3 varies at large t and, provided $D/t^3 \to \infty$ as $t \to \infty$ (which amounts to a requirement that the cloud of material is extended appreciably by convection in at least one direction), it is evident that these further terms do not have much effect on the value of the integral in Eq. (7.1.29). The numerical coefficient in Eq. (7.1.41) is 6% too small in the case of the pure straining motion and 10% too large in the case of the simple shearing motion . Aside from providing a fair estimate of the transfer for a wide class of linear ambient velocity distributions, Eq. (7.1.41) reveals E as the primary parameter determining the convective transfer.

1.3. The Partial Suppression of the Effect of Convection by Strong Rotation.

There is one circumstance in which the estimate Eq. (7.1.41) will be less accurate, viz. when the magnitude of the angular velocity vector $(\frac{1}{2}|\omega| = \Omega$, say) is large relative to the components of the rate-of-strain tensor . The effect of a strong rotation is to change the form of the streamline s so that instead of being open curves extending to infinity they become closed curves enclosing the rotation axis, and these closed curves become nearly circular as Ω still further increases, with a consequent suppression of the contribution to the transfer due to the convection associated with some of the components of the rate-of-strain tensor .

This may be seen clearly from the above consideration of the two-dimensional ambient motion in the (x_1, x_2) plane. The components of $\frac{1}{2}\omega$ are here (o, o, Ω), and the directions of the x_1 and x_2 axes are such that $E_{11} = E_{22} = 0$. The stream function describing this motion is

$$\psi = -\frac{1}{2}x_1^2(E_{12} + \Omega) + \frac{1}{2}x_2^2(E_{12} - \Omega),$$

and the streamline s are hyperbolic when $|\Omega/E_{12}| < 1$ and ellipses when $|\Omega/E_{12}| > 1$. The ratio of the principal diameters of the ellipses tends to unity as $E_{12}/\Omega \to 0$, and in this limit the increase in the transfer rate from the point source at the origin due to convection is zero. Analytically, we see from Eq. (7.1.29) and Eq. (7.1.39) that the transfer rate is given by

$$\frac{Nu - Nu^0}{Nu^0} = Nu^0 \frac{a(\Omega^2 - E_{12}^2)^{1/2}}{4\pi D_m)^{1/2}} \int_0^\infty \left\{ 1 - \frac{(\Omega^2 - E_{12}^2)^{1/2}s}{(\Omega^2 s^2 - E_{12}^2 \sin^2 s)^{1/2}} \right\} \frac{ds}{s^{3/2}}$$

when $|\Omega/E_{12}| > 1$. Hence, as $|\Omega/E_{12}| \to \infty$

$$\frac{Nu - Nu^0}{Nu^0} \sim Nu^0 \left(\frac{a^2 E_{12}}{D_m} \right)^{\frac{1}{2}} \left(\frac{E_{12}}{\Omega} \right)^{\frac{3}{2}} \frac{1}{4\sqrt{\pi}} \int_0^\infty \left(1 - \frac{\sin^2 s}{s^2} \right) \frac{ds}{s^{\frac{3}{2}}}$$

$$= \frac{2}{15} Nu^0 \left(\frac{a^2 E}{D_m} \right)^{\frac{1}{2}} \left(\frac{E}{\Omega} \right)^{\frac{3}{2}}, \qquad (7.1.42)$$

E being equal to $(2E_{12}^2)^{\frac{1}{2}}$ in this case. The suppression of the convective transfer by rotation in this case is thus quite strong.

But on the other hand it is clear that if an axisymmetric extensional motion, with the axis of symmetry coinciding with the x_3 axis, is combined with a rotation about the x_3 axis, the convective transfer due to this axisymmetric extensional motion is unaffected by the rotation.

Consider now the general linear ambient velocity distribution, with x_3 axis in the direction of ω and the directions of x_1 and x_2 axes such that $E_{11} = E_{22}, = -\frac{1}{2}E_{33}$. We shall write $\mathbf{E} = \mathbf{E}^{(0)} + \mathbf{E}^{(1)}$, where

$$\mathbf{E}^{(0)} = \begin{bmatrix} -\frac{1}{2}E_{33} & 0 & 0 \\ 0 & -\frac{1}{2}E_{33} & 0 \\ 0 & 0 & E_{33} \end{bmatrix}, \mathbf{E}^{(1)} = \begin{bmatrix} 0 & E_{12} & E_{13} \\ E_{21} & 0 & E_{23} \\ E_{31} & E_{32} & 0 \end{bmatrix},$$

$$\Omega = \begin{bmatrix} 0 & -\Omega & 0 \\ \Omega & 0 & 0 \\ 0 & 0 & 0 \end{bmatrix}, \tag{7.1.43}$$

and we suppose that all the components of $\mathbf{E}^{(1)}$ have small magnitude relative to Ω (there being no need for any restriction on E_{33}).

It appears that when the angular velocity is large the transfer rate from the point source is equal to that associated with an axisymmetric pure straining motion with a rate of extension E_{33} in the direction of the symmetry axis, that is, in terms of quantities which are independent of the axes of reference, with the rate of extension

$$E_\omega = \omega \cdot \mathbf{E} \cdot \omega / \omega^2$$

in the direction of the symmetry axis. The formula Eq. (7.1.37) then gives

$$\frac{Nu - Nu^0}{Nu^0} = 0.399 Nu^0 \left(\frac{a^2 |E_\omega|}{D_m} \right)^{\frac{1}{2}}. \tag{7.1.44}$$

The theoretical results for the transfer rate described in the above sections are asymptotically valid for small values of the Péclet number . We now restate the main results for a particle in various steady linear ambient flow fields.

For the additional rate of transfer from a particle due to convection at small Péclet number , we have the following three results which together cover virtually all kinds of linear ambient flow fields.

medskip (i) For any pure straining motion s,

$$\frac{Nu - Nu^0}{Nu^0} = 0.36 Nu^0 \left(\frac{a^2 E}{D_m} \right)^{\frac{1}{2}}, \tag{7.1.45}$$

where $E = E_{ij} E_{ij}$; the numerical coefficient is likely to be correct within a 3% error tolerance.

(ii) For a general linear ambient velocity distribution in which the ratio of the vorticity magnitude to E is not large compared to unity,

$$\frac{Nu - Nu^0}{Nu^0} = 0.34 Nu^0 \left(\frac{a^2 E}{D_m} \right)^{\frac{1}{2}}; \tag{7.1.46}$$

the numerical coefficient is approximate but is unlikely to be in an error more than 10%.

(iii) For a general linear ambient velocity distribution in which the vorticity magnitude is significantly larger than E (or, more precisely, significantly larger than $(E^2 - \frac{3}{2} E_\omega^2)^{1/2}$),

$$\frac{Nu - Nu^0}{Nu^0} = 0.40 Nu^0 \left(\frac{a^2 |E_\omega|}{D_m} \right)^{\frac{1}{2}}, \tag{7.1.47}$$

where $E_\omega = \omega \cdot \mathbf{E} \cdot \omega / \omega^2$.

The mass transfer theory for the case of large Péclet number is also available (see Batchelor 1979)[33]. However, the case in which the Reynolds number is small and the Péclet number is large only exists under certain circumstances in which the particles are suspended in liquid. On the other hand in the case of aerosol particles the Schmidt number is of order unity. Therefore, the Péclet number is always small when the Reynolds number is small in the case of aerosol particles. Thus, we will not introduce the theory here. It is enough to point out that when the Péclet number is large, there exists a thin concentration boundary layer and the effects of the convection are restricted in this thin boundary layer . We may thus expect that the suppression of the convective transfer which is induced by almost all the components of the ambient pure straining motion except E_ω will occur at any non-zero value of Ω, because within the boundary layer the contribution to the fluid velocity due to the ambient pure straining motion is proportional to the distance from the particle surface and so is small. This is the basis for Batchelor (1980) to construct his turbulent mass transfer theory at small Reynolds number and large Péclet number . Unfortunately, this theory can apply only to the case in which the particles are suspended in liquid. For the case of aerosol systems in which the Reynolds number and the Péclet number are both small and the suppression of the convective transfer will not occur at any non-zero value of Ω (it will occur only when the rotation velocity is large), the very important turbulent mass transfer problem is still unsolved.

2. The Effective Viscosity of a Dilute Suspension

We are concerned in this and next section with the bulk rheological properties of a suspension of particles in a Newtonian fluid (either liquid or gas) of uniform viscosity μ. The problem was not included in Fuchs' book[2] and the Chapter 12 of Prupppacher and Klett's book[3]. Thus, Fuchs' book and Pruppacher and Klett's Chapter 12 cannot be called "The mechanics of aerosols". It is better to call them "The dynamics of aerosol particles".

In this section, the suspension is assumed to be dilute. It means that the probability that two particle centres are simultaneously at the same place is extremely small. Thus, the isolated-particle approximation can be adopted in this section. It is of interest to know how such suspensions behave in response to applied forces and moving boundaries. Provided the characteristic length scale of the motion of the suspension is large compared to the average distance between the particles, we can regard the suspension as a homogeneous fluid, with mechanical properties different from those of the ambient fluid in which the particles are suspended. A random distribution of spherical particles does not convey any directional properties on the medium, so that provided the ambient fluid is 'Newtonian' the homogeneous fluid equivalent to a suspension of spherical particles is likewise 'Newtonian' and is characterized by a shear viscosity.

The background motion of the ambient fluid on which the disturbance flow due to the presence of one particular particle is superposed consists approximately of a uniform translation, a uniform rotation and a uniform pure straining motion. The particle translates and rotates with the surrounding fluid, so that only the background pure straining motion gives rise to a disturbance flow . It seems inevitable that the change in the straining motion due to the presence of one particle is accompanied by an increase in the total rate of dissipation and that the effective viscosity of the suspension is greater than the viscosity of the ambient flow ; we shall find this to be so.

In the first instance we assume the particles to be incompressible, so that the suspension is likewise incompressible and only the effective value of the shear viscosity is to be determined. Explicit knowledge of the disturbance flow due to a single incompressible particle is needed in determination of the effective viscosity of the suspension, and we therefore consider first the following problem of flow with negligible inertia forces.

2.1. The Flow due to a Sphere Embedded in a Pure Straining Motion

Fluid of viscosity μ and density ρ occupies the space outside a sphere of radius a, and far from the sphere the fluid is in pure straining motion specified by the rate-of-straining E_{ij}, where $E_{ii} = 0$. The velocity and pressure in the fluid may be written as

$$u_i = u_i' + E_{ij}x_j, \qquad p = p' + P, \tag{7.2.1}$$

where P is the pressure in the pure straining motion represented by E_{ij} in the absence of the sphere; u_i' and p' represent the changes due to the presence of the sphere and

$$u_i' \to 0, \qquad p' \to 0, \tag{7.2.2}$$

as $r(= |\mathbf{x}|) \to \infty$. We choose the centre of the sphere to be at the origin, so that by symmetry there is no tendency for the sphere to translate and the surface of the sphere is permanently located at $r = a$; thus

$$\mathbf{n} \cdot \mathbf{u} = 0 \qquad \text{at} \quad r = a. \tag{7.2.3}$$

There are further conditions at the particle surface which depend on the nature of the particle. We may include various kinds of particles within the scope of the analysis by supposing the spheres to contain incompressible fluid of viscosity $\bar{\mu}$ (the case of a rigid particle is a special one corresponding, as in section 1 of Chapter 3, to $\bar{\mu}/\mu \to \infty$). The velocity must be continuous across the interface, and so also is the tangential component of stress if we suppose, as in section 1 of Chapter 3, that the interface has no mechanical properties other than a uniform surface tension. Thus

$$\left. \begin{array}{rcl} u_i = u_i' + E_{ij}x_j &=& \bar{u}_i \\ \epsilon_{kli}n_l n_j(\sigma_{ij} - \bar{\sigma}_{ij}) &=& 0 \end{array} \right\} \qquad \text{at} \quad r = a, \tag{7.2.4}$$

where the over-bar indicates a quantity referring to the motion within the sphere and $\mathbf{n}$ is the normal to the interface. Also $\bar{p}$ and $\bar{u}_i$ are finite at $r = 0$.

When the particle radius a is small so that the Reynolds number of the disturbance flow is small, the flow s near the particle and inside the particle are governed by the Stokes equation , i.e.

$$\nabla p' = \mu \nabla^2 \mathbf{u}', \tag{7.2.5}$$

and

$$\nabla \overline{p} = \overline{\mu} \nabla^2 \overline{\mu}. \tag{7.2.6}$$

Finally, the mass conservation gives two equations

$$\nabla \cdot \mathbf{u}' = \nabla \cdot \overline{\mathbf{u}} = \nabla \cdot \mathbf{u} = 0. \tag{7.2.7}$$

The Eq. (7.2.5), Eq. (7.2.6), Eq. (7.2.7) and the boundary conditions Eq. (7.2.2), Eq. (7.2.3) and Eq. (7.2.4) governing the disturbance motion are linear and homogeneous in $\mathbf{u}', p', \overline{\mathbf{u}}, \overline{p}$ and E_{ij}. No vector occurs in the description of the interface, and, with an argument like that in section 1 of Chapter 3, the pressures (which are harmonic functions) and velocities are seen to be of the form[4]

$$\left. \begin{array}{rcl} p' & = & C\mu E_{ij}x_i x_j/r^5, \quad \overline{p} - \overline{p}_0 = \overline{C}\overline{\mu}E_{ij}x_i x_j, \\ u_i' & = & E_{ij}x_j M + E_{jk}x_i x_j x_k Q, \\ \overline{u}_i & = & E_{ij}x_j \overline{M} + E_{jk}x_i x_j x_k \overline{Q}, \end{array} \right\} \tag{7.2.8}$$

where M, Q, $\overline{M}$ and $\overline{Q}$ are functions of r alone, and C, $\overline{C}$ and $\overline{p}_0$ are constants. The forms of the functions satisfy the governing equations and the conditions far from the particle and at $r = 0$ are readily found to be

$$\left. \begin{array}{rcl} M & = & \frac{D}{r^5}, \quad Q = \frac{C}{2r^5} - \frac{5D}{2r^7}, \\ \overline{M} & = & \overline{D} + \frac{5}{21}\overline{C}r^2, \overline{Q} = -\frac{2}{21}\overline{C}, \end{array} \right\} \tag{7.2.9}$$

and the conditions at the interface $r = a$ are then satisfied provided

$$\frac{C}{(2\mu + 5\overline{\mu})a^3} = \frac{D}{\overline{\mu}a^5} = -\frac{2\overline{C}a^2}{21\mu} = \frac{2\overline{D}}{3\mu} = -\frac{1}{\mu + \overline{\mu}}. \tag{7.2.10}$$

We observe in passing that at large distance from the particle

$$u_i' = \frac{1}{2}C E_{jk}\frac{x_i x_j x_k}{r^5} + O(r^{-4}), \tag{7.2.11}$$

showing that the disturbance velocity is one order smaller than in the case of flow due to a sphere in a translational motion, as might have been expected from the 'dipole' nature of the condition on $\mathbf{u}'$ at the surface of the sphere (see Eq. (7.2.3) with Eq. (7.2.1)) in the present case.

2.2. The Increased Rate of Dissipation in an Incompressible Suspension

We proceed to use the foregoing results in the calculation of the effective shear viscosity of a suspension of small incompressible spherical particles undergoing a prescribed bulk motion. The precise manner in which the effective viscosity should be specified and determined is not obvious, so must be described with care.

A volume V_1 of the suspension will be supposed to be bound by a flexible surface A_1 on which the velocity is specified as a linear function of position; and in order to have a completely defined system we take this velocity to be exactly a linear function of $\mathbf{x}$. The rotation part of the motion at the boundary does not enter the analysis so for convenience we choose the (i-component of the) velocity at the boundary as $E_{ij}x_j$, where E_{ij} is a symmetrical tensor with $E_{ii} = 0$. The suspension takes up a motion compatible with that of the boundary, and if the suspension were a homogeneous fluid the velocity would be $E_{ij}x_j$ everywhere in the volume V_1; owing to the presence of the particles the velocity of the ambient fluid has this value only in average sense, and will be written as

$$E_{ij}x_j + u_i'.$$

Likewise the pressure in the suspension would have a certain value, P say, if the suspension were a homogeneous fluid of the same average density, whereas in reality the pressure in the ambient fluid has a more complicated dependence on position as

$$P + p'.$$

If the particles are far apart, each particle is embedded in a pure straining motion characterized by a rate-of-strain tensor E_{ij} and near one particle u_i' and p' have the forms given in Eq. (7.2.8); but there is no need for substitution.

The stress tensor at any point in the ambient fluid of viscosity μ is thus

$$\sigma_{ij} = -P\delta_{ij} + 2\mu E_{ij} + \sigma_{ij}', \tag{7.2.12}$$

where

$$\sigma_{ij}' = -p'\delta_{ij} + \mu\left(\frac{\partial u_j'}{\partial x_i} + \frac{\partial u_i'}{\partial x_j}\right). \tag{7.2.13}$$

On the other hand, if the suspension were a homogeneous fluid of the same average density and of viscosity μ^*, the stress tensor would be

$$\sigma_{ij} = -P\delta_{ij} + 2\mu^* E_{ij}. \tag{7.2.14}$$

We wish to choose the value of μ^* so that in a physically significant way it represents the net effect of the disturbance flow s due to the presence of all the particles in the suspension. An appropriate quantity which when calculated in two different ways yields a value of the effective viscosity of the suspension is the rate of dissipation of mechanical energy in V_1; this rate of dissipation is a direct consequence of internal friction and it has the desired property of including the effect of the disturbance flow s due to all the particles within the volume V_1.

Now the additional rate of dissipation per unit volume in the fluid at distance r from a single particle varies asymptotically as r^{-3} (for see Eq. (7.2.1) and Eq. (7.2.11)), and this frustrates direct evaluation of the total additional rate of dissipation by integration over the fluid in V_1; the integral is not actually divergent, but is found to have a value dependent on the shape of the distant outer boundary of the region of integration for one particle. We must therefore proceed in a different way. The rate at which work is being done by forces at the boundary A_1 is

$$\int_{A_1} E_{ik} x_k \sigma_{ij} n_j \mathrm{d}A, = E_{ik} \int_{A_1} (-P\delta_{ij} + 2\mu E_{ij} + \sigma'_{ij}) x_k n_j \mathrm{d}A;$$

and if the suspension were a homogeneous fluid of the same average density and of viscosity μ^* the power at the boundary would be

$$E_{ik} \int_{A_1} (-P\delta_{ij} + 2\mu^* E_{ij}) x_k n_j \mathrm{d}A.$$

The term involving P is common to the two expressions and accounts for any increase in kinetic energy associated with the linear velocity field. The remaining part of the two expressions represent the rates of dissipation within V_1, and the effective viscosity μ^* will be defined to have a value such that they are equal; that is, after use of the divergence theorem,

$$2\mu^* E_{ij} E_{ij} V_1 = 2\mu E_{ij} E_{ij} V_1 + e_{ik} \int_{A_1} \sigma'_{ij} x_k n_j \mathrm{d}A. \qquad (7.2.15)$$

The last term of Eq. (7.2.15) representing the additional rate of dissipation in V_1 due to the presence of particles, can be transformed to an integral over the surface of the particles. For we have

$$2(\mu^* - \mu) E_{ij} E_{ij} V_1 = E_{ik} \int_{V_1 - NV_0} \left(\frac{\partial \sigma'_{ij}}{\partial x_j} x_k + \sigma'_{ik} \right) \mathrm{d}V + E_{ik} N \int_{A_0} \sigma'_{ij} x_k n_j \mathrm{d}A, \qquad (7.2.16)$$

where A_0 and V_0 are the surface and volume of one particle, with n as the outward normal to A_0, and N denotes the total number of the particles in V_1. The disturbance motion due to the presence of each particle is governed by Eq. (7.2.5) when the condition Eq. (7.2.4) is satisfied, as we shall assume to be the case, so that $\sigma'_{ij}/x_j = 0$. Also

$$E_{ik} \int_{V_1 - NV_0} \sigma'_{ik} \mathrm{d}V = E_{ik} \int_{V_1 - NV_0} 2\mu \frac{\partial u'_i}{\partial x_k} \mathrm{d}V,$$

$$= -E_{ik} N \int_{A_0} 2\mu u'_i n_k \mathrm{d}A,$$

since $\mathbf{u}' = 0$ on the boundary A_1. Thus the relation Eq. (7.2.15) becomes

$$2(\mu^* - \mu) E_{ij} E_{ij} = \frac{N}{V_1} E_{ik} \int_{A_0} (\sigma'_{ij} x_k n_j - 2\mu u'_i n_k) \mathrm{d}A, \qquad (7.2.17)$$

the right-hand-side of which represents the average additional rate of dissipation per unit volume due to the presence of the particles. This expression for the effective viscosity of an isotropic suspension is valid for any spacing of the particles in V_1.

If now we assume the suspension to contain spherical particles at distances from each other which is large compared to their diameters, the disturbance flow near one particle is approximately independent of the existence of other particles and we may use the results obtained earlier to evaluate the integral in Eq. (7.2.7). It follows readily from Eq. (7.2.13), Eq. (7.2.8) and Eq. (7.2.9) that

$$\sigma'_{ij}x_jx_k - 2\mu u'_i x_k = \mu E_{ij}x_jx_k \left(\frac{C}{r^3} - \frac{10D}{r^5}\right) + \mu E_{jl}x_i x_j x_k x_l \left(-\frac{5C}{r^5} + \frac{25D}{r^7}\right). \quad (7.2.18)$$

The surface A_0 is a sphere of radius a, and with the well-known identities

$$\int n_j n_k d\Omega = \frac{4}{3}\pi\delta_{jk}, \quad \int n_i n_j n_k n_l d\Omega = \frac{4}{15}\pi(\delta_{ij}\delta_{kl} + \delta_{ik}\delta_{jl} + \delta_{il}\delta_{jk}), \quad (7.2.19)$$

in which the integration is over the complete solid angle subtended at the centre of the sphere, we find

$$\int_{A_0} (\sigma'_{ij}x_k n_j - 2\mu u'_i n_k)dA = -\frac{4}{3}\pi\mu C E_{ik}.$$

Thus

$$\frac{\mu^*}{\mu} = 1 - \frac{2\pi}{3V_1}NC$$

and, from Eq. (7.2.10) we see that $C = -(2\mu + 5\overline{\mu})a^3/(\mu + \overline{\mu})$.

If all the particles have the same internal viscosity, then (Batchel or 1967)[4]

$$\frac{\mu^*}{\mu} = 1 + \left[\frac{\mu + \frac{5}{2}\overline{\mu}}{\mu + \overline{\mu}}\right]\varphi, \quad (7.2.20)$$

where $\varphi = NV_0/V_1$ is the volume concentration of particles. For a suspension of rigid particles the effective viscosity is greater than the viscosity of the ambient fluid by a factor $\frac{5}{2}\varphi$ (a result first obtained by Einstein (1906,1911)), while for gas bubbles suspended in liquid corresponding fraction is φ.

The formulae Eq. (7.2.19) and Eq. (7.2.20) are subject to the restriction $\varphi \ll 1$; when the concentration is not small compared to unity, the presence of neighbouring particles affects the disturbance flow due to one particle and the expression Eq. (7.2.18) then needs a modification. The experimental evidence for the validity of the relation Eq. (7.2.20) appears not to be decisive, but the viscosity of a suspension of small rigid spheres is believed to be represented by the 'Einstein formula' $(\mu(1 + \frac{5}{2}\varphi))$ for values of φ less than about 0.02.

3. The Effective Viscosity of a Sub-dilute Suspension

The relative bulk velocity field in a small neighbourhood of **x** can always be represented as the superposition of a pure straining motion , characterized by the rate-of-strain tensor E_{ij}, and a rigid body rotation with angular velocity Ω. These two quantities E_{ij} and Ω are continuous functions of **x** in general, and there will be a region about **x**, of volume V say, in which E_{ij} and Ω are approximately uniform. In such a region the bulk stress σ_{ij} will likewise be approximately uniform; and the local velocity gradient and stress will be stationary random functions of position within V. If now the linear dimensions of V (the macroscopic scale of the suspension) are large compared with the average distance between the particles (the microscopic scale of the suspension), and ensemble average at the point $\mathbf{x}$ is identical with a spatial average over V for one realization, the latter sometimes being more convenient for analytical purpose.

It is known that, under the conditions described above, and for any concentration of particles, there is a relation between the deviatoric part of the bulk stress, and the conditions at the surfaces of individual particles. This relation is

$$\sigma_{ij} - \frac{1}{3}\delta_{ij}\sigma_{kk} = 2\mu E_{ij} + \sigma_{ij}^{(p)}, \tag{7.3.1}$$

where the first term on the right-hand-side is the deviatoric stress that would be generated in the ambient fluid in the absence of the particles, and the particle stress $\sigma_{ij}^{(p)}$ is given, for particles which are force-free and couple-free, by

$$\sigma_{ij}^{(p)} = \frac{N}{V}S_{ij}, \tag{7.3.2}$$

and

$$S_{ij} = \int_{A_0} \{(\sigma_{ik}x_j - \frac{1}{3}\delta_{ij}\sigma_{lk}x_l)n_k - \mu(u_in_j + u_jn_i)\}\mathrm{d}A. \tag{7.3.3}$$

The summation in Eq. (7.3.2) is over all the particles in V, A_0 is the surface of one of the particles, the unit (outward) normal there being $\mathbf{n}$, and u_i and σ_{ik} are the local velocity and stress.

In the case of a suspension which is so dilute that the fluid motion near one particle is independent of the presence of the other particles, it is a simple matter to determine the force dipole strength S_{ij}. The isolated-particle results given in the above section can be used to calculate the integral in Eq. (7.3.3). Then we have

$$S_{ij} = \frac{20}{3}\pi a^3 \alpha \mu E_{ij}, \tag{7.3.4}$$

and

$$\sigma_{ij}^{(p)} = 5\varphi\alpha\mu E_{ij}, \tag{7.3.5}$$

where

$$\alpha = \left(\overline{u} + \frac{2}{5}\mu\right)/(\overline{\mu} + \mu), \tag{7.3.6}$$

and thus Eq. (7.3.1) becomes

$$\sigma_{ij} - \frac{1}{3}\delta_{ij}\sigma_{kk} = 2\mu\left(1 + \frac{5}{2}\alpha\varphi\right) + O(\varphi^2). \tag{7.3.7}$$

Eq. (7.3.7) reduces to the Einstein in formula Eq. (7.2.21) for the effective viscosity as $\overline{\mu}/\mu \to \infty$ and $\alpha \to 1$. Measurements with suspension of rigid spheres suggest that the term of order φ^2 in Eq. (7.3.7) is negligible only for φ being less than about 0.02. When the suspension is not so dilute that φ is larger than 0.02, but still smaller than unity, the multi-particle hydrodynamic interaction must be considered. We shall come across the familiar problem, of non-absolutely convergent integrals representing the effect of all pair interactions in which one specified particle takes part. This difficulty remained unovercome until Batchelor and Green's work (1972), and we will introduce their work [41] in this section.

When the volume V contains $N(\gg 1)$ spherical particles all of radius a, the summation in Eq. (7.3.2) is equivalent to N times the average value of S_{ij} and we may write

$$< \sigma_{ij}^{(p)} >= n < S_{ij} >$$

$$= 5\varphi\alpha\mu E_{ij} + 5\varphi\alpha\mu[\frac{< S_{ij} >}{\frac{20}{3}\pi a^3 \alpha\mu} - E_{ij}]. \tag{7.3.8}$$

The first term on the right-hand-side of Eq. (7.3.8) is the particle stress obtained when hydrodynamic interaction s between the particles are completely neglected, and so the second term represents the effects of those interactions.

Here, the definition of the ensemble average is the same as that given in Chapter 5. Thus, we have

$$< S_{ij} >= \frac{1}{N!} \int S_{ij}(\mathbf{x}_0, \mathcal{C}_N) P(\mathcal{C}_N|\mathbf{x}_0) d\mathcal{C}_N, \tag{7.3.9}$$

where $S_{ij}(\mathbf{x}_0, \mathcal{C}_N)$ denotes the value of the surface integral in Eq. (7.3.3) for a spherical particle centred at $\mathbf{x}_0$ in the presence of a configuration of N other particles with locations specified by $\mathcal{C}_N$, $P(\mathcal{C}_N)$ is the probability density of this configuration, and $P(\mathcal{C}_N|\mathbf{x}_0)$ is the conditional probability density which applies when the presence of an additional particle centred at $\mathbf{x}_0$ is given. Then, the second term on the right-hand-side of Eq. (7.3.8), representing the effect of particle interactions on the bulk stress, may likewise be written as

$$\frac{5\varphi\alpha\mu}{N!} \int \left\{ \frac{S_{ij}(\mathbf{x}_0, \mathcal{C}_N)}{\frac{20}{3}\pi a^3 \alpha\mu} - E_{ij} \right\} P(\mathcal{C}_N|\mathbf{x}_0) d\mathcal{C}_N. \tag{7.3.10}$$

Now the probability that the centre of one particle of the configuration $\mathcal{C}_N$ lies within a distance of a few sphere radii from the reference particle centred at $\mathbf{x}_0$ is of order φ, which is small for a dilute suspension. And the probability that two particles of $\mathcal{C}_N$ simultaneously lie in the region surrounding the reference particle is of order φ^2. We also observe that the quantity within the curly bracket in Eq. (7.3.10) falls to zero as the distance of all the surrounding particles from $\mathbf{x}_0$ tend to infinity . It is therefore natural to think that we could obtain the first approximation to the effect of hydrodynamic interaction s on the bulk stress by supposing that only one particle of the configuration $\mathcal{C}_N$ has a significant effect on the value of S_{ij} for the reference particle. This would correspond to replacing

$$\frac{1}{N!} \int \cdots P(\mathcal{C}_N|\mathbf{x}_0) d\mathcal{C}_N \quad \text{by} \quad \int \cdots P(\mathbf{x}_0 + \mathbf{r}|\mathbf{x}_0) d\mathbf{r}$$

in Eq. (7.3.10), and so a necessary condition for the 'natural' supposition to be correct is that the quantity $S_{ij}(\mathbf{x_0}, \mathbf{x_0} + \mathbf{r})/(\frac{20}{3}\pi a^3\alpha\mu) - E_{ij}$ should be integrable with respect to $\mathbf{r}$, that is, that it should be of smaller order than a^3/r^3 when $a/r \ll 1$. Here $S_{ij}(\mathbf{x_0}, \mathbf{x_0} + \mathbf{r})$ stands for the force dipole strength of the reference particle in the presence of a second particle centred at $\mathbf{x_0} + \mathbf{r}$.

To see whether this condition is satisfied we must consider the way in which two spherical particles interact when their separation is large. Each particle exerts zero resultant force on the fluid and acts as a force dipole, of strength given approximately by Eq. (7.3.4), which generates at distance r in unbound fluid a disturbance velocity of order $\alpha E_{ij}a^3/r^2$,(see Eq. (7.2.11)), a disturbance velocity gradient of order $\alpha E_{ij}a^3/r^3$, etc. Such a disturbance motion gives it an additional translational velocity, and also requires a change of the distribution of stress at the surface of the second particle, to ensure satisfaction of the no-slip condition there in the presence of the disturbance rate of strain. The corresponding addition to the value of S_{ij} for each particle may be seen from Eq. (7.3.4) to be proportional to the disturbance rate of strain, and so to be of order $\alpha^2\mu E_{ij}a^6/r^3$. It appears that the quantity $S_{ij}(\mathbf{x_0}, \mathbf{x_0} + \mathbf{r})/(\frac{20}{3}\pi a^3\mu) - E_{ij}$ is (only just) not integrable. Consideration of two-particle interactions alone is not yet sufficient for the determination of the φ^2–term and we must return to Eq. (7.3.10).

The device that we adopt here, following the general procedure proposed earlier in Chapter 5, is to look for a quantity, Θ_{ij}, whose mean value is known exactly and which has the same dependence on the position of one distant particle as $S_{ij}/(\frac{20}{3}\pi a^3\alpha\mu) - E_{ij}$, and then to simplify the expression for the difference between the mean value s of

$$S_{ij}/ \left(\frac{20}{3}\pi a^3\alpha\mu \right) - E_{ij}$$

and of Θ_{ij} by considering only the effect of one particle of the configuration C_N.

The appropriate choice of the function Θ_{ij} is made evidently by the fact that the contribution to the value of S_{ij} for a particle centred at $\mathbf{x_0}$ due to the presence of the second particle centred at $\mathbf{x_0} + \mathbf{r}$ is proportional, when $a/r \ll 1$, to the rate of strain induced at $\mathbf{x_0}$ by that second particle. Noting the proportionality constant in Eq. (7.3.4), we choose

$$\Theta_{ij} = e_{ij}(\mathbf{x_0}, C_N) - E_{ij}$$

where $e_{ij}(\mathbf{x_0}, C_N)$ is the rate-of-strain tensor at point $\mathbf{x_0}$ in the presence of the N particles specified by C_N. We have the exact result

$$E_{ij} = \frac{1}{N!} \int e_{ij}(\mathbf{x_0}, C_N) P(C_N) \mathrm{d}C_N, \qquad (7.3.11a)$$

and

$$< \Theta_{ij} >= \frac{1}{N!} \int \{e_{ij}(\mathbf{x_0}, C_N) - E_{ij})\} P(C_N) \mathrm{d}C_N = 0, \qquad (7.3.11b)$$

which allows us to write

$$\sigma_{ij} - 5\varphi\mu E_{ij} = \frac{5\varphi\alpha\mu}{N!} \int \left[\left\{ \frac{S_{ij}(\mathbf{x_0}, C_N)}{\frac{20}{3}\pi a^3\alpha\mu} - E_{ij} \right\} P(C_N|\mathbf{x_0}) \right.$$

$$- \{e_{ij}(\mathbf{x}_0, \mathcal{C}_N) - E_{ij}\} P(\mathcal{C}_N)] \, \mathrm{d}\mathcal{C}_N. \tag{7.3.12}$$

We can now apply the dilute approximation to Eq. (7.3.12). Then equation Eq. (7.3.12) reduces to (Batchelor and Green 1972)[41]

$$\sigma_{ij}^{(p)} - 5\varphi\alpha\mu E_{ij} = 5\varphi\alpha\mu \int \left[\left\{ \frac{S_{ij}(\mathbf{x}_0, \mathbf{x}_0 + \mathbf{r})}{\frac{20}{3}\pi a^3 \alpha\mu} - E_{ij} \right\} P(\mathbf{x}_0 + \mathbf{r}|\mathbf{x}_0) \right.$$

$$\left. - \{e_{ij}(\mathbf{x}_0, \mathbf{x}_0 + \mathbf{r}) - E_{ij}\} P(\mathbf{x}_0 + \mathbf{r})\right] \, \mathrm{d}\mathbf{r} + O(\varphi^2). \tag{7.3.13}$$

The force dipole strength $S_{ij}(\mathbf{x}_0, \mathbf{x}_0 + \mathbf{r})$ in the integrand of Eq. (7.3.13) may be written as (Batchelor and Green 1972)[34]

$$\frac{S_{ij}(\mathbf{x}_0, \mathbf{x}_0 + \mathbf{r})}{\frac{20}{3}\pi a^3 \alpha\mu} = E_{ij}(1 + K) + E_{kl}\left(\frac{r_i r_k \delta_{jl} + r_j r_k \delta_{il}}{r^2} - \frac{r_k r_l}{r^2}\frac{2}{3}\delta_{ij} \right) L$$

$$+ E_{kl}\frac{r_k r_l}{r^2}\left(\frac{r_i r_j}{r^2} - \frac{1}{3}\delta_{ij} \right) M, \tag{7.3.14}$$

where the dimensionless scalar functions K, L, M are functions of the dimensionless distance r/a, and the asymptotic forms of them are

$$K = O(\frac{a^3}{r^3}), \qquad L \sim -\frac{5\alpha a^3}{2r^3}, \qquad M \sim \frac{25\alpha a^3}{2r^3}, \tag{7.3.15}$$

as $r/a \to \infty$. Later we shall encounter the average value of S_{ij} over all directions of $\mathbf{r}$, which is

$$\frac{1}{4\pi r^2}\int_{r=const} \frac{S_{ij}(\mathbf{x}_0, \mathbf{x}_0 + \mathbf{r})}{\frac{20}{3}\pi a^3 \alpha\mu}\mathrm{d}A = E_{ij}\left(1 + K + \frac{2}{3}L + \frac{2}{15}M \right), = E_{ij}(1 + J), \tag{7.3.16}$$

say, where J is also a function of r/a. The asymptotic form of J is

$$J = K + \frac{2}{3}L + \frac{2}{15}M \sim \frac{15a^5}{2r^6} \qquad \text{as} \quad r/a \to \infty, \tag{7.3.17}$$

and

$$J = 0.2214 \qquad \text{at} \quad r/a = 2. \tag{7.3.18}$$

The expression for the rate of strain in Eq. (7.3.12) is given by (Batchelor and Green 1972)[41]

$$e_{ij}(\mathbf{x}_0, \mathbf{x}_0 + \mathbf{r}) = E_{ij}\left(1 - \frac{\beta a^5}{r^5} \right) + E_{kl}\left(\frac{r_i r_k \delta_{jl} + r_j r_k \delta_{il}}{r^2} - \frac{r_k r_l}{r^2}\frac{2}{3}\delta_{ij} \right)\left(-\frac{5\alpha a^3}{2r^3} + \frac{5\beta a^5}{r^5} \right)$$

$$+ E_{kl}\frac{r_k r_l}{r^2}\left(\frac{r_i r_j}{r^2} - \frac{1}{3}\delta_{ij} \right)\left(\frac{25\alpha a^3}{2r^3} - \frac{35\beta a^3}{2r^5} \right), \tag{7.3.19}$$

where the parameter β is[41]

$$\beta = \frac{\bar{\mu}}{\bar{\mu} + \mu}. \tag{7.3.20}$$

The solution of the governing equation for the pair-distribution function $p(\mathbf{r}, t)$ at large Péclet number , to which the conditional probability density $P(\mathbf{x}_0 + \mathbf{r}|\mathbf{x}_0)$ is related by $P(\mathbf{x}_0 + \mathbf{r}|\mathbf{x}_0) = np(\mathbf{r}, t)$ is given by Batchelor and Green (1972)[40]

$$p(\mathbf{r}, t) = q(r) = \frac{1}{1 - A} \exp\left\{ \int_r^\infty \frac{3(B - A)}{r(1 - A)} dr \right\}, \tag{7.3.21}$$

where A and B are dimensionless scalar functions of the scalar variable r/a in the expression of the velocity $\mathbf{V}(\mathbf{r})$ which is described by (Batchelor and Green 1972)[40]

$$V_i(\mathbf{r}) = \epsilon_{ijk}\Omega_j r_k + r_j E_{ij} - \left\{ A\frac{r_i r_j}{r^2} + B\left(\delta_{ij} - \frac{r_i r_j}{r^2} \right) \right\} r_k E_{jk}. \tag{7.3.22}$$

Note that the solution Eq. (7.3.21) is valid only for the case of a material point in $\mathbf{r}-$space which comes or goes to infinity. This is true for the particular case of a steady pure straining motion , since in that case every point in the accessible parts of the $\mathbf{r}-$ space lies on a trajectory of a material point in $\mathbf{r}-$ space which has come from infinity. It is necessary to point out that at this point not all the trajectories of bulk flow come from infinity. It may happen that some of the trajectories are closed and occupy a region into which open trajectories from infinity do not penetrate; for instance, in the important case of steady simple shearing motion both experimental observation and theoretical analysis show the existence of 'bound' pairs of rigid spheres which move relative to each other along closed paths. The equation for the pair-distribution function in that case remains unsolved and in this section we shall not consider further the problem of determining the probability density function $p(\mathbf{r}, t)$ in the region of $\mathbf{r}-$space occupied by closed trajectories which have not come from infinity.

The general expression for the particle stress in a suspension of spherical particles, correct to order φ^2, is given in Eq. (7.3.13), with Eq. (7.3.14), Eq. (7.3.19), Eq. (7.3.21) as auxiliary relations. The expression Eq. (7.3.13) takes an illuminating form if we carry out the integration with respect to $\mathbf{r}$ explicitly over the region $r < 2a$, where $P(\mathbf{x}_0 + \mathbf{r}|\mathbf{x}_0) = 0$, since the two particles cannot be overlapped. Thus, we find

$$\sigma_{ij}^{(p)} - 5\varphi\alpha\mu E_{ij} = 5\varphi^2\alpha^2\mu E_{ij}$$

$$+ 5n^2\alpha\mu \int_{r \geq 2a} \frac{4}{3}\pi a^3 \left[\left\{ \frac{S_{ij}(\mathbf{x}_0, \mathbf{x}_0 + \mathbf{r})}{\frac{20}{3}\pi a^3 \alpha\mu} - E_{ij} \right\} p(\mathbf{r}, t) \right.$$

$$\left. - \{ e_{ij}(\mathbf{x}_0, \mathbf{x}_0 + \mathbf{r}) - E_{ij} \} \right] d\mathbf{r} + o(\varphi^2). \tag{7.3.23}$$

For the case of a bulk steady pure straining motion , then, the probability density function $p(\mathbf{r}, t)$ has the form $q(r)$ (see Eq. (7.3.21)), and Eq. (7.3.23) may be rewritten as

$$\sigma_{ij}^{(p)} - 5\varphi\alpha\mu E_{ij} = 5\varphi^2\alpha^2\mu E_{ij}$$

$$+20\pi n^2 \alpha \mu E_{ij} \left(\frac{4}{3\pi a^3}\right)^2 \int_{2a}^{\infty} J\left(\frac{r}{a}\right) q\left(\frac{r}{a}\right) r^2 dr + o(\varphi^2). \qquad (7.3.24)$$

Before going further with the manipulation of Eq. (7.3.24), we observe that the term of order φ^2 in the expression for the particle stress now has the Newtonian form. This important conclusion has been established only for a steady bulk pure straining motion , and is of course a direct consequence of the isotropy of the two–particle structure of the suspension which in turn is a consequence of the open form of all trajectories of one particle relative to another. The Newtonian stress form Eq. (7.3.24) implies that, to order φ^2, the suspension can be characterized by an effective viscosity μ^* given by

$$\frac{\mu^*}{\mu} = 1 + \frac{5}{2}\varphi\alpha + \varphi^2 \left\{\frac{5}{2}\alpha^2 + \frac{15\alpha n^2}{2\varphi^2} \left(\frac{4}{3}\pi a^3\right)^2 \left(\frac{1}{a^3}\right) \int_{2a}^{\infty} J\left(\frac{r}{a}\right) q\left(\frac{r}{a}\right) r^2 dr\right\}. \qquad (7.3.25)$$

Enough is known about the function J and q for the case of $\overline{\mu}/\mu = \infty$ to estimate μ^*/μ for a suspension of identical rigid spheres, then Eq. (7.3.25) reduces to[41]

$$\frac{\mu^*}{\mu} = 1 + \frac{5}{2}\varphi + 7.6\varphi^2. \qquad (7.3.26)$$

The quadratic term in the expression is sometimes written as $k(2.5\varphi)^2$, with k being termed the Huggins constant, and the estimate Eq. (7.3.26) for rigid spheres of uniform size corresponds to $k = 1.22$.

We now turn to another extreme when the Péclet number , Pe, is small. ($Pe \ll 1$). In that case the effect of convection on the pair-distribution function $p(\mathbf{r},t)$ is a perturbation of the uniform distribution. Thus we may write $p(\mathbf{r},t)$ as (see expression (5.2.18))

$$p(\mathbf{r},t) = 1 + p_1(\mathbf{r},t) + O(Pe^2), \qquad (7.3.27)$$

and substituting to Eq. (4.4.27), we obtain the equation for the perturbation function p^1 i.e.

$$\nabla \cdot \mathbf{D} \cdot \nabla p_1 = \nabla \cdot \mathbf{V}, \qquad (7.3.28)$$

where the divergence term $\nabla \cdot \mathbf{V}$ is given by

$$\nabla \cdot \mathbf{V} = \frac{\mathbf{r} \cdot \mathbf{E} \cdot \mathbf{r}}{r^2} W(s), \qquad (7.3.29)$$

The dimensionless scalar function $W(s)$ is a function of dimensionless distance s ($s = r/a$), and may be written as

$$W(r) = -3(A - B) - r\frac{dA}{dr}. \qquad (7.3.30)$$

From Eq. (7.3.28), Eq. (7.3.29) we may see that p_1 is evidently linear in $\mathbf{E}$, and must therefore be of the form

$$p_1 = -\frac{a^2}{D_{ij}^{(0)}} \frac{\mathbf{r} \cdot \mathbf{E} \cdot \mathbf{r}}{r^2} Q(s). \qquad (7.3.31)$$

Substituting the expressions for p_1 and $\mathbf{D}$ in Eq. (7.3.31) and Eq. (7.4.18) into Eq. (7.3.28) and Eq. (7.3.29) one gets

$$\frac{d}{ds}\left(s^2 Q \frac{dQ}{ds}\right) - 6HQ = -s^2 W. \qquad (7.3.32)$$

The boundary conditions on Q are

$$G\frac{dQ}{ds} = 0 \qquad \text{at} \qquad s = 2 \quad \text{and} \quad Q \to 0 \quad \text{as} \quad s \to \infty. \qquad (7.3.33)$$

As $s \to \infty$ both G and H tend to unity, corresponding to an isotropic relative Brownian diffusivity tensor , and it follows from the asymptotic form for $W(s)(\propto s^{-6})$ that the particular integral of Eq. (7.3.32) behaves as s^{-4} for $s \gg 1$, and that the complementary function behaves as s^{-3} and is dominant. This asymptotic variation of $Q(s)$ as s^{-3} is a consequence of the quadrupole character of the source term $\nabla \cdot \mathbf{V}$ (see Eq. (7.3.29)) in the diffusion Eq. (7.3.28).

In order to be able to make use of the inner boundary condition we first investigate the form of the solution near $s = 2$, which is a regular point of the equation. It is known that (Batchelor and Green 1972)[34]

$$A(s) \sim 1 - 4.077\xi, \qquad B(s) \to 0.406, \quad \text{as} \quad \xi \to 0$$

where $\xi = s - 2$, whence it follows from Eq. (7.3.30), Eq. (4.4.23), Eq. (4.4.26) that

$$W(s) \to 6.372, \quad G \sim 2\xi, \quad H \to 0.401, \quad \text{as} \quad \xi \to 0.$$

Near the point $\xi = 0$ the equation can therefore be written as

$$\xi\frac{d^2Q}{d\xi^2} + \frac{dQ}{d\xi} - 0.301Q = -3.186. \qquad (7.3.34)$$

The complementary function for this equation is

$$Q(\xi) = \alpha I_0(\eta) + \beta K_0(\eta), \qquad (7.3.35)$$

where $\eta = 2(0.301\xi)^{1/2}$ and I_0 and K_0 are the modified Bessel functions, α and β being arbitrary constants. A particular integral of Eq. (7.3.34) is $Q=3.186/0.301=10.58$. The requirement that $2\xi(dQ/d\xi)$ be zero at $\xi =0$ can be satisfied only if $\beta = 0$. A solution of Eq. (7.3.34) which satisfies the boundary condition at $\xi = 0$ is therefore

$$Q = 10.58 + \alpha I_0(\eta). \qquad (7.3.36)$$

But Eq. (7.3.36) is a valid approximation to Eq. (7.3.32) only if contributions to W which are smaller than ξ^0 are neglected. Hence what we learn from Eq. (7.3.36) is that

$$Q_0 = 10.58 + \alpha, \quad Q_0' = 0.301\alpha. \qquad (7.3.37)$$

It is also useful for numerical purposes to note the asymptotic form of the solution of Eq. (7.3.32) as $s \to \infty$. From the asymptotic forms G, H, we find that

$$Q(s) \sim \frac{\gamma}{s^3} + \frac{3\gamma - 25}{2s^4}. \tag{7.3.38}$$

Numerical trials have shown that the constant γ is 9.41. Figure 2 of Chapter 7 shows

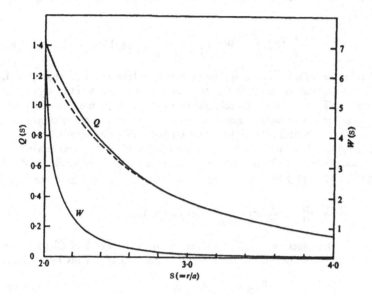

Fig. 2. The scalar functions $W(s)$ and $Q(s)$ specifying $\nabla \cdot \mathbf{V}$ and the pair-distribution function $p(\mathbf{r})$ respectively. The broken line is the asymptotic form $Q = (\gamma/s^3) + (3\gamma - 25)/2s^4$ with γ being 9.41 (From Batchelor 1977)[42]

both W and Q as functions of s (Batchelor 1977[41]); it appears that the asymptotic form Eq. (7.3.38), with $\gamma = 9.41$, is in fact a fair approximation to Q over the whole range. The pair-distribution function for the case $Ea^2/D_{ij}^{(0)} \ll 1$ is then given, correct to order $Ea^2/D_{ij}^{(0)}$, by Eq. (7.3.27) and Eq. (7.3.31) with these values of Q.

The complete and exact expression for the bulk stress has been given in Eq. (7.3.1). An approximate form of the term $\sigma_{ij}^{(p)}$ in Eq. (7.3.1) correct to order φ^2 is provided by Eq. (7.3.13). This term represents the hydrodynamic effects of the particles in the given bulk flow but is not completely independent of the existence of the Brownian motion because the Brownian diffusion affects the form of the pair-distribution function. In the case of relatively strong Brownian motions (i.e. $Pe = Ea^2/D_{ij}^{(0)} \ll 1$), the pair-distribution function is approximately uniform (see Eq. (7.3.27) and Eq. (7.3.31)) regardless of the nature of the bulk flow, and so the value of $\sigma_{ij}^{(p)}$ correct to the orders of φ^2 and $(Ea^2/D_{ij}^{(0)})^0$ in the small quantities φ and $Ea^2/D_{ij}^{(0)}$ is obtained from the expression Eq. (7.3.13) with $p(\mathbf{r}) = 1$. The integral term in Eq. (7.3.13) has been evaluated with $p(\mathbf{r}) = 1$ (Batchelor and Green 1972)[40], and the result for the effective viscosity is

$$\frac{\mu^*}{\mu} = 1 + \frac{5}{2}\varphi + 5.2\varphi^2. \qquad (7.3.39)$$

The direct contribution to the bulk stress due to the Brownian motion is given in (Batchelor 1977)[41]

$$\sigma_{ij}^{(B)} = -\frac{1}{2}n^2 kT \int_{r \geq 2a} W(r)\left(\frac{r_i r_j}{r^2} - \frac{1}{3}\delta_{ij}\right)p(\mathbf{r})\mathrm{d}\mathbf{r}. \qquad (7.3.40)$$

Here we need a better approximation to the pair-distribution function than $p = 1$, because with a uniform pair-distribution function the deviatoric part of the expression Eq. (7.3.40) is zero. Furthermore, the integral in Eq. (7.3.40) is multiplied by kT, which is large compared with $\mu a^3 E$ when Brownian motion effects are dominant, so the small departure from uniformity of the pair-distribution function leads to a direct contribution to the bulk stress which is independent of kT and is comparable with the hydrodynamic contribution. Substituting the expression for p given by Eq. (7.3.27) and Eq. (7.3.31) into Eq. (7.3.40), and remembering that $D_{ij}^{(0)} = kT/3\pi\mu a$, we find

$$\sigma_{ij}^{(B)} = \frac{9}{20}\varphi^2 \mu E_{ij} \int_2^\infty s^2 W(s)Q(s)\mathrm{d}s. \qquad (7.3.41)$$

Using the values of the function $W(s)$ and $Q(s)$ shown in Figure 2 of Chapter 7 a numerical calculation gives the value of the integral in Eq. (7.3.41). Thus Eq. (7.3.41) becomes

$$\sigma_{ij}^{(B)} = 2\mu(0.97\varphi^2)E_{ij}. \qquad (7.3.42)$$

It appears from Eq. (7.3.42) that, when $Ea^2/D_{ij}^{(0)} \ll 1$, the deviatoric part of the bulk stress to order φ^2 is of the Newtonian form, regardless of the form of the bulk flow , and is the same as that for a fluid with the viscosity[42]

$$\mu^* = \mu\left(1 + \frac{5}{2}\varphi + 6.2\varphi^2\right). \qquad (7.3.43)$$

The Newtonian behaviour of the stress to leading order of the small quantity $Ea^2/D_{ij}^{(0)}$ is of course a general property of a system subjected to a small departure from the state of thermodynamic equilibrium, and is to be expected for all values of φ.

Measurements of the values of the effective viscosity of a suspension of spheres at small concentration have not yet given an empirical value of the coefficient of φ^2. One of the best available sets of measurements at low rates of strain was made by Krieger (1972). His observations and the relation Eq. (7.3.43) are shown in Figure

3 of Chapter 7 but there are too few observations to draw any conclusions from the comparison.

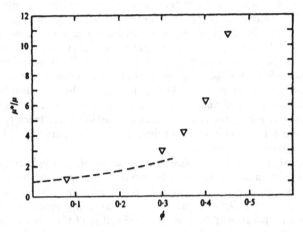

Fig. 3. Measurements by Krieger (1972) of the effective viscosity of a suspension of rigid spherical particles in simple shearing flow at low rates of shear ($Ea^2/D_{ij}^{(0)} \ll 1$). The broken line is the theoretical relation Eq. (7.3.43) correct to order φ^2 (Form Batchelor 1977[42])

4. The Evolution of the Size Distribution of Aerosol Particles

The evolution of the aerosol size distribution is important in a wide variety of environmental and technological applications, including the growth and evolution of atmospheric aerosols, the behavior of particles generated in nuclear reactor accidents, the controlled production of fine particles for industrial applications, and the formation of particles in combustion systems. The aerosol size distribution is described by a master equation usually referred to as the General Dynamic Equation (GDE), which includes contributions from mass transfer (condensation/evaporation) processes, coagulation, particle inception by nucleation, and particle loss by various removal routes. If chemically there is only a single species, then the aerosol simulation is characterized by a single size-variable such as particle radius. On the other hand, a multicomponent aerosol is characterized by both size and chemical components.

In its most general form, the spatially homogeneous multicomponent GDE which governs the size-composition distribution function of an aerosol, $n(m,t)$, is (Seinfeld 1990)[43]

$$\frac{\partial n(m,t)}{\partial t} + \sum_{i=1}^{s} \frac{\partial}{\partial m_i}[I_i(m,t)n(m,t)]$$

$$= \frac{1}{2}\int_0^{m_1} \cdots \int_0^{m_s} K(u, m-u)n(m-u,t)n(u,t)du$$

$$-n(m,t) \int_0^\infty \cdots \int_0^\infty K(u,m)n(u,t)du, \qquad (7.4.1)$$

where m_i is the mass of the i-th component in a particle, and m is a vector of compositions $(m_1, \cdots, m_s)$, where s is the total number of components. $n(m,t)dm$ is the number of particles having mass of component i, in the range $[m_i, m_i+dm_i]$ at time t. I_i is the time rate change of the mass concentration of the i-th component from mass transfer (condensation/evaporation) processes. $K(u,m) = K(m,u)$ is the binary coagulation coefficient ususlly referred to as the collectional kernel, and $K(u,m)n(u)du$ is the probability that a given particle of mass m will encounter a particle of mass u per unit time interval. The first term on the right-hand-side of Eq. (7.4.1) gives the rate of appearance of new particles by coalescences where the factor of $1/2$ is included to avoid counting the same interaction event twice $(u+m = m+u)$, and the second term gives the rate of disappearance of particles of mass m by coagulation with a particle of mass u.

A number of approaches exist for the numerical solution of the single-component aerosol GDE, whether for mass transfer (condensation/evaporation) processes only, coagulation only, or simultaneous mass transfer and coagulation. Numerical solution of the multi-component GDE presents a more demanding problem than the size distribution only case because of the multidimensionality of the size-composition distribution function. However, we shall only briefly introduce the case of the single-component aerosols for coagulation in this section. Exact solutions exist in some idealized cases. Thus, the General Dynamic Equation (GDE) Eq. (7.4.1) reduces to the Kinetic Coagulation Equation , i.e.

$$\frac{\partial n(v,t)}{\partial t} = \frac{1}{2} \int_0^v K(u,v-u)n(v-u)n(u)du - n(v,t) \int_0^\infty K(u,v)n(u)du. \qquad (7.4.2)$$

We have here transformed Eq. (7.4.1) directly into Eq. (7.4.2) for $n(v,t)$ by noting that $n(v,t)dv = n(m,t)dm$ where v is the particle volume, and $K(u,v)n(u)du$ is the probability that a given particle of volume v encounters a particle of volume u per unit time interval. Actually this probability is the rate of coagulation obtained in Chapter 5. Thus from equations (6.1.26) and (1.1.1) the collectional kernel $K_B(a,b)$ for the Brownian coagulation is

$$K_B(a,b) = \frac{2C_\varphi kT}{3\mu}(a+b)\left[\left(\frac{1}{a}+\frac{1}{b}\right) + \alpha\lambda_m\left(\frac{1}{a^2}+\frac{1}{b^2}\right)\right], \qquad (7.4.3)$$

where a is the radius of particle A and b is the radius of particle B, λ_m is the mean free path of the air molecule and α is a constant given in (1.1.1).

We may also obtain the collectional kernel s $K_G(a,b)$ for gravitational coagulation, $K_{AP}(a,b)$ for axisymmetric pure strain coagulation, $K_{SS}(a,b)$ for simple shear coagulation from (6.1.4) with the coefficient in (6.1.4) being $1, 8/3\sqrt{3}$ and $8/3\pi$ respectively for the corresponding coagulation cases. Then,

$$K_G(a,b) = \pi(a+b)^2 E_{ij} V_{ij}^{(0)}, \qquad (7.4.4)$$

$$K_{AP}(a, b) = \frac{4}{3\sqrt{3}}\pi(a + b)^3 E_{ij}E, \tag{7.4.5}$$

$$K_{SS}(a, b) = \frac{4}{3}(a + b)^3 E_{ij}\Gamma, \tag{7.4.6}$$

where $V_{ij}^{(0)}$ is the gravitational velocity of particle A relative to particle B when they are far apart, E is the characteristic rate-of-strain of the ambient axisymmetric pure straining motion and Γ is the velocity gradient of the ambient simple shear flow .

In an idealized case Smoluchowski (1916, 1917) obtained an approximate solution to Eq. (7.4.2) for the Brownian coagulation. Suppose initially we have a homogeneous aerosol of particles of volume v_1 and concentration $n_1(0) = n_0$. Coagulation sets in and soon there appear particles of volume $v_2 = 2v_1$ with concentration $n_2(n_2(0) = 0), v_3 = 3v_1$ with concentration $n_3(n_3(0) = 0)$, etc. Then to an excellent approximation the rate of the Brownian coagulation of v_i and v_j particles per unit volume of aerosol to form particles of volume $v_{i+j} = (i + j)v_1$ is given by $4\pi D_{ij}^{(0)}C_\varphi(a_i + a_j)n_i n_j$, with $a_i + a_j = (3/4\pi)^{1/3}(v_i^{1/3} + v_j^{1/3})$, where n_i and n_j are the ambient concentrations of the v_i and v_j particles at time t. Then the Kinetic Coagulation Eq. (7.4.2) reduces to

$$\frac{\partial n}{\partial t} = \frac{1}{2}\sum_{i+j=k} 4\pi D_{ij}^{(0)}C_\varphi(a_i + a_j)n_i n_j - \sum_i 4\pi D_{ij}^{(0)}C_\varphi(a_i + a_k)n_i n_k. \tag{7.4.7}$$

This is the discrete form of the Kinetic Coagulation Equation for the Brownian coagulation with collectional kernel K_{ij} being $4\pi D_{ij}^{(0)}C_\varphi(a_i + a_j)$. Suppose that the collectional kernel is independent of particle size , then, $K_{ij} = K =$ constant. If we introduce $f_k(\tau) = n_k/n_0$ and $\tau = \frac{1}{2}Kn_0t$, Eq. (7.4.7) becomes

$$\frac{\mathrm{d}f_k}{\mathrm{d}\tau} = \sum_{i+j=k} f_i f_j - 2f_k \sum_i f_j. \tag{7.4.8}$$

Summing over k, we obtain

$$\frac{\mathrm{d}}{\mathrm{d}\tau}\left(\sum_k f_k\right) = \sum_{i=1}\sum_{j=1} -2\sum_{k=1}\sum_{i=1} f_i f_k = -\left(\sum_k f_k\right)^2, \tag{7.4.9}$$

which upon integration yields

$$\sum_k f_k = \frac{1}{1 + \tau}. \tag{7.4.10}$$

With this result, the solution for f_k may be obtained successively. Thus the equation for f_1 is

$$\frac{\mathrm{d}f_1}{\mathrm{d}\tau} = -2f_1 \sum_k f_k = \frac{-2f_1}{1 + \tau}, \tag{7.4.11}$$

which leads to $f_1 = (1 + \tau)^{-2}$. By induction one may easily show that for arbitrary k,

$$f_k = \frac{\tau^{k-1}}{(\tau + 1)^{k+1}}, \tag{7.4.12}$$

which is Smoluchowski 's solution.

For the Brownian coagulation of submicron aerosol particles in the approximate range $0.1\mu m < a < 1\mu m$ with $0.5 \leq a_i/a_j \leq 2$, the approximation $K_{ij} = $ constant is reasonably good, and the reciprocal dependence of the total particle concentration on time predicted by Eq. (7.4.10) has been verified by extensive experiments over the past years. The total particle concentration is reduced by a factor two at the time $t_{1/2}$ given by

$$t_{1/2} = (Kn_0)^{-1} = (8\pi D a n_0)^{-1} = \frac{3\mu}{4kTn_0(1 + \alpha Kn)} \qquad (7.4.13)$$

including the slip correction and assumption $C_\varphi = 1$. As would be expected, the coagulation time $t_{1/2}$ decreases with decreasing particle size , increasing concentration, and increasing temperature of the aerosol. One may use $t_{1/2}$ to estimate the size range of atmospheric aerosols for which the Brownian coagulation is important. As an example, we may assume that a typical concentration for Aitken particles with $0.01\mu m$ radius is $n_0 = 10^5 cm^{-3}$, which means $t_{1/2} = 2.6 \times 10^3 sec$ at $15^0 C$ and $1000mb$, the sufficiently short time indicates that the Brownian coagulation is the dominant loss mechanism for such particles; for particles of $1\mu m$ radius with concentration $n_0 = 10^3 cm^{-3}$, the coagulation time is 10^3 longer.

The most serious limitation of the Smoluchowski 's solution is that it assumes an initially homogeneous aerosol. In order to investigate the effect of the Brownian coagulation on realistic aerosol spectra, Junge (1955, 1957) and Junge and Abel (1965) carried out numerical calculations of the Kinetic Coagulation Equation, using measured tropospheric aerosol spectra for the initial conditions and allowing for the dependence of the collectional kernel on particle size . For this purpose they used a continuous function $n(a, t)$. where $n(a, t)da$ is the number of particles at time t with radius between a and $a + da$, per unit aerosol volume. Then Eq. (7.4.2) is transformed into an equation for $n(a, t)$, viz.

$$\frac{\partial n(a, t)}{\partial t} = \frac{1}{2} \int_0^a K(b, a) \left(1 - \frac{b^3}{a^3}\right)^{-2/3} n(b, t)n(a, t)db$$

$$-n(a, t) \int_0^\infty K(b, a)n(b, t)db, \qquad (7.4.14)$$

with the collectional kernel $K(b, a)$ given by Eq. (7.4.3).

Eq. (7.4.14) and Eq. (7.4.3) were solved numerically by Junge (1957, 1963) and Junge and Abel (1965) for the case of an initial particle spectrum representing the average tropospheric conditions, and assuming that the colliding aerosol particles have a sticking efficiency of unity and behave as droplets. The result of this calculation is shown in Figure 4 of Chapter 7 in terms of the concentration $N(a, t)$ of particles with radii $\geq a$. It is seen that the modification of the size distribution due to the Brownian

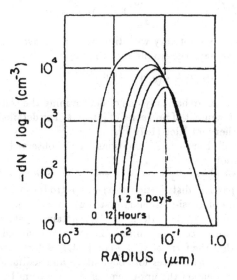

Fig. 4. Numerical calculation of the change of an average atmospheric aerosol particle distribution due to Brownian coagulation (Junge and Abel 1965).

coagulation is confined mainly to particles with radii less than 0.1μm, i.e., the Aitken particle size range. With increasing time, particles of radii less than 0.1μm rapidly disappear, while the maximum of the original size distribution shifts to larger sizes, but the shift becomes less and less, finally the maximum is centering over $a \approx 0.1\mu$m after a few days. From the computed corresponding particle volume changes Junge, and Junge and Abel also concluded that the coagulation causes a steady flux of aerosol material from Aitken size range into the 'large' particle size range, forming 'mixed' particles in the size range of 'large' particles.

Another problem which the Kinetic Coagulation Equation can be used to solve is scavenging of aerosols. This is the 'wet removal processes' by which aerosol particles may be attached to, or 'scavenged' by cloud drops and ice crystals, some fraction of which may subsequently fall to earth as precipitation. The aerosol particles are thus lost from the atmosphere.

From Eq. (7.4.14) we see that the loss rate of aerosol particles per unit volume of air due to scavenging process by cloud drops is given by

$$-\frac{\partial n(a,t)}{\partial t} = n(a,t) \int_0^\infty K(a,b)n_d(b,t)db, \qquad (7.4.15)$$

where b denotes the cloud drop radius, and $n_d(b,t)db$ is the cloud drop number per unit volume of air at time t in the size interval b to $b + db$. The fractional depletion rate of the aerosol concentration by scavenging is called the scavenging coefficient, Λ. From Eq. (7.4.15) we thus have

$$-\frac{1}{n}\frac{\partial n(a,t)}{\partial t} \equiv \Lambda(a,t) = \int_0^\infty K(a,b)n_d(b,t)\mathrm{d}b. \qquad (7.4.16)$$

If the cloud drop distribution does not vary with time then the scavenging coefficient Λ becomes a constant, and we have the simple result that[3]

$$n(a,t) = n(a,0)\mathrm{e}^{-\Lambda(a)t}. \qquad (7.4.17)$$

The scavenging problem is therefore basically one of determining the collectional kernel $K(a,b)$ for the various attachment of processes of interest. For detailed discussions of the problem, see Pruppacher and Klett(1978)[3].

The third problem we shall discuss is explanations for the observed size distribution of the atmospheric aerosol.

A particularly simple and physically appealing way of dealing with the problem of the steady state aerosol particle distribution $n(a)$ was introduced by Friedlander (1960), who proposed a theory of quasi-stationary distribution (QSD). This theory is based on an assumption that for $a \geq 0.1\mu$m the aerosol has attained a state of dynamic equilibrium between the Brownian coagulation and the gravitational sedimentation . The theory further assumes that the form of $n(a)$ is completely determined by the two parameters characterizing the process rates for coagulation and sedimentation , and by the rate ϵ_m at which matter enters the upper end of the spectrum by coagulation of smaller particles. The QSD theory is somewhat analogous to the Kolmogorov's theory of turbulence in the inertial subrange. There the assumption was made that $v_\lambda = v_\lambda(\lambda, \epsilon)$, where ϵ measures the energy flow rate down the turbulent eddy size spectrum. Similarly, according to QSD an aerosol size spectrum in dynamic equilibrium is characterized principally by ϵ_m, the flow rate of matter passing up the aerosol size spectrum.

Friedlander delineates two subranges for $n(a)$. At the lower end of the equilibrium range, the 'coagulation subrange' for which $0.1 \leq a \leq 0.5\mu$m, he assumes sedimentation to be negligible. Since the concentration of particles in this subrange is much larger than in the range $a > 1\mu$m, practically all the matter being transferred up the spectrum over the coagulation subrange will do so by the Brownian coagulation rather than by inertial impact with larger sedimentation particles. Therefore, for the coagulation subrange it is reasonable to assume that $n = n(a, C, \epsilon_m)$, where $C = 2kT/3\mu$ is the characteristic coagulation parameter (cf. Eq. (7.4.3)). If L^3 denotes a characteristic air volume and 1 denotes a characteristic spectral length (particle radius), then on dimensional grounds one finds that the units of n, a, C, and $\epsilon_m : [n] = L^{-3}l^{-1}, [a] = l, [C] = L^3t^{-1}, [\epsilon_m] = l^3L^{-3}t^{-1}$, where t represents time. It therefore follows from the dimensional analysis that

$$n(a) = B_1\left(\frac{\epsilon_m}{C}\right)^{1/2}a^{-5/2}, \qquad 1 \leq a \leq 0.5\mu\text{m}, \qquad (7.4.18)$$

where B_1 is a dimensionless constant.

The second subrange is considered to be confined to the upper end of the spectrum, $a > 5\mu$m, where it may be assumed that the Brownian coagulation is negligible. Thus, matter entering this subrange is lost by sedimentation without significant further transfer with the range by coagulation. Therefore, for this subrange n is assumed to be a function of a, ϵ_m and the characteristic sedimentation parameter $A = (2/9)(g/\mu)(\rho_p - \rho_a)$ (cf. Eq. (3.1.21)). Since $[A] = Ll^{-2}t^{-1}$, by dimensional analysis we find

$$n(a) = B_2 \left(\frac{\epsilon_m}{A}\right)^{3/4} a^{-19/4}, \quad a \geq 5\mu\text{m}, \qquad (7.4.19)$$

where B_2 is another dimensionless constant.

The physical basis of this second spectrum form appears not be as sound as that for the coagulation subrange. Particle loss by sedimentation from an aerosol volume element requires a vertical gradient of particle concentration, and the characteristic length for this gradient will depend in part on some additional transport mechanism such as the turbulent diffusion. The model leading to Eq. (7.4.19) misrepresents the physics of this situation by ignoring the possibility of such an additional independent characteristic length.

Although the power laws obtained for the two particle subranges agree qualitatively with the atmospheric aerosol spectra, it should be recalled from section 2 of Chapter 1 that in the lower troposphere the observed particle size spectra are usually better represented by $n \sim a^{-4}$, so called the 'Junge distribution ' over a range as large as $0.1 < a < 100\mu$m (see section 2 of Chapter 1).

Another plausible approach for explaining the observed regularities of size distributions for tropospheric aerosols involves the notion that they represent asymptotic solutions to the Kinetic Coagulation Equation , rather than equilibrium solutions as in the QSD theory. It seems reasonable to expect that as time proceeds, a coagulating aerosol might lose its 'birth marks' and acquire a size distribution independent of its initial form. It is also reasonable to anticipate that such asymptotic solutions should have relatively simple forms, which might be investigated by the use of similarity transformations, i.e., a transformation of variables which will reduce the coagulation equation to an equation with only one independent variable. The single variable would then suffice to describe the form of the asymptotic distribution. Following the example of the self-preserving hypothesis used in the theory of turbulence (see, for example, Townsend, 1956), Friedlander (1961) introduced a similarity transformation which forms the basis of his theory of self-preserving distribution(SPD).

The similarity transformation for $n(v, t)$ in the SPD theory is as follows:

$$n(v, t) = g(t)\psi_1 \left(\frac{v}{v_+(t)}\right), \qquad (7.4.20)$$

where g and v_+ are functions of time, and it is assumed that ψ_1, the dimensionless 'shape' of the distribution, does not change with time. The functions g and v_+ can be evaluated to within a constant from any two integral functions of $n(v, t)$, such as the zeroth and first moments of the distribution (i.e., the total number of particles per unit volume of aerosol, N, and the volume fraction of dispersed phase, φ). Thus we have

$$N = \int_0^\infty n(v,t)\mathrm{d}v = v_+ g c_1 \qquad (7.4.21)$$

and

$$\varphi = \int_0^\infty vn(v,t)\mathrm{d}v = v_+^2 g c_2, \qquad (7.4.22)$$

where c_1 and c_2 are constants since ψ_1 is assumed to be independent of time. Substituting the resultant expressions for v_+ and g into Eq. (7.4.20), we obtain

$$n(v,t) = \frac{c_2 N^2}{c_1^2 \varphi} \psi_1 \left(\frac{c_2 Nv}{c_1 \varphi} \right) = \frac{N^2}{\varphi} \psi(\eta), \qquad (7.4.23a)$$

where

$$\eta = Nv/\varphi, \qquad (7.4.23b)$$

and ψ is another suitable dimensionless distribution function. This representation also makes sense on a dimensional ground, since $\frac{\varphi}{N} \sim l^3$ is a characteristic spectral volume, namely the average particle volume.

The conjecture that Eq. (7.4.20) and Eq. (7.4.23) constitutes a solution can only be tested for a specific kernel K by substituting it into the coagulation equation. The SPD theory was developed to apply to situations in which there is no addition or removal of aerosol mass in any volume element (as, for example, by sedimentation), so that a basic condition for the validity of Eq. (7.4.20) is that φ be constant in time. The pure coagulating process is a mass conserving process.

Physically, it is obvious that Eq. (7.4.2) must satisfy the condition $\varphi = $ constant. It can also be easily demonstrated by integration Eq. (7.4.2) multiplied by v:

$$\frac{\mathrm{d}\varphi}{\mathrm{d}t} = \frac{\mathrm{d}}{\mathrm{d}t} \int_0^\infty \mathrm{d}v\, vn(v;t) = \frac{1}{2} \int_0^\infty \mathrm{d}v\, v \int_0^v K(u, v-u)n(u,t)n(v-u,t)\mathrm{d}u$$

$$- \int_0^\infty \mathrm{d}v\, vn(v,t) \int_0^\infty K(u,v)n(u,t)\mathrm{d}u. \qquad (7.4.24)$$

If the order of integration is interchanged in the first term on the right-hand-side of Eq. (7.4.24) (i.e., $\int_0^\infty \mathrm{d}v \int_0^v \mathrm{d}u F(u,v) = \int_0^\infty \mathrm{d}u \int_0^\infty \mathrm{d}v F(u,v)$), and this is followed by a variable substitution $v' = v - u$, the two integrals on the right-hand-side of Eq. (7.4.24) cancel each other, so that $\varphi = $ constant as expected.

We would like now to show that Eq. (7.4.23) represents a solution to Eq. (7.4.2) for the case that K is a homogeneous function of its arguments, i,e., $K(au, av) = a^b K(u,v)$. This includes the important cases of the Brownian coagulation without the slip correction, and shear coagulation. In order to facilitate our discussion here, let us first introduce a dimensionless form of the coagulation equation. Following Scott(1968) and Drake (1972), we write

$$u \equiv v_0 y, \quad v \equiv v_0 x, \quad \tau \equiv K_0 N(0)t, \quad K(u,v) \equiv K_0 \alpha(y,x),$$

$$v_0 n(v,t) \equiv N(0)f(x,\tau) \quad v_0 n(v,0) = N(0)f(x,0) \equiv N(0)f_0(x), \qquad (7.4.25)$$

where $N(0)$ is the initial total number density, v_0 the initial mean particle volume ($v_0 = \varphi/N(0)$), K_0 a normalization factor with a dimension of volume per unit time, τ the dimensionless time, $f(x, \tau)$ the dimensionless concentration, $\alpha(x, y)(= \alpha(y, x))$ the dimeneionless collectional kernel , and x and y are dimensionless particle volume. Substituting Eq. (7.4.25) into Eq. (7.4.2) gives for the dimensionless coagulation equation

$$\frac{\partial f(x, \tau)}{\partial \tau} = \frac{1}{2} \int_0^x \alpha(x - y, y) f(x - y, \tau) f(y, \tau) \mathrm{d}y$$

$$-f(x, \tau) \int_0^\infty \alpha(x, y) f(y, \tau) \mathrm{d}y. \tag{7.4.26}$$

If we denote the moments of f by $M_n(\tau)$, i.e.,

$$M_n(\tau) \equiv \int_0^\infty x^n f(x, \tau) \mathrm{d}x, \tag{7.4.27}$$

we find from the condition of aerosol mass conservation ($\varphi = $const.$= v_0 N(0)$) that $M_1(\tau) = 1$. Therefore, a similarity transformation for $f(x, \tau)$ which is completely equivalent to Eq. (7.4.23) is as follows:

$$f(x, \tau) = M_0^2(\tau) \psi(\eta), \tag{7.4.28a}$$

where

$$\eta = M_0(\tau) x. \tag{7.4.28b}$$

Now let us proceed to show that this transformation is indeed a solution to the coagulation equation for homogeneous kernels. From Eq. (7.4.28a) we have

$$\frac{\partial f}{\partial \tau} = \left(2\psi + \eta \frac{\partial \psi}{\partial \eta}\right) M_0 \frac{\mathrm{d}M_0}{\mathrm{d}\tau}. \tag{7.4.29}$$

Evidently, an expression for $\mathrm{d}M_0/\mathrm{d}\tau$ is needed. An ordinary integral-differential equation for any of the moments is easily obtained by integration of the coagulation equation in the manner discussed above. The result is

$$\frac{\mathrm{d}M_n}{\mathrm{d}\tau} = \frac{1}{2} \int_0^\infty \int_0^\infty [(x + y)^n - x^n - y^n] \alpha(x, y) f(x, \tau) f(y, \tau) \mathrm{d}x \mathrm{d}y. \tag{7.4.30}$$

Therefore, we have

$$\frac{\mathrm{d}M_0}{\mathrm{d}\tau} = -\frac{1}{2} \int_0^\infty \int_0^\infty \alpha(x, y) f(x, \tau) f(y, \tau) \mathrm{d}x \mathrm{d}y$$

$$= -\frac{M_0^{2-b}}{2} \int_0^\infty \int_0^\infty \alpha(\eta, \xi) \psi(\eta) \psi(\xi) \mathrm{d}\eta \mathrm{d}\xi, \tag{7.4.31}$$

in view of the homogeneity of α. Similarly, the right-hand-side of Eq. (7.4.26) in similarity variable is

$$M_0^{3-b}\{0.5 \int_0^\eta \alpha(\eta - \xi, \xi) \psi(\eta - \xi) \psi(\xi) \mathrm{d}\xi - \psi(\eta) \int_0^\infty \alpha(\eta, \xi) \psi(\xi) \mathrm{d}\xi\}.$$

Combining these expressions we see that the similarity transformation does work and reduces the coagulation equation to the following integral-differential equation:

$$(\eta\frac{\mathrm{d}\psi}{\mathrm{d}\eta} + 2\psi)\int_0^\infty \int_0^\infty \alpha(\eta,\xi)\psi(\eta)\psi(\xi)\mathrm{d}\eta\mathrm{d}\xi$$

$$= 2\int_0^\infty \alpha(\eta,\xi)\psi(\eta)\psi(\xi)\mathrm{d}\xi - \int_0^\eta \alpha(\eta-\xi,\xi)\psi(\eta-\xi)\psi(\xi)\mathrm{d}\xi, \qquad (7.4.32)$$

(Friedlander and Wang, 1966). Two accompanying integral constraints, corresponding to $M_0(0) = M_1(\tau) = 1$, are as follows:

$$\int_0^\infty \eta\psi(\eta)\mathrm{d}\eta = \int_0^\infty \psi(\eta)\mathrm{d}\eta = 1. \qquad (7.4.33)$$

In order to gain some idea of the spectral shapes predicted by the SPD theory, let us now solve Eq. (7.4.32) for the simplest case of the Brownian coagulation with a constant collision kernel. For $\alpha = $ constant, Eq. (7.4.32) reduces to

$$\eta\frac{\mathrm{d}\psi}{\mathrm{d}\eta} = -\int_0^\eta \psi(\eta-\xi)\psi(\xi)\mathrm{d}\xi. \qquad (7.4.34)$$

The convolution form suggests the use of Laplace transforms, by which the following solution is readily obtained:

$$\psi = \mathrm{e}^{-\eta}. \qquad (7.4.35)$$

This also satisfies Eq. (7.4.33). The corresponding spectrum function is

$$f(x,\tau) = M_0^2(\tau)\mathrm{e}^{-M_0(\tau)x}. \qquad (7.4.36)$$

To proceed further we must choose an explicit representation for τ, or, in other words, K_0 (see Eq. (7.4.25)). In order to obtain the best correspondence with the notion for Smoluchowski 's discrete solution Eq. (7.4.12), we choose $K_0 = K/2$, and $\alpha = 2$. From the form Eq. (7.4.30) we have $\mathrm{d}M_0/\mathrm{d}\tau = -M_0^2$ or $M_0 = (1+\tau)^{-1}$, so that the spectral function is

$$f(x,\tau) = \frac{\mathrm{e}^{-\frac{x}{1+\tau}}}{(1+\tau)^2}. \qquad (7.4.37)$$

This result represents a particular asymptotic solution to the coagulation equation. It corresponds to the initial spectrum $f(x,0) = \exp(-x)$, but it is expected that any initial distribution would approach to Eq. (7.4.37) for large τ.

An interesting feature of Eq. (7.4.37) is that SPD due to the Brownian coagulation declines faster at large sizes than observed tropospheric spectra. The inclusion of size dependence and a slip correction for the Brownian kernel does not alter this conclusion. Evidently then, the shape of the Brownian coagulation SPD does not conform well with tropospheric aerosol distributions. Therefore, although there remains a need for research into the possibilities of SPD for other mechanism, we may conclude that the occurrence of SPD for the Brownian coagulation is unlikely in the tropospheric aerosol.

Let us now suppose that the coagulation equation is extended to include possible sources and sinks of particles, so that in general φ is no longer a constant. Liu and Whitby (1968) investigated the spectrum which could result in this situation if the following restrictions were to apply: (1) dynamic equilibrium exists within some subrange, so that $\partial n/\partial t = 0$; (2) in that subrange a self–similar form ψ for n exists, so that φ is a constant or nearly so over the time interval of interest.

These strong constraints suffice to specify ψ and n to within a constant. Thus, the governing equation for ψ is simply

$$2\psi + \eta\frac{\partial\psi}{\partial\eta} = 0, \tag{7.4.38}$$

as can be seen from Eq. (7.4.29). The solution is then $\psi = c_1\eta^{-2}$, where c_1 is a dimensionless constant. The corresponding solution for n is obtained from Eq. (7.4.23), $n(v) = c_1\eta^{-2}$, or[3]

$$n(a) = \frac{c_1\varphi}{v^2}\frac{dv}{da} = \frac{c_2}{a^4}, \tag{7.4.39}$$

where c_2 is another dimensionless constant (Liu and Whitby, 1968).

Of course, this simple result is in good agreement with the observed 'Junge distribution', and it is tempting to regard this as a demonstration that the tropospheric aerosol is a quasi-stationary self-preserving distribution. However, it must be remembered that the a^{-4} law is not rigorously experimental, and that there is no proof of the existence of SPD for general coagulating mechanisms. The simple expression Eq. (7.4.39) is just the envelopping curve of the assumed SPD, and is a result of an almost overdetermined formulation.

REFERENCES

1. G. K. Batchelor, Developments in Microhydrodynamics, General Lecture to IUTAM, Delft; *Advances in Mechanics* **7** (1977) 105 (in Chinese).
2. N. A. Fuchs, *The Mechanics of Aerosols* (Science Press, Beijing, 1960)(in Chiese); (Perg amon Press, Oxford, 1964).
3. H. R. Pruppacher and J. D. Klett, *Microphysics of Clouds and Precipitation* (D. Reidel Publishing Company, 1978).
4. G. K. Batchelor, *An Introduction to Fluid Dynamics* (Cambridge University Press, Cambrige, 1967).
5. V. G. Levich, *Physicochemical Hydrodynamics* (Prentice-Hall, 1962).
6. M. E. O'Neill and K. Stewartson, On the Slow Motion of a Sphere Parallel to a Nearby Plane Wall, *J.Fluid Mech.* **27** (1967) 705.
7. M. E. O'Neill and S. R. Majumdar, Asymmetrical Slow Viscous Fluid Motions Caused by the Translation or Rotation of Two Spheres. Part II: Asymptotic Forms of the Solutions when the Minimum Clearance between the Spheres Approaches Zero, *Z. Angew. Math. Phys.* **21** (1970) 180.
8. M. E. O'Neill and S. R. Majumda, Asymmetrical Slow Viscous Fluid Motions Caused by the Translation or Rotation of Two Spheres. Part I: The Determination of Exact Solutions for Any Values of the Ratio of Radii and Separation Parameters, *Z. Angew. Math. Phys.* **21** (1970) 164.
9. D. J. Jeffrey and Y. Onishi, Calculation of the Resistance and Mobility Functions for Two Unequal Rigid Spheres in Low-Reynolds- Number Flow, *J.Fluid Mech.* **139**(1984) 261.
10. J. Happel and H. Brenner, *Low Reynolds Number Hydrodynamics* (Prentice-Hall,1965).
11. G. K. Batchelor, Sedimentation in a Dilute Polydisperse System of Interacting Spheres. Part 1. General Theory, *J. Fluid Mech.* **119**(1982) 379; *J.Fluid Mech.* **137** (1983) 467.
12. G. K. Batchelor and C. S. Wen, Sedimentation in a Dilute Polydisperse System of Interacting Spheres. Part 2. Numerical Results, *J.Fluid Mech.* **124** (1982) 495; *J.Fluid Mech.* **137** (1983) 467.
13. C. S. Wen and G. K. Batchelor, The Rate of Coagulation in a Dilute Suspen sion of Small Particles, *Proc. 2nd Asian Cong. Fluid Mech.* (Beijing, 1983, 705); *Scientia Sinica A* **11** (1984) 996 (in Chinese); *Scientia Sinica A* **18** (1985) 172.
14. Y. G. Wang and C. S. Wen, The Effect of Weak Gravitational Force on Brownian Coagulation of Small Particles, *J.Fluid Mech.* **214** (1990) 599
15. C. S. Wen, L. Zhang and H. Lin, The Rate of Coagulation of Particles in a Sedimenting Dispersion at Large Péclet Number, *J.Colloid and Interface Sci.* **142** (1991) 257.

16. A. Nir and A. Acrivos, On the Creeping Motion of Two Arbitrary-Sized Touching Spheres in a Linear Shear Field, *J.Fluid Mech.* **59** (1973) 209.

17. H. C. Hamaker, The London–van der Waals Attraction between Spherical Particles, Physica **4** (1937) 1058.

18. G. K. Batchelor, Brownian Diffusion of Particles with Hydrodynamic Interaction, *J.Fluid Mech.* **74** (1976) 1.

19. S. Chandrasekhar, Stochastic Problems in Physics and Astronomy, *Rev. Modern Phys.* **15** (1943) 1.

20. G. K. Batchelor, Diffusion in a Dilute Polydisperse System of Interacting Spheres, *J.Fluid Mech.* **131** (1983) 155.

21. G. K. Batchelor, Sedimentation in a Dilute Dispersion of Spheres, *J.Fluid Mech.* **52** (1972) 245.

22. P. Y. Cheng and H. K. Schachman, Studies on the Validity of the Einstein Viscosity Law and Stokes's Law of Sedimentation, *J.Polymer Sci.* **16** (1955) 19.

23. R. Buscall, J. W. Goodwin, R. H. Ottewill and T. F. Tadros, The Settling of Particles through Newtonian and Non-Newtonian Media, *J.Colloid and Interface Sci.* **85** (1982) 78.

24. R. H. Davis and A. Acrivos, Sedimentation of Non-Colloidal Particles at Low Reynolds Numbers, *Ann. Rev. Fluid Mech.* **17** (1985) 91.

25. R. H. Davis and K. H. Birdsell, Hindered Settling of Semidilute Monodisperse and Polydisperse Suspensions, *AIChE J.* **34** (1988) 123.

26. W. B. Russel, D. A. Saville and W. R. Schowalter, *Colloidal Dispersions* (Cambridge University Press, Cambrige,1989.)

27. B. V. Derjaguin and V. M. Muller, Slow Coagulation of Hydrosols, *Dokl. Akad. Nauk. U. S. S. R.* **176** (1967) 869. (in Russia).

28. B. V. Derjaguin, Theory of Coagulation of Colloidal Dispersions, *Dokl. Akad. Nauk. U. S. S. R.* **109** (1956) 967. (in Russia).

29. T. G. M. van de Ven and S. G. Mason, The Microrheology of Colloidal Dispersions. VIII. Effect of Shear on Perikinetic Doublet Formation, *Colloid and Polymer Sci.* **255** (1977) 794.

30. L. M. Hocking, The Effect of Slip on the Motion of a Sphere Close to a Wall and of Two Adjacent Spheres, *J.Eng Math.* **7** (1973) 207.

31. R. H. Davis, The Rate of Coagulation of a Dilute Polydisperse System of Sedimenting Spheres, *J.Fluid Mech.* **145** (1984) 179.

32. G. R. Zeichner and W. R. Schowalter, Use of Trajectory Analysis to Study Stability of Colloidal Dispersicons in Flow Fields, *AIChE J.* **23** (1977) 243.

33. G. K. Batchelor, Mass Transfer from a Particle Suspended in Fluid with a Steady Linear Ambient Velocity Distribution, *J.Fluid Mech.* **95** (1979) 369.

34. G. K. Batchelor and J. T. Green, The Hydrodynamic Interaction of Two Small Freely-Moving Spheres in a Linear Flow Field, *J.Fluid Mech.* **56** (1972) 375.

35. C. S. Wen and H. Lin, Numerical Solution of the Boundary Layer Equation for the Pair-Distribution Function and the Rate of Gravitational Coagulation in a Dilute Polydisperse Suspension, *Proc. Intern. Conf. Fluid Mech.* (Beijing,

1987, 774.)

36. D. L. Feke and W. R. Schowalter, The Effect of Brownian Diffusion on Shear-Induced Coagulation of Colloidal Dispersions, *J.Fluid Mech.* **133** (1983) 117.

37. C. S. Wen and L. Zhang, The Effect of Brownian Diffusion on Gravitational Coagulation of Aerosol Particles at Large Péclet Number, *Proc. 3rd Intern. Aerosol Conf.* (Kyoto, Pergamon Press, 1990, 450.)

38. D. H. Melik and H. S. Fogler, Effect of gravity on Brownian flocculation, *J. Colloid and Interface Sci.* **101** (1984) 84.

39. Y. G. Wang and C. S. Wen, The Effect of Gravitational Force on Brownian Coagulation of Colloidal Dispersions at Small Péclet Number, *Proc. 4th Asian Cong. Fluid Mech.* (Hong Kong, 1989, 23.)

40. A. Acrivos and T. D. Taylor, Heat and Mass Transfer from Single Sphere in Stokes Flow, *Phys. Fluids* **5** (1962) 387.

41. G. K. Batchelor and J. T. Green, The Determination of the Bulk Stress in a Suspension of Spherical Particles to Order c^2, *J.Fluid Mech.* **56** (1972) 401.

42. G. K. Batchelor, The Effect of Brownian Motion on the Bulk Stress in a Suspension of Spherical Partcles, *J.Fluid Mech.* **83** (1977) 97.

43. J. H. Seinfeld, 1990, Aerosol dynamics, *Proc. 3rd Intern. Aerosol Conf.* (Pergamon Press, Kyoto, 55.)

44. R. L. Qiao and C. S. Wen, The Coagulation Rate of Small Particles in a Sedimentating Dispersion, *J.Colloid and Interface Sci.* (to be published in 1996)

INDEX

A

additivity assumption, 180,193,
    195–197
Aitken nucleus, 2,3,5,16,91,92,124
asymptotic expansion, 60,70,71,161,
    182–184,190,195,
attractive potential, 77,152,154,192
axisymmetric pure straining motion,
    161,163,171
    extensional flow, 171
    compressible motion, 171

B

binary coagulation coefficient, 228
bispherical coordinates, 82
Boltzmann distribution, 92,95,140,122,
    124,125
boundary layer, 161–168,170–175,212
boundary layer equation for the pair-
    distribution function, 163,
Brownian Diffusion, 4,5,35,90,91,93,
    106,118,132,133,163,165,179,
    180,197,
Brownian diffusivity, 5,35,92,96,98,
    100,121,132,150–153,197,
Brownian motion, 11,91–94,96,98,
    100,101,116,119,120,132,133,
    150,151,154,174–176,179,180,
    182,194,225,226

C

capture eiffciency, 155–160,165,
    167–169,171,173–175,177,
    179,195,
cell model, 107, 114,115,129
chemical potential, 93,100
classification of aerosols, 1,3,6
cloud physics, 156,161
coagulation, 9,38,60,77,78,99,149,
    154–156,179–181,187–197

Brownian, 149–152,159,174,
    182,188,190,192
gravitational, 84,149,150,154,
    158,159,174
shear-induced, 179,182,192
collectioal kernel, 228–230,232,235,
configuration, 39,40,41,94,95,
of N spheres, 107
conservation equation, 17,19
    of Fokker–Planck type for the pair-
    distribution function, 98
    mass-conservation equation, 57,65,
    73,75
contact surface, 150–152,165
convection–diffusion equation, 133,199
Coutte flow, 24,26,40

D

Debye–Hükel screening length, 142
deviatoric stress tensor, 111
DLVO theory, 191
double layer, 90,123,184
doublets, 151,
drag, 30,31,40,43,46,112,113
    coefficient, 47
dynamic similarity, 3,5,30

E

Einstein, 91–94,137,145,217,218
electrostatic repulsion, 163
ensemble average, 92,218,219
evolution of the size distribution
    of aerosol particles,
    78,227

F

Faxen theorem, 109
first order of approximation, 34,57,
    66,70,79
flow, 3,10,11,22–25,29–33,37,41–43,